Evolution—the Extended Synthesi

EVOLUTION—THE EXTENDED SYNTHESIS

edited by Massimo Pigliucci and Gerd B. Müller

The MIT Press
Cambridge, Massachusetts
London, England

MIT Press books may be purchased at special quantity discounts for business or sales promotional use. For information, please email special_sales@mitpress.mit.edu or write to Special Sales Department, The MIT Press, 55 Hayward Street, Cambridge, MA 02142.

This book was set in Syntax and Times Roman by Toppan Best-set Premedia Limited. Printed and bound in the United States of America.

Library of Congress Cataloging-in-Publication Data

Evolution—the extended synthesis / edited by Massimo Pigliucci and Gerd B. Müller.
   p. cm.
Includes bibliographical references and index.
ISBN 978-0-262-51367-8 (pbk. : alk. paper) 1. Evolution (Biology) 2. Evolutionary genetics. 3. Developmental biology. I. Pigliucci, Massimo, 1964– II. Müller, Gerd B.
QH366.2.E8627 2010
576.8—dc22

2009024587

10 9 8 7 6 5 4 3

Contents

Preface

The biological sciences evolve at a perplexing pace. Since the mid-twentieth century we have witnessed the molecular revolution, a dramatic technical turn in all fields of biology, and the rise and spread of computation, jointly leading to vast amounts of new data, concepts, and models about the organic world. Fundamentally different kinds of information are now available as compared with the time of the last major conceptual integration in the biosciences, the Modern Synthesis of the 1930s and 1940s—memorably summarized by Julian Huxley and newly accessible in the companion to this volume. As a consequence, but less noted than many of the spectacular empirical advances, the core theoretical framework underlying the biological sciences is undergoing ferment. Evolutionary theory, as practiced today, includes a considerable number of concepts that were not part of the foundational structure of the Modern Synthesis. Which of these will actually coalesce into a new kind of synthesis, augmenting the traditional framework in a substantial fashion, is a major challenge for the theorists of today.

To begin to meet that challenge, a group of 16 prominent evolutionary biologists and philosophers of science convened at the Konrad Lorenz Institute for Evolution and Cognition Research in Altenberg, Austria, in July 2008. The "Altenberg 16," as the group was labeled by the media, met over three days to discuss the new information, both empirical and theoretical, from a large number of different fields. Conceptual change was seen to emerge from traditional domains of evolutionary biology, such as quantitative genetics, as well as from entirely new fields of research, such as genomics or EvoDevo. The structure of the present volume reflects the areas in which the conceptual progress was perceived to be most significant.

The modifications and additions to the Modern Synthesis presented in this volume are combined under the term Extended Synthesis, not because anyone calls for a radically new theory, but because the current scope and practice of evolutionary biology clearly extend beyond the boundaries of the classical framework. The Altenberg group jointly concluded that by incorporating the new results and insights into our understanding of evolution, the explanatory power of evolutionary theory is greatly expanded within biology and beyond. As is the nature of science, some of the new ideas will stand the test of time, while others will be substantially modified over the course of the next few years. Nonetheless, the authors agree that there is much justified excitement in evolutionary biology today. This is a propitious time to engage the scientific community in a vast interdisciplinary effort to further our understanding of how life evolves. An extended evolutionary framework will be key for this endeavor.

The editors wish to thank all those who made this work possible. Foremost, we express our gratitude to the workshop participants and authors who contributed their expertise in so many areas of evolutionary biology. We are grateful to the staff of the KLI for their dedicated assistance in preparing and running the workshop, and we are equally grateful to the devoted editors at the MIT Press—Katherine Almeida, Susan Buckley, and Bob Prior—without whose experience, patience, and encouragement this volume and its companion would not have happened.

the central tenets of the current framework, while wanting to relax some of its assumptions and to introduce what they see as significant conceptual augmentations of the basic MS structure— just as the architects of the Modern Synthesis themselves had done with previous versions of Darwinism and the ideas that had been debated by biologists around the turn of the twentieth century.

Whenever we talk to colleagues who are inclined toward a conservative position about the status of evolutionary theory, we are confronted with the question "So, what exactly is so new that we may speak of an Extended Synthesis?" This volume is the beginning of a response to that question, and we shall provide the reader with an overview below. The commonest reaction to our explanations is something along the lines of "But that is already understood as part of the Modern Synthesis anyway." We beg to differ. Many of the empirical findings and ideas discussed in this volume are simply too recent and distinct from the framework of the MS to be reasonably attributed to it without falling into blatant anachronism. Concepts such as evolvability (Wagner and Altenberg 1996; R. L. Carroll 2002; Love 2003b; Wagner 2005; Hansen 2006; Hendrikse et al. 2007; Colegrave and Collins 2008; Pigliucci 2008), for instance, did not exist in the literature before the early 1990s; phenotypic plasticity (West-Eberhard 1989; Scheiner 1993; Pigliucci 2001; Schlichting and Smith 2002; West-Eberhard 2003; Borenstein et al. 2006) was known, but consistently rejected as a source of nuisance, not of significant micro- and macro-evolutionary change. Or consider EvoDevo, an entirely new field of evolutionary research that has emerged in full only since the late 1980s, precisely because of the perceived explanatory deficits of the MS in the realm of phenotypic evolution (Laubichler and Maienschein 2007; Müller 2007, 2008; Sansom and Brandon 2008). Yet another common retort to our arguments is that the new ideas are "not inconsistent" with the framework of the Modern Synthesis; this may very well be true—and most of us would gladly agree—but being consistent with the MS is not at all the same thing as being a part of the MS!

Much of the confusion and resistance to new ideas may derive from the fact that evolutionary biologists of course have gradually updated their thinking beyond the Modern Synthesis, without necessarily paying too much attention to the fact that in so doing, they have stepped well outside of its original boundaries. Also part of the problem is that most practicing biologists do not have the time to read the papers, and especially the books, that shaped and solidified the MS during the 1930s and 1940s, and therefore may not be too familiar with its actual claims.

Therefore, before plunging into this book's examination of how the Synthesis is already extended in the current usage of evolutionary biologists, it may be useful to briefly summarize the conceptual history of evolutionary thought as well as the basic tenets of the Modern Synthesis. This will put us in a better position to judge how the ideas advanced in this volume relate to the central corpus of the discipline, and how much of an extension is really warranted or provided.

Modern evolutionary thought, of course, began with Charles Darwin and Alfred Russel Wallace's paper to the Linnean Society (1858), although the idea of biological change over time had been around since ancient Greek philosophy. The original Darwinism, as it was soon to be known, was based on two fundamental ideas: the common descent of all living organisms, and the claim that natural selection is the major agent of evolutionary change, as well as the only one that can bring about adaptation. The first idea was quickly accepted (indeed, it had been advanced by others before Darwin, though nobody had done the hard work of systematically collecting evidence in its favor). Natural selection, on the other hand, was more controversial, and Darwinism underwent a period of "eclipse" (Bowler 1983) in the scientific community toward the end of the nineteenth century. According to Julian Huxley, who actually coined the term "eclipse" for that period, several alternative evolutionary mechanisms were proposed at the time, including a revival of Lamarckian-type inheritance (which Darwin himself had flirted with), so-called orthogenesis (macroevolutionary trends directed by internal forces), and saltationism, the idea that evolutionary change is not gradual, as assumed by Darwin, but proceeds in major leaps.

During the same period, by contrast, Wallace and August Weismann pushed a view of evolution that again situated natural selection at the forefront, but eliminated any vestiges of Lamarckism. The physiologist George Romanes famously coined the term neo-Darwinism (not to be confused, as it so often is, with the Modern Synthesis) to mock Wallace and Weismann's pan-selectionist theories. Things did not look any better for the Darwinian view of evolution at the onset of the twentieth century, when the rediscovery of Mendel's work and the beginnings of genetics appeared to deal a blow to the theory. The problem was that Mendel's laws, as well as the newly discovered phenomenon of mutations, implied that there could not be any "blending of inheritance" (the second idea about inheritance, after Lamarckism, to which Darwin had turned in a halfhearted attempt to complete his theory). Mendelian traits seemed to be inherited as discrete units, which would imply the impossibility of gradual evolutionary change: Lamarck may have been definitely abandoned by then, but saltationism was alive and well.

It is this perceived contrast between Mendelism and neo-Darwinism that set up the conditions for what was to become the Modern Synthesis. A group of mathematically oriented biologists, including Ronald Fisher, J. B. S. Haldane, and Sewall Wright, began work that eventually showed that there was no contradiction between Mendelian genetics, the observation of mutations, and the more continuous variation in so-called quantitative characters that was the focus of neo-Darwinism. Fisher's 1918 paper, "The Correlation between Relatives on the Supposition of Mendelian Inheritance," which in particular was the milestone that demonstrated that Mendelian traits affected by several genes would produce the phenotypic distribution typical of quantitative characters (a bell curve), the smoothing of the distribution being ensured by the combination of the additive effects of several loci and the blurring effect of environmental variation. Fisher's 1930 book, as well as seminal papers by Haldane (1932) and Wright (1932), established the field of population genetics, today still considered the theoretical-mathematical backbone of evolutionary biology.

That work by itself, however, did not constitute the Modern Synthesis. The process took another several decades to complete (essentially ranging from 1918, when Fisher's paper appeared, to 1950, the year of publication of Stebbins's seminal book on plant evolution). What was needed was the understanding and acceptance of population genetics by the majority of practicing evolutionary biologists, as well as an articulation of the new ideas in direct reference to the classic disciplines of natural history, systematics, paleontology, zoology, and botany. This happened thanks to the now classic books by Dobzhansky (1937), Mayr (1942), Simpson (1944), Stebbins (1950), and Rensch (1959), although it was the 1942 volume by Julian Huxley that introduced the term Modern Synthesis into evolutionary jargon.

The major contributions of these authors are now well known, but need to be briefly revisited in order to build our case that the MS did not, in fact, include most of the concepts examined in this volume. Dobzhansky was one of the first geneticists to work with natural populations, and contributed mounting evidence (especially with his famous "GNP" series; Provine 1981) that there is much more variation for quantitative traits than previously suspected; since this variation is the fuel for natural selection, and hence essential to the neo-Darwinian process, Dobzhansky helped to establish population genetics as the empirical field that provided the long-missing piece to the original Darwinian puzzle.

Up to this point one could consider the MS as, in fact, a synthesis: from Fisher to Dobzhansky, it was a fusion of neo-Darwinism and Mendelism achieved through the theory and practice of the new population-statistical genetics. The other major contributions, however, went beyond synthesis to actually adding new concepts to the neo-Darwinian edifice, and in some cases to even contradicting some of Darwin's own positions. Take, for instance, Mayr's related ideas of the so-called biological species concept and of allopatric speciation. Despite the fact that they are both still controversial today (Mishler and Donoghue 1982; Templeton 1989; Grant 1994; Sterelny 1994; Barraclough and Nee 2001; Hey 2001; Schluter 2001; Pigliucci 2003; Coyne and Orr 2004; Gavrilets 2004; de Queiroz 2005), they were a direct rebuttal of Darwin's conception of species as arbitrary demarcation lines imposed by the human mind on an otherwise continuous process of diversification (which is why The Origin of Species does not really deal with, well, the origin of species). Mayr made species into the fundamental unit of the biological hierarchy, a move that implied that the study of the process of speciation is part and parcel of what an evolutionary biologist ought to do. This was definitely going beyond, not just synthesizing, neo-Darwinism and Mendelism.

The contributions of Simpson (paleontology) and Stebbins (botany), while historically important, are much less clear from a conceptual perspective. Simpson's *Tempo and Mode in Evolution* (1944) reflects the author's own ambiguity about what is often referred to as "the hardening" of the synthesis in the late 1940s and beyond, when more heterodox ideas were purged or marginalized to yield the core that still characterizes evolutionary theory today. Here the "synthesis" was achieved as much through exclusion as it was through integration. For instance, Simpson initially defended—on the basis of paleontological evidence—the idea that macroevolutionary change may occur very rapidly, on a geological scale. While his concept of tachytelic (fast) evolution may not have been intrinsically incompatible with the MS emphasis on gradualism, it was eventually dropped in favor of a more prosaic view in which microevolutionary processes directly extrapolate to macroevolutionary time scales, rendering paleontology little more than an appendage to the population genetic view of things. We had to wait until Eldredge and Gould's (1972) challenge of punctuated equilibria for that debate to be reignited.

The situation is similar for Stebbins: it is hard to see why, exactly, botany needed to be "brought in" with the Modern Synthesis, unless the plant world offered a perspective significantly different from the typically animal-centric (mostly, in the case of population genetics, Drosophila-centric) view. The work of Stebbins and others did have the potential to challenge central concepts of the MS, beginning with Mayr's insistence on the "biological" species concept and allopatric speciation. Plants are known to exhibit a variety of isolating mechanisms that make it difficult to fit one simple criterion to the reality of plant species, and instantaneous, sympatric speciation is common in plants through hybridization and both allo- and auto-polyploidization. But none of this came to be considered anything other than a set of curious "exceptions" by the central architects of the Modern Synthesis, particularly Mayr, or by their modern intellectual heirs (Coyne and Orr 2004).

As is well known, other branches of biology were left entirely out of the Modern Synthesis, which is a major reason why so many authors in recent years have clamored for its expansion. Most famously, embryology and developmental biology were not incorporated, despite a long tradition of research that had yielded tantalizing insights into the evolution of organismal form (Gould 1977). This may very well have been at least in part a result, as Mayr often claimed, of the lack of interest on the part of developmental biologists, or perhaps it happened because there was at the time no figure in that field comparable to the likes of Simpson or Stebbins. In any case, the need for the growing field of EvoDevo to be explicitly and organically incorporated into evolutionary theory is obvious and largely undisputed (Robert 2004; Müller and Newman 2005; Müller 2007; Love 2006; S.B. Carroll 2008).

Curiously largely unnoticed is the fact that ecology also missed out on the Modern Synthesis. The closest the MS came to ecology was the establishment of the field of evolutionary ecology, almost single-handedly started by Ford (1964). But Ford and his intellectual descendants worked very much within the standard Fisherian model of almost exclusive emphasis on natural selection and population genetics, and ecology as a field kept developing with little to add to, or import from, evolutionary biology (a situation that persists today, despite the widespread existence of departments of "ecology and evolutionary biology" throughout the world). This lacuna began to be addressed only much later by unorthodox researchers such as Van Valen (1973) with his "Red Queen Hypothesis" and Odling-Smee (2003) with his work on the concept of "niche construction." Much still needs to be done in this area.

Be that as it may, the Modern Synthesis became the established framework in evolutionary biology, which has been summarized by Douglas Futuyma (1986: 12) in the following fashion:

The major tenets of the evolutionary synthesis, then, were that populations contain genetic variation that arises by random (i.e., not adaptively directed) mutation and recombination; that populations evolve by changes in gene frequency brought about by random genetic drift, gene flow, and especially natural selection; that most adaptive genetic variants have individually slight phenotypic effects so that phenotypic changes are gradual (although some alleles with discrete effects may be advantageous, as in certain color polymorphisms); that diversification comes about by speciation, which normally entails the gradual evolution of reproductive isolation among populations; and that these processes, continued for sufficiently long, give rise to changes of such great magnitude as to warrant the designation of higher taxonomic levels (genera, families, and so forth).

As we will see in the rest of this volume, several of these tenets are being challenged as either inaccurate or incomplete. It is important, however, to understand the kind of challenge being posed here, in order to avoid wasting time on unproductive discussions that miss the point of an extended evolutionary synthesis. Perhaps a parallel with another branch of biology will be helpful. After Watson and Crick discovered the double-helix structure of DNA, and the molecular revolution got started in earnest, one of the first principles to emerge from the new discipline was the unfortunately named "central dogma" of molecular biology. The dogma (a word that arguably should never be used in science) stated that the flow of information in biological systems is always one way, from DNA to RNA to proteins. Later on, however, it was discovered that the DNA > RNA flow can be reversed by the appropriately named process of reverse transcription, which takes place in a variety of organisms, including some viruses and eukaryotes (through retrotransposons). Moreover, we now know that some viruses replicate their RNA directly by means of RNA-dependent RNA polymerases, enzymes also found in eukaryotes, where they mediate RNA silencing. Prions have shown us how some proteins can catalyze conformational changes in similar proteins, a phenomenon that is not a case of replication, but certainly qualifies as information transfer. Finally, we also have examples of direct DNA translation to protein in cell-free experimental systems in the presence of ribosomes but not of mRNA. All of these molecular processes clearly demolish the alleged central dogma, and yet do not call for the rejection of any of the empirical discoveries or conceptual advances made in molecular biology since the 1950s. Similarly, we argue, individual tenets of the Modern Synthesis can be modified, or even rejected, without generating a fundamental crisis in the structure of evolutionary theory–just as the Modern Synthesis itself improved upon but did not cause the rejection of either Darwinism or neo-Darwinism.

Specifically, this book presents six sections aimed at outlining the directions that contribute to an Extended Synthesis, with the goal of stimulating further conceptual discussion and empirical work to move the field forward. Part II concerns significant advances in our understanding of the tightly linked ideas (in the MS) of natural selection and adaptation. John Beatty explores the relative roles of contingency and chance variation in evolutionary theory, important because the balance between these and determining processes such as natural selection seems to be perennially shifting in the minds of biologists. Sergey Gavrilets writes about what happens when one takes seriously the idea that adaptive landscapes—introduced as a powerful metaphor by Sewall Wright in the 1930s—are mathematical constructs of very high dimensionality and surprising properties. David Sloan Wilson revisits the debate on group selection, a concept that has experienced vertiginous ups and downs since the 1960s, and brings it up to modern standards within the broader context of multilevel selection theory.

Part III deals with the new information from molecular genetics and genomics that brings significant new issues to evolutionary theory. It begins with Gregory Wray's characterization of the consequences of a shift of focus from individual genes to gene networks. Michael Purugganan continues with his analysis of the revolutionary impact of genomic science on the study of evolution.

Part IV tackles the kinds of hereditary and replicatory mechanisms that are not considered within the framework of the Modern Synthesis, and that pose theoretical and empirical hurdles for it: Eva Jablonka and Marion J. Lamb present the case of transgenerational epigenetic inheritance, John Odling-Smee treats niche inheritance, and Chrisantha Fernando and Eörs Szathmáry discuss replication in systems chemistry and extend the replicator principle to neuronal evolution, brains, and language.

Part V is dedicated to the ongoing revisions of the MS that come from the new field of evolutionary developmental biology (EvoDevo), and contains chapters by Marc Kirschner and John Gerhart on the contribution of developmental systems to evolutionary variation, by Stuart Newman on the developmental-genetic mobilization of physical forces in the origin of animal body plans, and by Gerd Müller on the roles of development in the generation of phenotypic innovation. Part VI considers principles of macroevolution and evolvability that are outside the scope of the traditional Modern Synthesis. David Jablonski writes about the large-scale evolutionary processes underlying patterns of innovation, and Massimo Pigliucci considers the potential role of phenotypic plasticity in macroevolution. Günter P. Wagner and Jeremy Draghi present an essay on the evolution of evolvability.

We conclude with part VII, which consists of two essays, by the philosophers of science Alan Love and Werner Callebaut, which explicitly discuss the conceptual structure and theoretical implications of both the Modern Synthesis and the ongoing expansions, presented in this volume and elsewhere, that we collectively term the Extended Synthesis (figure 1.1).

Though part VII provides detailed considerations of the theoretical and philosophical implications of an extended evolutionary framework, we will briefly reflect here on whether the extended approach differs in any principal aspects from the traditional account. Couldn't it be argued, for instance, that the new views introduced through molecular genetics

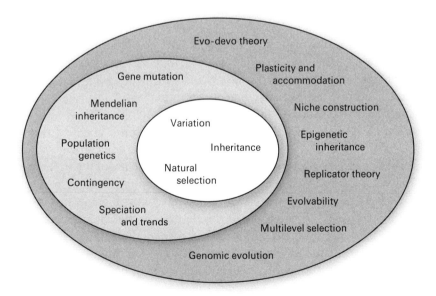

Figure 1.1
Schematic representation of key concepts of Darwinism (center field), the Modern Synthesis (intermediate field), and the Extended Synthesis (outer field). The scheme is meant to depict the broad steps in the continuous expansion of evolutionary theory, not to enumerate all concepts belonging to each of these steps.

and genomics merely add more detail to the classical concepts of variation and selection? Or that non-DNA based mechanisms of inheritance are still part of the inheritance component implicit in the theory anyway? Isn't the environment given merely a little more weight as one effective factor of change, with EvoDevo only adding new mechanistic detail explaining how development itself evolves? That is, besides the obvious disciplinary expansions, one may ask whether any of the general principles of the population-dynamical core of the classical theory are compromised by these new views. Well, no. The concepts we bring together in this volume for the most part do not concern population dynamics, our understanding of which is improved but not fundamentally altered by the new results. Rather, the majority of the new work concerns problems of evolution that had been sidelined in the MS and are now coming to the fore ever more strongly, such as the specific mechanisms responsible for major changes of organismal form, the role of plasticity and environmental factors, or the importance of epigenetic modes of inheritance. This shift of emphasis from statistical correlation to mechanistic causation arguably represents the most critical change in evolutionary theory today.

Underlying the shift toward a causal-mechanistic approach in evolutionary theory is a hugely expanded knowledge base consisting of large data sets in genetics, development, plasticity, inheritance, and other empirical domains. While the MS, in the absence of such data, had to contend with black-boxing all mechanistic aspects, and thus was unable to explain how organismal change is realized at the phenotypic level, the organism as an explanandum has returned through the extended accounts. The traditional theory treated phenotypic traits as abstract quantitative measures, but, as shown in this volume, it is now possible to establish naturalized models of evolutionary variation and innovation based, for instance, on gene regulatory and cell behavioral parameters. Similar functional and dynamical models of evolving processes are used in genetics, development, ecology, behavior, systems biology, and other fields of biology (Laubichler and Müller 2007). The predictions that follow from such qualitative models can be experimentally tested, and the outcomes can be compared with natural patterns of organismal change, a new feature entirely beyond the scope of the MS.

The ongoing shift from a population-dynamic account to a causal-mechanistic theory of phenotypic evolution brings with it a significantly expanded explanatory capacity of evolutionary theory. It has become possible to address phenomena of evolution that were untreatable by the MS, and to cast them as "how" questions, such as How did body plans originate? How did homoplasies arise? How did novelties evolve? How do organisms change phenotypes in response to different environments? Whereas in the classical theory the traits used for quantitative studies were taken as given, the extended accounts can address the conditions of trait generation, fixation, and variation. That is, evolutionary theory is no longer confined to the explanation of the increase in frequency and maintenance of favorable variants, but also becomes a theory of the mechanistic conditions for the origin and innovation of traits.

The extended framework overcomes several basic restrictions and methodological commitments that had been necessary for the correlational approach of the MS to work. One is gradualism. Because the population-dynamic formalism operated on the assumption of continuous and incremental genetic variation, all nongradualist forms of evolutionary change were excluded. Several approaches discussed in this volume show that nongradual change is a property of complex dynamical systems, including biological organisms, and that various kinds of mechanisms for discontinuous change are now known from the domains of genome evolution, phenotypic plasticity, epigenetic development, and nongenetic inheritance. The dynamics of biological systems illuminates the capacity of continuous selectional regimes to produce the nongradual phenotypic change frequently observed in the paleontological record. Accounting for these forms of discontinuous change amounts to a significant extension of the evolutionary synthesis.

A second restriction overcome by the new approach is externalism. The nearly exclusive concentration of the Modern Synthesis on natural selection gave priority to all external factors that realize adaptation through differential reproduction, a fundamental feature of Darwinism not rooted solely in scientific considerations (Hull 2005). Organismal shape and structure were interpreted as products uniquely of external selection regimes. All directionality of the evolutionary process was assumed to result from natural selection alone. The inclusion of EvoDevo in particular, as shown in section five of this volume, represents a major change of this paradigm by taking the contributions of the generative processes into account, as entrenched properties of the organism promote particular forms of change rather than others. On this view, natural selection becomes a constantly operating background condition, but the specificity of its phenotypic outcome is provided by the developmental systems it operates on. Hence the organisms themselves represent the determinants of selectable variation and innovation. At the theoretical level, this shifts a significant portion of the explanatory weight from the external conditions of selection to the internal generative properties of evolving phenotypes.

A third restriction of the Modern Synthesis is its gene centrism. The focus on the gene as the sole agent of variation and unit of inheritance, and the dogmatic insistence on this stance by the popularizers of the Synthesis, quelled all calls for more comprehensive attitudes. Although gene centrism has been a major point of contention, including strong criticism from philosophy of science (Keller 2000; Moss 2003; Neumann-Held 2006), this aspect could not be changed from within the paradigm of the MS, which rested on it both explicitly and implicitly. But gene centrism necessarily disappears in an extended account that provides for multicausal evolutionary factors acting on organismal systems' properties, including the non-programmed components of environment, development, and inheritance. Far from denying the importance of genes in organismal evolution, the extended theory gives less overall weight to genetic variation as a generative force. Rather, the opinions expressed in several contributions to this volume converge on the view of "genes as followers" in the evolutionary process, ensuring the routinization of developmental interactions, the faithfulness of their inheritance, and the progressive fixation of phenotypic traits that were initially mobilized through plastic responses of adaptive developmental systems to changing environmental conditions. In this way, evolution progresses through the capture of emergent interactions into genetic-epigenetic circuits, which are passed to and elaborated on in subsequent generations.

The overcoming of gradualism, externalism, and gene centrism are general hallmarks of the Extended Synthesis, whether in the forms presented here or in various other accounts to a similar effect published since the late 1990s. The editors and authors of this volume offer this extended view of evolutionary theory to the scientific community as, we hope, much food for thought and a stimulus for constructive discussions. It took almost four decades for the Modern Synthesis to take shape, and we certainly do not expect to achieve an equivalent result with a single edited volume. Others, including many not represented here, have advanced along similar intellectual lines, and more will undoubtedly do so in the near future. No matter what the final outcome, 150 years after the publication of the *Origin of Species*, evolutionary theory is still making enormous progress in its capacity to explain the world we live in.

References

Barraclough TG, Nee S (2001) Phylogenetics and speciation. Trends in Ecology and Evolution 16: 391–399.

Borenstein E, Meilijson I, Ruppin E (2006) The effect of phenotypic plasticity on evolution in multipeaked fitness landscapes. Journal of Evolutionary Biology 19: 1555–1570.

Bowler PJ (1983) The Eclipse of Darwinism. Baltimore: Johns Hopkins University Press.

Carroll RL (2000) Towards a new evolutionary synthesis. Trends in Ecology and Evolution 15: 27–32.

Carroll RL (2002) Evolution of the capacity to evolve. Journal of Evolutionary Biology 15: 911–921.

Carroll SB (2008) EvoDevo and an expanding evolutionary synthesis: A genetic theory of morphological evolution. Cell 134: 25–36.

Colegrave N, Collins S (2008) Experimental evolution: Experimental evolution and evolvability. Heredity 100: 464–470.

Coyne JA, Orr HA (2004) Speciation. Sunderland, MA: Sinauer.

Darwin C, Wallace AR (1858) Laws which affect the production of varieties, races, and species. Communication to the Linnean Society. http://www.linnean.org/index.php ?id=380.

de Queiroz K (2005) Different species problems and their resolutions. BioEssays 27: 1263–1269.

Dobzhansky T (1937) Genetics and the Origin of Species. New York: Columbia University Press.

Eldredge N, Gould SJ (1972) Punctuated equilibria: An alternative to phyletic gradualism. In Models in Paleobiology. TJM Schopf, ed.: 82–115. San Francisco: Freeman, Cooper.

Fisher R (1918) The correlation between relatives on the supposition of Mendelian inheritance. Transactions of the Royal Society of Edinburgh 52: 399–433.

Fisher RA (1930) The Genetical Theory of Natural Selection. Oxford: Clarendon Press.

Ford EB (1964) Ecological Genetics. London: Chapman and Hall.

Futuyma D (1986) Evolutionary Biology. Sunderland, MA: Sinauer.

Gavrilets S (2004) Fitness Landscapes and the Origin of Species. Princeton, NJ: Princeton University Press.

Gould SJ (1977) Ontogeny and Phylogeny. Cambridge, MA: Harvard University Press.

Gould SJ (2002) The Structure of Evolutionary Theory. Cambridge, MA: Harvard University Press.

Grant V (1994) Evolution of the species concept. Biologische Zentralblatt 113: 401–415.

Haldane JBS (1932) The time of action of genes, and its bearing on some evolutionary problems. American Naturalist 66: 5–24.

Hansen TF (2006) The evolution of genetic architecture. Annual Review of Ecology and Systematics 37: 123–157.

Hendrikse JLT, Parsons E, Hallgrimsson B (2007). Evolvability as the proper focus of evolutionary developmental biology. Evolution and Development 9: 393–401.

Hey J (2001) Genes, Categories and Species. Oxford: Oxford University Press.

Hull DL (2005) Deconstructing Darwin: Evolutionary theory in context. Journal of the History of Biology 38: 137–152.

Huxley JS (1942) Evolution: The Modern Synthesis. London: Allen & Unwin.

Jablonka E, Lamb MJ (1995) Epigenetic Inheritance and Evolution. Oxford: Oxford University Press.

Keller EF (2000) The Century of the Gene. Cambridge, MA: Harvard University Press.

Kirschner M, Gerhart J (2005) The Plausibility of Life: Resolving Darwin's Dilemma. New Haven, CT: Yale University Press.

Kutschera U, Niklas KJ (2004) The modern theory of biological evolution: An expanded synthesis. Naturwissenschaften 91: 255–276.

Laubichler M, Maienschein J, eds. (2007) From Embryology to Evo-Devo: A History of Developmental Evolution. Cambridge, MA: MIT Press.

Laubichler MD, Müller GB, eds. (2007) Modeling Biology: Structures, Behaviors, Evolution. Cambridge, MA: MIT Press.

Love AC (2003a) Evolutionary morphology, innovation, and the synthesis of evolutionary and developmental biology. Biology and Philosophy 18: 309–345.

Love AC (2003b) Evolvability, dispositions, and intrinsicality. Philosophy of Science 70: 1015–1027.

Love AC (2006) Evolutionary morphology and EvoDevo: Hierarchy and novelty. Theory in Biosciences 124: 317–333.

Maynard Smith J, Szathmáry E (1995) The Major Transitions in Evolution. Oxford: W.H. Freeman.

Mayr E (1942) Systematics and the Origin of Species. New York: Columbia University Press.

Mishler BD, Donoghue MJ (1982) Species concepts: A case for pluralism. Systematic Zoology 31: 491–503.

Moss L (2003) What Genes Can't Do. Cambridge, MA: MIT Press.

Müller GB (2007) EvoDevo: Extending the evolutionary synthesis. Nature Reviews Genetics 8: 943–949.

Müller GB (2008) EvoDevo as a discipline. In Evolving Pathways: Key Themes in Evolutionary Developmental Biology. A Minelli, G Fusco, eds.: 3–29. Cambridge: Cambridge University Press.

Müller GB, Newman SA, eds. (2003) Origination of Organismal Form. Cambridge, MA: MIT Press.

Müller GB, Newman SA (2005) The innovation triad: An EvoDevo agenda. Journal of Experimental Zoology 304B: 487–503.

Neumann-Held E, Rehmann-Sutter C, eds. (2006) Genes in Development: Re-reading the Molecular Paradigm. Durham, NC: Duke University Press.

Odling-Smee FJ, Laland KN, Feldman MW (2003) Comments on Niche Construction: The Neglected Process in Evolution. Princeton, NJ: Princeton University Press.

Pennisi E. (2008) Modernizing the Modern Synthesis. Science 321: 196–197.

Pigliucci M (2001) Phenotypic Plasticity: Beyond Nature and Nurture. Baltimore: Johns Hopkins University Press.

Pigliucci M (2003) Species as family resemblance concepts: The (dis)solution of the species problem? BioEssays 25: 596–602.

Pigliucci M (2007) Do we need an extended evolutionary synthesis? Evolution 61: 2743–2749.

Pigliucci M (2008) Is evolvability evolvable? Nature Reviews Genetics 9: 75–82.

Provine WB (1981) Origins of the GNP series. In Dobzhansky's Genetics of Natural Populations. RC Lewontin, JA Moore, WB Provine, eds.: 5–83. New York: Columbia University Press.

Rensch B (1959) Evolution Above the Species Level. New York: Columbia University Press.

Robert JS (2004) Embryology, Epigenesis, and Evolution: Taking Development Seriously. Cambridge: Cambridge University Press.

Rose MR, Oakley TH (2007) The new biology: Beyond the Modern Synthesis. Biology Direct 2:30.

Sansom R, Brandon RN, eds. (2007) Integrating Evolution and Development: From Theory to Practice. Cambridge, MA: MIT Press.

Scheiner SM (1993) The genetics and evolution of phenotypic plasticity. Annual Review of Ecology and Systematics 24: 35–68.

Schlichting CD, Pigliucci M (1998) Phenotypic Evolution: A Reaction Norm Perspective. Sunderland, MA: Sinauer.

Schlichting CD, Smith H (2002) Phenotypic plasticity: Linking molecular mechanisms with evolutionary outcomes. Evolutionary Ecology 16: 189–211.

Schluter D (2001) Ecology and the origin of species. Trends in Ecology and Evolution 16: 372–380.

Simpson GG (1944) Tempo and Mode in Evolution. New York: Columbia University Press.

Stebbins GL (1950) Variation and Evolution in Plants. New York: Columbia University Press.

Sterelny K (1994) The Nature of Species. Philosophical Books 35: 9–20.

Templeton AR (1989) The meaning of species and speciation: A genetic perspective. In Speciation and Its Consequences. D Otte, JA Endler, eds.: 3–27. Sunderand, MA: Sinauer.

Van Valen L (1973) A new evolutionary law. Evolutionary Theory 1: 1–30.

Wagner A (2005) Robustness and Evolvability in Living Systems. Princeton, NJ: Princeton University Press.

Wagner GP, Altenberg L (1996) Complex adaptations and the evolution of evolvability. Evolution 50: 967–976.

West-Eberhard MJ (1989) Phenotypic plasticity and the origins of diversity. Annual Review of Ecology and Systematics 20: 249–278.

West-Eberhard MJ (2003) Developmental Plasticity and Evolution. Oxford: Oxford University Press.

Whitfield J (2008) Postmodern evolution? Nature 455: 281–284.

Wright S (1932) The roles of mutation, inbreeding, crossbreeding and selection in evolution. In Proceedings of the Sixth International Congress of Genetics. Vol. 1: 356–366.

II VARIATION AND SELECTION

2 Reconsidering the Importance of Chance Variation

John Beatty

In his book *Wonderful Life*, Stephen Gould offered the following thought experiment in order to express what he took to be the highly contingent nature of evolutionary outcomes:

> I call this experiment "replaying life's tape." You press the rewind button and, making sure you thoroughly erase everything that actually happened, go back to any time and place in the past....Then let the tape run again and see if the repetition looks at all like the original. (Gould 1989: 48)

His expectation was that "any replay of the tape would lead evolution down a pathway radically different from the road actually taken" (p. 51).

I will focus here on one particular source of contingency, namely, chance variation and the order in which it appears. Gould's emphasis on contingency was part of his case against the all-importance of natural selection. This chapter is also in part about how chance variation—as a source of contingency—undercuts the all-importance of selection. This may sound odd. In discussions of relative importance, natural selection and chance variation generally go hand in hand: *natural selection of chance variations*, conceived as one process, is usually said to be a more important cause of evolution than any other putative cause, for example, directed variation.

To the extent that the importance of chance variation has been distinguished from, and compared with, that of natural selection, the latter usually comes out the overwhelming winner. I will discuss the tradition and line of reasoning, from Darwin throughout the Modern Synthesis, that completely subordinates the importance of chance variation to that of natural selection. This will set the stage for introducing what Gould considered to be the very "essence of Darwinism": the conviction that selection, not variation—certainly not chance variation—gives direction to evolution and is the source of whatever creativity we might attribute to the evolutionary process. And yet, as I will discuss, there are in Darwin's early work some very nice examples of how the particular pathways and outcomes of evolution may be due—and are perhaps often due—to the chance order in which variation appears. But Darwin subsequently changed his mind about this. The very strong position that he ultimately adopted concerning the sole direction-giving and creative role of natural selection, and the imagery he employed in this regard, were in turn adopted and reinforced by twentieth-century Darwinians. Recently, though, there have been a number of studies that effectively "replay life's tape," and suggest that the significance of chance variation and the order of variation, relative to natural selection, may need to be reassessed or even reconceived.

Lurking in the background throughout this chapter is the bogey of "directed variation." There is, I think, a common perception that the generation of variation is important *only* to the extent that it is predictably ordered (or biased or constrained, etc.). This seems to be assumed by proponents of the importance of natural selection, as well as by proponents of the importance of directed variation. But the post facto order of chance variation can be important as well. Variation (per se) is thus even more important than proponents of directed variation have recognized.

Darwin's Invention and Subordination of Chance Variation

That chance variation is a source of evolutionary contingency is hardly a recent discovery. Darwin provided striking illustrations, especially in his work on orchids. But Darwin hardly championed the importance of chance variation relative to natural selection, and in fact accorded it less and less significance in ways and for reasons that are worth recounting. What importance it had for Darwin was always its role in championing natural selection, especially relative to *directed* variation.

I would even say that Darwin *came* to the notion of chance variation, as we now understand it, in large part through the realization that such a conception of variation would enhance the role of natural selection, again especially relative to directed variation. In the *Origin*, Darwin explained that he used the term "chance" merely to signify his (and others') "ignorance" of the causes of variation. He was prompted to give a more positive characterization of chance variation largely in response to the suggestion that he had overemphasized the importance of selection, and had failed to consider the possibility that the generation of variation itself gives direction to evolution and is the ultimate source of evolutionary creativity.

Two influential proponents of this view, Asa Gray and Charles Lyell, suggested further that it was God who, directly or indirectly, caused just the right variations to appear, at just the right times, to be accumulated by natural selection, thus leading to all the specific outcomes that He intended (including, most importantly of course, humans in His image). Lyell made what has since become the familiar case that natural selection can only preserve or eliminate, it cannot create:

> If we take the three attributes of the deity of the Hindoo Triad, the Creator, Brahma, the preserver or sustainer, Vishnu, & the destroyer, Siva, Natural Selection will be a combination of the two last but without the first, or the creative power, we cannot conceive the others having any function.
>
> The destroy[ing] force is selection, the sustaining [force] preserves things, . . . but in order that life shd. exist where there was none before, . . . & mind in the course of time, . . . this is not selection, but creation, the variety-making not the destroying, or continuing by inheritance, power. Nothing new wd. appear if there were not the creative force. (Lyell in Wilson 1970: 369)

Lyell accused Darwin of "deifying" natural selection by attributing to it the sort of creativity that should be reserved for the Creator (Lyell to Darwin, 15 June 1860, in Darwin 1993: 255).

Gray also recommended a creative role for God in the evolutionary process. It should not be surprising that He would have some way of "[bringing] to pass . . . new and fitting events at fitting times" (Gray [1860] 1963: 47–48). One way in which God could direct the course of evolution would be by bringing to pass new and fitting variations at fitting times. After all, Darwin had admitted his ignorance concerning the causes of variation, thus leaving a gap that could be both conveniently and appropriately filled by God.

> [A]t least while the physical cause of variation is utterly unknown and mysterious, we should advise Mr. Darwin to assume, in the philosophy of his hypothesis, that variation has been led [by God] along certain beneficial lines. (Gray [1860] 1963: 121–122)

Darwin had some theological qualms about saddling God with that much responsibility for the pathways and outcomes of evolution (Beatty 2009). But even leaving aside God's role in all this, Darwin could not concede that the direction of evolution was set by the direction of variation, and merely endorsed by natural selection. Selection had a much greater responsibility and deserved much more credit. Darwin began to express his view of the matter in terms of an analogy between natural selection and an architect:

> As squared stone, or bricks, or timber, are the indispensable materials for a building, and influence its character, so is variability not only indispensable but influential. Yet in the same manner as the architect is the all important person in a building, so is selection with organic bodies. (Darwin to Lyell, 14 June 1860, in Darwin 1993: 254)

And

> [I]n admiring a well-contrived or splendid building one speaks of the architect alone & not of the brick-maker." (Darwin to Hooker, 12 June 1860, in Darwin 1993: 252)

In other words, bricks and timber do not guarantee any particular style of building, nor any building at all. It is the architect who puts them all together. And similarly, the variation within a species at any one time does not itself guarantee that the species will be shaped or reshaped in any particular direction. That is the job of natural selection.

It was a later version of this analogy that Darwin used to articulate the notion of chance variation as we know it. This time around, the building materials were not cut to order, and the architect accordingly assumed even more responsibility for the outcome:

> Let an architect be compelled to build an edifice with uncut stones, fallen from a precipice. The shape of each fragment may be called accidental; yet the shape of each has been determined by the force of gravity, the nature of the rock, and the slope of the precipice,—events and circumstances, all of which depend on natural laws; *but there is no relation between these laws and the purpose for which each fragment is used by the builder.* In the same manner the variations of each creature are determined by fixed and immutable laws; but *these bear no relation to the living structure which is slowly built up through the power of selection,* whether this be natural or artificial selection.
>
> If our architect succeeded in rearing a noble edifice, using the rough wedge-shaped fragments for the arches, the longer stones for the lintels, and so forth, we should admire his skill even in a higher degree than if he had used stones shaped for the purpose. So it is with selection, whether applied by man or by nature; for though variability is indispensably necessary, yet, when we look at some highly complex and excellently adapted organism, variability sinks to a quite subordinate position in importance in comparison with selection, in the same manner as the shape of each fragment used by our supposed architect is unimportant in comparison with his skill. (Darwin 1868: 2, 248–249; emphases added)

In this version of the analogy, there is no conceivable directionality to be had from the process of variation itself. Direction is imposed entirely by natural selection, and thus to the extent that evolution by natural selection is a creative process, it is selection alone that does the creating.

But there is another important component of the architecture analogy, besides the chance formation of the building materials, which bears on the supposedly "paramount" importance of natural selection relative to variation (and which, as we will see, bears on the issue of evolutionary contingency). And that has to do with the abundance of the building materials/variations present at any one time. To see this, imagine an alternative analogy in which the architect works with rocks as they fall *sequentially*, rather than a pile of rocks provided all at once. This would correspond to Gray's (and I suppose Lyell's) view of things if stones of particular sizes and shapes appeared in the order ordained by God for the architect to employ precisely in the sequence provided—perhaps large rectangular foundation stones at the beginning, thin and wide shingle stones later on, and maybe a Star of David or cross-shaped stone at the very end. In this case, the building would owe its form as much or more to the process by which the materials were generated, as to the architect. Of course, this is precisely the scenario that Darwin wanted to rule out.

But the building materials could appear sequentially in no predictable order, and still influence the outcome. Since the architect builds—of necessity—from the ground up, the base of the building will be constructed from among the rocks that fall first, and the roof from rocks that fall later. In this case, the particular outcome will be much more likely to reflect the order in which rocks of various sizes and shapes *happened* to be made available. The architect would still deserve considerable credit for the outcome, but would also be considerably more constrained than if rocks of all sizes and shapes had been available at every stage of the building process.

Darwin himself offered correspondingly parallel, alternative narratives to illustrate evolution by natural selection in early editions of the *Origin*. Consider, for instance, the two hypothetical wolf pack examples that follow his elaboration of the concept of natural selection in chapter 4 (1859: 90–91). In the first example there is, at the start, considerable variation among the wolves with regard to body proportions and speed, but no evolution. Evolution commences only when there is a change in the environment—specifically, a reduction in the number of prey (deer)—that leaves the faster and slimmer wolves much better off.

In the second narrative, evolution commences with the origin of a favorable variation (in this case a new dietary preference) that was not previously present. The new variation confers greater survival ability and is subsequently accumulated by natural selection. It is this second narrative, combined with chance variation, that figures prominently in stories of evolutionary contingency. I will illustrate this with a brief discussion of Darwin's work on orchid evolution.

In his book *On the Various Contrivances by Which Orchids Are Fertilised by Insects*, published in 1862, shortly after the *Origin*, Darwin focused on what he took to be a common sort of phenomenon—namely, the existence among closely related lineages of multiple solutions to the same adaptive problem. This sort of phenomenon, he argued, was best explained in terms of the natural selection of whatever variations happen to arise in each group, and in whatever order. Orchid floral morphology illustrated the general point quite well. As diverse as orchid flowers are, Darwin argued, they all serve basically the same function, namely, to enlist flying insects in their cross-pollination, and thus avoid inbreeding. These otherwise very different "contrivances" for intercrossing had evolved, he believed, under virtually the same environmental circumstances, such as the same range of available insects (small flies, large flies; small bees, large bees; etc.). Sometimes one part of the flower had been modified to entice insects in the vicinity, by mimicry or by scent; sometimes another part had been modified to do the same job. Once the insects had arrived, the pollen had to be attached to them. Some flowers were so constructed as to catapult pollen at the visiting insects; some catapult the insects against the pollen; some simply induce the visitors to travel past and brush up against the pollen. Et cetera, et cetera. Thus cross-pollination is accomplished in very different ways, the different outcomes being due in large measure, Darwin argued, to natural selection acting on the different variations that happened to arise in the different lineages.

An example that particularly struck him involved the position of the "labellum" petal, which in most fully formed orchid flowers is the lowermost of the three petals. In that position, it often serves as a landing pad for pollinators. But interestingly, the labellum actually arrives at that position through a 180-degree twisting of the flower's stem as the flower develops. Darwin reckoned that the position of the labellum in the ancestral orchid had been uppermost, presumably on the grounds that this is also the original position in development, and assuming more generally that the order of development reflects the order of ancestry.

He understood the now typical, lowermost position of the labellum to be an outcome of evolution by natural selection of the more twisted variations that had happened, by chance, to arise (Darwin 1877: 284–285).

But Darwin was especially intrigued by cases where the labellum had resumed its uppermost position, which in some cases had resulted from the selection of less and less twisted variations, and in other cases had come about as the result of selection for more and more twisted forms. Flowers of the latter sort twist a full 360 degrees to resume their starting position (figure 2.1)! As Darwin described the situation,

... in many Orchids the ovarium (but sometimes the foot-stalk) becomes for a period twisted, causing the labellum to assume the position of a lower petal, so that insects can easily visit the flower; but ... it might be advantageous to the plant that the labellum should resume its normal position on the upper side of the flower, as is actually the case with *Malaxis paludosa*, and some species of *Catasetum*, &c. This change, it is obvious, might be simply effected by the continued selection of varieties which had their ovaria less and less twisted, but if the plant [for whatever reason, by chance] only afforded variations with the ovarium more twisted, the same end could be attained by the selection of such variations, until the flower was turned completely on its axis. This seems to have

Figure 2.1
The 360-degree twist of the flower stalk of *Malaxis paludosa*. The labellum is labeled "l" (Darwin 1877: 130).

actually occurred with *Malaxis paludosa*, for the labellum has acquired its present upward position by the ovarium being twisted twice as much as is usual. (1877: 284–285)

So it had apparently become advantageous for *Malaxis paludosa* and some species of *Catasetum* to have their labella uppermost. But due to differences in the variations that happened to occur in the two lineages, evolution by natural selection had resulted in very different means of accomplishing this end: a 360-degree twist in the first case, and no twist in the second (see also Lennox 1993). Here we have two replays from the labellum-down starting position, under presumably similar selective environments (favoring the labellum up), with two different outcomes. Selection alone cannot be responsible for the difference; it is also a matter of the chance differences in variation that occurred. NB: it is worth emphasizing that it is crucial to Darwin's account that the variations in question appeared sequentially, and in no particular direction overall.

Darwin invoked basically the same process—evolution by natural selection of chance differences in sequentially appearing variations— to account for the endless diversity of floral morphologies among orchids:

In my examination of Orchids, hardly any fact has struck me so much as the endless diversities of structure ... for gaining the very same end, namely, the fertilization of one flower by pollen from another plant. This fact is to a large extent intelligible on the principle of natural selection. As all the parts of a flower are co-ordinated, if slight variations in any one part were preserved from being beneficial to the plant, then the other parts would generally have to be modified in some corresponding manner. But *these latter parts might not vary at all* [who knows, it is a matter of chance], *or they might not vary in a fitting manner* [again, who knows], *and these other variations, whatever their nature might be* [ditto], which tended to bring all the parts into more harmonious action with one another, would be preserved by natural selection. (1877: 284; emphases added)

But Darwin did not pursue such contingencies for long. One major reason was the intervention of the Scottish engineer Fleeming Jenkin, whose extremely critical review of the *Origin*—appearing rather late, and coming from out of the blue—forced serious rethinking on Darwin's part (Jenkin 1867). Among other things, Jenkin convinced Darwin that natural selection of sequentially arising variations simply would not work. The problem was Darwin's "blending" theory of inheritance, according to which parents who differ with respect to some trait (e.g., height) would give rise to offspring intermediate between them. The possessor of a single, new advantageous variation would thus not pass the same trait to its offspring; rather, the trait would be diluted through mating with an organism of the previously prevailing type. And accordingly, the advantage associated with the new variation would also be reduced. The new trait, along with its associated advantage, would be further diluted in subsequent generations until the trait and its advantage had almost entirely disappeared. Darwin responded in part by jettisoning the idea that selection acts on sequentially appearing variations, and by supposing instead that there is always considerable variation present for natural selection to act on, and that the variations that ultimately prove advantageous are sufficiently numerous that there is a good chance that their possessors will mate with each other, so that blending will not prevent their accumulation (see the excellent, nuanced discussion of this historical episode in Gayon 1998: 85–102).

In subsequent editions of the *Origin*, Darwin deleted the second illustrative narrative concerning wolf evolution (in which evolutionary change begins with the appearance of a favorable variation that did not previously exist in the population). In its place, he substituted a paragraph in which he acknowledged having previously envisioned natural selection sometimes acting on copious, preexisting variation, and sometimes on single, sequentially appearing variations. But Jenkin's review had convinced him to bet everything on the former:

In former editions of this work I sometimes spoke as if this latter alternative had frequently occurred. . . . Nevertheless, until reading an able and valuable article in the "North British Review" (1867), I did not appreciate how rarely single variations, whether slight or strongly-marked, could be perpetuated. (Darwin 1872: 71)

This was a capitulation of sorts, but the effect was to make natural selection an even more important determinant of the pathways and outcomes of evolution than would be the case if variations arose sequentially. There is thus a double meaning to the appreciation that Darwin expressed in correspondence to Joseph Hooker: "Fleeming Jenkin has given me much trouble, but has been of more real use than any other essay or review" (F. Darwin and Seward 1903: 2, 379). Jenkin had, to be sure, pointed out a serious problem with regard to Darwin's views on variation (or, to be more precise, with regard to his views on variation combined with his blending theory of inheritance). But Darwin's solution only made natural selection that much more directive and creative relative to the input of variation.

The Continued Subordination of Chance Variation throughout the Evolutionary Synthesis

By way of transition to twentieth-century Darwinism and the Modern Synthesis, it helps first to backtrack to Asa Gray. Gray had initially congratulated Darwin for his book on orchids. Darwin was appreciative, but perplexed. Had Gray really understood it? Darwin questioned him over and over again as to what he thought about the last chapter, which reiterated the lesson illustrated by the twisted orchid. Finally, Gray admitted to Darwin that the book gave him a "cold chill" (Gray to Darwin, 7 July 1863; in Darwin 1999: 525–526). Such whimsy hardly bore witness to God's direction. Gray reacted more vigorously to what became the preferred case against the importance of directed variation, namely, the assumption that there was at all times ubiquitous variation, in all directions, for selection to act upon. As it was later put to Gray by the arch-Darwinian George Romanes, "The theory [of evolution by natural selection] merely supposes that variations *of all kinds and in all directions* are constantly taking place, and that natural selection seizes upon the more advantageous" (Romanes 1883: 529).

"Merely supposes?!" Gray retorted (in effect). *Were* it the case that variations *of all kinds and in all directions* were ever present for natural selection to act upon, then of course it would make no sense to propose a role for God in directing the course of variation in order to guarantee the evolutionary outcomes He favored. But to Gray, the accomplished naturalist, the facts were simply otherwise:

[O]mnifarous variation is no fact of observation, nor a demonstrable or, in my opinion, even a warrantable inference from observation and experiment. I am curious to know how far the observations and impressions of the most experienced naturalists and cultivators conform to my own, which favour the idea that variations occur, in every degree indeed, but along comparatively few lines. (Gray 1883).

Mendelism would shortly put an end to concerns about blending inheritance, and to the need to assume "omnifarous" variation to shore up the accumulative power of natural selection. But twentieth-century geneticists and evolutionary biologists had new reasons to believe that there would always be plenty of variation for natural selection to act upon. A number of theoretical and empirical studies could be cited in this regard. I will consider just three.

1. On purely theoretical grounds, R. A. Fisher argued that even a small number of polyallelic loci would give rise to a staggering amount of genetic variation. A mere 100 loci, with two alleles each, can recombine to form 10^{47} different genotypes. There is so much variation present at any one time that new mutations could play no conceivable role in the direction of evolution by natural selection:

> There are, moreover, millions of different directions in which such modification may take place, so that without the occurrence of further mutations all ordinary species must already possess within themselves the potentialities of the most varied evolutionary modifications. It has often been remarked, and truly, that without mutation evolutionary progress, whatever direction it may take, will ultimately come to a standstill for lack of further possible improvements. It has not so often been realized how very far most existing species must be from such a state of stagnation, or how easily with no more than one hundred factors a species may be modified to a condition considerably outside the range of its previous variation, and this in a large number of different characteristics. (Fisher [1930] 1958: 103)

2. Despite the possibility (on Mendelian grounds) of such extensive variation, there was very little variation to be *seen*. One need only recall T.H. Morgan's excitement at finally discovering an observable *Drosophila* mutant whose pattern of inheritance could be studied. Subsequently, variations were found in ever greater numbers in Morgan's laboratory. But it was initially believed that these variations were artifacts; *Drosophila* in the wild seemed to have little, if any, intraspecific variation, and hence little material for future adaptive evolution. Theodosius Dobzhansky popularized the suggestion of Sergei Chetverikov, who proposed that most mutations were recessive and would result in observable differences only when doubled up in the homozygous state (as when inbred in the lab). In this way, natural populations soak up variations "like a sponge," all the while remaining phenotypically uniform:

> The mutation process constantly and unremittingly generates new hereditary variants—gene mutations and chromosomal changes. These variants accumulate in populations of sexually reproducing and cross-fertilizing organisms, and form a great store of potential variability, mostly concealed as a mass of recessive mutants carried in heterozygous condition. According to the succinct metaphor of Chetverikov, a sexual species is "like a sponge" which absorbs and stores the genetic variability. (Dobzhansky 1951: 73)

Various means of exposing this hidden variation, such as the manner proposed by Chetverikov himself, turned up more and more. But nothing like the amount of variation exposed by Richard Lewontin and Jack Hubby, who employed gel electrophoresis to detect amino acid differences in proteins—differences attributable to nucleotide differences in the genes coding for those proteins. Their studies of natural populations of *Drosophila pseudoobscura* suggested—given the inherent biases of their methods—that nearly every gene locus has at least two alleles, and that a third of all loci are heterozygous (Hubby and Lewontin 1966; Lewontin and Hubby 1966; Lewontin 1974).

3. Hidden genetic variation was also revealed by one artificial selection experiment after another. It seemed possible to bring about evolutionary change in most any direction that selection was applied. As Lewontin summarized the results among *Drosophila* workers:

There appears to be no character—morphogenetic, behavioral, physiological, or cytological—that cannot be selected in *Drosophila*. The only known failure is the attempt of Maynard Smith and Sondhi ... to select for left-handed flies. (Lewontin 1974: 92; see also Dobzhansky 1970: 201–209)

This was a somewhat more indirect demonstration of standing variation than, say, the methods employed by Lewontin and Hubby. But it bore more directly on the question of whether selection was the all-important determinant of the direction of evolutionary change, or whether the direction of evolutionary change was significantly constrained by the input of variation.

These and other theoretical and empirical considerations made a strong case for the "omnifarous" variation required by Darwin's first evolutionary narrative, and undermined the second narrative. It came to be believed by major architects of the Modern Synthesis that natural selection does not wait for variation to appear. Nothing hinges on input (in this regard see also the insightful article by Stoltzfus 2006). As Dobzhansky summarized, long after Fisher had suggested basically the same thing:

That selection can work only with raw materials arisen ultimately by mutation is manifestly true. But it is also true that populations, particularly those of diploid, outbreeding species, have stored in them a profusion of genetic variability. A temporary suppression of the mutation process, even if it could be brought about, would have no immediate effect on evolutionary plasticity. (Dobzhansky 1970: 201)

And as Richard Dawkins more recently assured his many readers:

For simplicity we speak of mutation as the first stage in the Darwinian process, natural selection as the second stage. But this is misleading if it suggests that natural selection hangs about waiting for a mutation which is then either rejected or snapped up and the waiting begins again. It could have been like that: natural selection of that kind would probably work, and maybe does work somewhere in the universe. But as a matter of fact on this planet it usually isn't like that. (Dawkins 1997: 87)

All of this contributed further to the idea that natural selection is the all-important cause of the direction of evolutionary change and, to the extent that evolution by natural selection is a "creative" process, that natural selection is *the* creative element. Rather than multiply quotations to this effect from Darwinian evolutionary biologists throughout the twentieth century, I will quote Gould's characterization of the "essence of Darwinism" and a key component of the Modern Synthesis:

[W]hy was natural selection compared to a composer by Dobzhansky; to a poet by Simpson; to a sculptor by Mayr; and to, of all people, Mr. Shakespeare by Julian Huxley? [Gould could and should have asked why natural selection was originally compared to an architect by Darwin.] I won't defend the choice of metaphors, but I will uphold the intent, namely, to illustrate the essence of Darwinism—the creativity of natural selection. The essence of Darwinism lies in its claim that natural selection creates the fit. Variation is ubiquitous and random in direction. It supplies the raw material only. Natural selection directs the course of evolutionary change. (Gould 1977: 44; and again I highly recommend Stoltzfus 2006)

Chance Variation Redux

But if variation is so extensive that selection alone determines the direction of evolutionary change, then why would we find, among closely related species, multiple solutions to the same adaptive problems? Why would initially very similar species, inhabiting similar environments, diverge evolutionarily if indeed they have much the same store of variation for natural selection to act on? Must we always assume in such cases that the environments are quite different after all, and pose quite dissimilar adaptive challenges, so that the divergence can be explained solely in terms of differences in the directionality of selection? In their influential critique of the state of evolutionary biology circa 1979, Gould and Lewontin attacked what they took to be the all-too-common view that natural selection is "omnipotent" to explain all evolutionary outcomes (Gould and Lewontin 1979: 584–585). As one of their main objections, they pointed to the existence of multiple adaptive solutions to the same problem among populations of the same species and closely related species. This shows, they argued, that adaptive outcomes are "the result of history; the first steps went in one direction, though others would have led to adequate prosperity as well" (p. 593).

They did not elaborate on the process by which related populations or species would take their "first steps" in different directions on their way to alternative adaptive outcomes. That process might involve the introduction of different mutations in different lineages, and/or in a different order. But there is another possibility that was, I suspect, more prevalent in the minds of evolutionary biologists at the time. This alternative involves, as a first step, the random drifting of the frequencies of already present alleles, leading to the fixation (or near fixation) of different alleles in different groups. That initial divergence due to random drift might then result in the establishment of different genetic backgrounds against which the rest of the existing variation would be tested. For example, at a particular locus, A_1 might drift to fixation in one population while A_2 drifts to fixation in another. This might then influence the outcome of selection at a second locus, where B_1 might be favored in the first population, in combination with A_1, while B_2 might be favored in the second. If the fitness of an allele at one locus depends on which particular alleles are present at other loci—a situation known as epistasis—then initial differences due to drift (or any other cause) can have cascading effects and lead to considerable divergence.

The latter scenario, relying on existing variation—rather than the first scenario, relying on chance differences in variational input—would surely have been first on the minds of those evolutionary biologists who pondered Gould and Lewontin's suggestion at the time it was published. It would have reminded them of the first "phase" of Sewall Wright's "shifting balance" theory of evolution, which had been a controversial issue off and on throughout the century (Wright 1932; Provine 1986; Coyne et al. 1997; Wade and Goodnight 1998). If one had been given the exam assignment "Discuss the relative roles of chance and selection in evolutionary divergence" at this time, one would surely have been expected to focus on *random drift* and selection, and the instructor would not have deducted a single point for failing to mention chance variation and the order of variation.

One important turn of events in this regard was the commencement of Richard Lenski and collaborators' long-term (and still ongoing) study of 12 initially identical (cloned) lines of *E. coli*, evolving in identical (and identically altered) environments. The experimental system was designed to allow investigators to distinguish between the roles of selection and "chance" in adaptive evolution and divergence. Given that the 12 lines were genetically identical at the start, there was a clear need and an excellent opportunity to include differences in variational input, and not just random drift, under the category "chance." And this was an explicit aim of the Lenski group. The study began shortly after the publication of *Wonderful Life* (1989), in which Gould had mused, regretfully, that of course the tape of life cannot actually be replayed; there can never be a controlled experiment to confirm or disconfirm the contingency of an evolutionary outcome. Nonetheless, that is what the Lenski group aimed to do, and Gould's replay analogy came to serve as a framing device in many of their publications.

I will focus on several of the results obtained from this experimental setup in order to indicate the very real role that chance variation and the order of variation play in directing the course of evolution, but also the difficulties of interpreting the overall importance of chance variation in this regard.

To begin, the investigators grew the 12 lines on a novel medium (i.e., one to which the strain of *E. coli* in question was not previously adapted) for 2,000 generations. The phenotype they monitored was fitness (measured as the ratio of rate of increase of the derived form in competition with the ancestral form, populations of which were kept frozen and then revived for the purpose of comparison). Not surprisingly, each line increased in fitness, but there was some variance among the fitnesses achieved. The investigators somewhat tentatively interpreted these differences as "transient divergence" due to "stochastic variation in time of origin of particular classes of beneficial alleles among replicate populations" (Lenski et al. 1991: 1334). In other words, they expected (again, tentatively) that with time the same beneficial variations would appear in each lineage and spread throughout, and that the lineages would thereby converge on the same fitness. This expectation would be realized only if there were no significant epistatic interactions involving the loci under selection.

In another experiment, lines were drawn from the 2,000th generation of each of the 12 lineages and grown separately on yet another novel medium (Travisano, Mongold, et al. 1995). Initially the fitnesses of these new lines differed considerably; the differences were much greater than those between the donor lineages, although over the course of 8,000 generations in the new medium, the differences were reduced. The initially large differences in the second environment suggested to investigators that the genetic differences at the end of 2,000 generations in the first environment were much greater than previously thought. It seemed that the various lines had evolved genetically quite different ways of being fit in the first environment:

Thus, we agree with Bull and Molineux (1992, p. 892) that "elementary models of selection predict the outcome of evolution with respect to the phenotype under direct selection, but the models are not successful at predicting ... the multiplicity of genetic states satisfying the selected phenotypic criterion." (Travisano, Vasi, and Lenski 1995: 196).

But what about the lines continuing to evolve in the first environment? Interestingly, the differences in fitness that had arisen during the first 2,000 generations in that environment were maintained, and even increased somewhat during the subsequent 8,000 generations. The divergence was not transient after all (figure 2.2). The investigators continued to argue that the genetic differences between the lineages had arisen because of differences in timing and order of mutations, but now added that the initial differences had substantial epistatic implications for other loci, so that strong selection in identical environments would lead to different evolutionary outcomes:

Evolutionary biologists usually regard diversification as being caused by either (i) adaptation to different environments, which often produces conspicuous

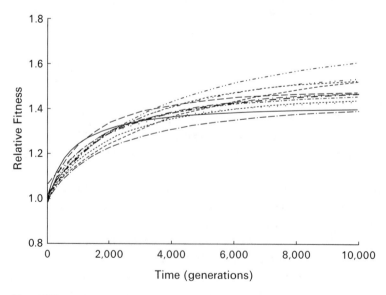

Figure 2.2
Increases in fitness of twelve initially identical lines of *E. coli*, evolving in identical environments (Lenski and Travisano 1994: 6811).

phenotypic variation, or (ii) random genetic drift, which is usually seen in molecular genetic variation. Yet our experiments demonstrate diversification, in identical environments and with very large populations, of no less selected a trait than fitness itself. Someone confronted with the variability among our derived populations (and unaware of the experimental design) might attribute this diversity to environmental heterogeneity ... but any such "just-so story" would clearly be misguided in this case. Instead, our experiment demonstrates the crucial role of chance events [the order of mutations] in adaptive evolution....

Sustained divergence in mean fitness supports a Wrightian model of evolution, in which replicate populations found their way onto different fitness peaks. Although the experimental populations were so large that the same mutations occurred in all of them, the order in which various mutations arose would have been different. As a consequence, some populations may have incorporated mutations that were beneficial over the short-term but led to evolutionary dead ends. (Lenski and Travisano 1994: 6813)

A third experiment was based on an adaptive opportunity that was built into the experimental setup from the beginning. The novel environment on which the various lineages of bacteria were grown had included citrate, which *E. coli* was known not to metabolize. But the investigators considered it within the realm of possibility that *E. coli might* evolve the ability to make use of citrate as a carbon source. As of 30,000 generations, none of the 12 populations had done so. But by 31,500 generations, one lineage had succeeded.

This left the question whether the one population had experienced an extremely rare mutation that would ultimately occur in the other populations as well, rendering the contingent outcome only temporary. Or whether the lucky population had by that time, through a *series* of contingencies, evolved to become *uniquely* capable of taking the final evolutionary steps in the direction of citrate metabolism. The investigators were able to discriminate between the possibilities by employing the frozen "fossil record" of evolution up to that point. That is, after every 500 generations, samples of each lineage had been frozen. So the researchers were able to back up to a point in time in the history of the lucky lineage, prior to the evolution of citrate metabolism, and replay the evolution of that lineage multiple times from that point. And what they found was that the ability to metabolize citrate arose over and over again, suggesting that, by this point, the lineage in question had—again through a series of contingencies—become uniquely capable of making this evolutionary breakthrough. They argued, in particular, that the lucky lineage had evolved a genetic background that substantially increased the mutability of the citrate metabolism locus in question.

Their very literary conclusion drew from the last passage of the *Origin* and from Robert Frost's most famous poem:

[O]ur study shows that historical contingency can have a profound and lasting impact under the simplest, and thus most stringent, conditions in which initially identical populations evolve in identical environments. Even from so simple a beginning, small happenstances of history may lead populations along different evolutionary paths. A potentiated cell took the one less traveled by, and that has made all the difference. (Blount et al. 2008: 7905)

Other selection experiments involving genetically identical starting populations evolving in identical environments have similarly demonstrated that the pathways and outcomes of evolution depend in part on chance differences in the order of mutation. In some of these, it has been possible to characterize the divergence at the genetic, and even at the nucleotide, level. In one study by Holly Wichman and colleagues, two large (10^7) replicate lineages of a DNA bacteriophage were cultured in identical, novel environments for approximately 1,000 population doublings. On the basis of entire genome sequencing, Wichman and her collaborators detected 13 nucleotide substitutions in one lineage, and 14 in the other (all of these nucleotide substitutions resulted in amino acid differences and were thus also phenotypically significant in this sense). Of these 13/14 substitutions, 7 were common between the two lineages. Thus there was a fair degree of parallel evolution. But the other half of the detected substitutions occurred only in one or the other lineage.

On the basis of the very large population sizes that they maintained, over the period of time in question, the investigators reasoned that all of the mutations that occurred in one lineage would probably occur in the other. But whereas one group of mutations had spread to fixation in each lineage, another group had not. This must be due, they reasoned, to the fact that this group of mutations excluded each other. That is, if one of them were to appear early in the experiment, and spread throughout the population, its prevalence would render the other mutations in that group adaptively unfavorable (again, epistasis figures prominently). This interpretation fit with the fact that the seven parallel substitutions occurred in different orders in the two lineages, suggesting that this group of mutations did not exclude each other. None of them, upon spreading, rendered the others unfit.

The degree of parallel evolution that would have signified the all-importance of natural selection in this case (13/14 out of 13/14 substitutions) was not met. Why?

Why is parallel evolution not complete? Given the population size in the chemostat, all these substitutions would have arisen multiple times during the course of these selections. This variation among replicates suggests that stochastic features, such as the identity of the earliest change to sweep through the population and the order in which substitutions arise, may influence the pattern of adaptation even in systems where parallelism is the rule rather than the exception. (Wichman et al. 1999: 424)

Another very interesting study focuses on identifiable mutations at a small number of loci, and demonstrates that the selectively most advantageous combination can be reached only if the mutations occur in a particular order (or in a small subset of all the possible orderings in which those mutations might arise). Daniel Weinrich and colleagues studied five point mutations in *E. coli* that individually contribute, in varying degrees, to antibiotic resistance (specifically to β-lactam antibiotics, such as penicillin; Weinrich et al. 2006). There are 32 combinations of these 5 mutations (where, for example, AB, BCD, and ABCDE are 3 combinations). The investigators experimentally determined the resistances of all 32 combinations. The optimal combination was all 5 mutations together (ABCDE), which increased resistance (to cefotaxime) 100,000-fold.

Then they considered all 120 ordered sequences by which all 5 mutations could be acquired (where, for example, A to AB to ABC to ABCD to ABCDE is one such sequence, and B to BD to BCD to ABCDC to ABCDE is another). Many of these sequences, they calculated, would not be achievable by natural selection for increased resistance alone, because one or more steps in each sequence would involve decreases in resistance. They concluded that 18 of the 120 possible trajectories to optimal resistance were selectively traversable.

So, starting from identical lineages in a particular antibiotic environment, the sequence of mutations would have to be one of the appropriate 18, out of the possible 120, in order to guarantee parallel or convergent evolution to the same optimal outcome. Mutations arising in any other order would result in one of the many alternative, suboptimal combinations.

What to Make of It All

Evolutionary divergence is sometimes due to differences in the order of appearance of chance variations, *and not* to differences in the direction of selection. Variation does not have to be predictably directed (or biased or constrained) in order to influence evolutionary outcomes. This much we can confidently say. But what do studies such as those just described reveal about the *overall* significance of chance variation and the order of variation? Surely they demonstrate that the importance of chance variation should not be subordinated completely to the importance of natural selection. But while studies such as these count against the *unimportance* (or the "triviality") of chance variation, I am not sure how they bear on its overall importance.

First, there is the issue of whether the studies in question are biased toward the production (and subsequent detection) of order-of-mutation effects. I will not address this issue. Even if these studies are not particularly biased, it would still not be clear how one would extrapolate from the various positive instances generated by them, to the importance of chance variation and order of mutation in general.

Even if we knew something like the true proportion of cases in which identical selection pressures would lead from identical starting points to different endpoints (depending on order of mutation, over some specified number of generations, etc.), there would still be the question of whether the proportion is "significant" or "important." This can be seen by considering the latter two studies, which I chose in part because Lewontin earlier employed them as textbook demonstrations that "The order of occurrence of mutations is of critical importance in determining whether evolution by natural selection will or will not actually reach the most advantageous state [or will instead reach one of possibly many alternative, suboptimal states]" (Griffiths et al. 2004). How interesting, though, that Wichman and her coauthors titled their paper "Different Trajectories of Parallel Evolution During Viral Adaptation," as if to emphasize the parallel substitutions that occurred, rather than the equally frequent divergent substitutions. And how interesting that Weinrich and his coauthors concluded their paper, "It now appears that intramolecular interactions render many mutational trajectories selectively inaccessible, which implies that replaying the protein tape of life might be surprisingly repetitive" (Weinrich et al. 2006: 113). One might have thought they had demonstrated something closer to Gould's thesis, that life's tape is surprisingly *not* repetitive.

A rather different challenge to making sense of the overall importance of chance variation and the order of variation has to do with where, along the continuum from genotype through development to phenotype, one chooses to focus. Take the case of the twisted orchid. Darwin was struck by the different ways in which *Malaxis* and *Catasetum* evolved an uppermost labellum petal: in the former case through the selection of more and more twisted flower stems, and in the latter case through the selection of less and less twisted forms. But someone else might be struck by the fact that selection accomplished roughly the same result in both cases, namely, an uppermost labellum. Similarly, some might be struck by the genotypic differences between the 10,000- generation-old *E. coli* lines in the Lenski group studies, while others might be struck by the overall similarity in fitness gained (figure 2.2). And so on.

There is another, altogether different, sort of issue that is involved in assessing the relative roles of chance variation and natural selection. This one has to do with the difficulty of disentangling causal processes. Darwinians often trivialize the importance of variation by saying things like "though variability is indispensably necessary, yet . . . ," as Darwin did in the extended passage comparing natural selection to an architect (quoted above). Or, as Dobzhansky put it, "The statement that [some instance of evolution by natural selection] required a supply of 'chance' mutations is true but trivial. What is far more interesting is . . ." natural selection, of course (Dobzhansky 1974: 323). How can variation be "necessary" and "required," and at the same time "trivial?" Consider the most charitable case for the Darwinian, where variation is truly "omnifarous," as Romanes would have had it, and there is strong directional selection. To be sure, there is an important sense in which the particular direction taken must be due to selection and not to the availability of variation, which is in this case omnidirectional. But on the other hand, the direction-giving power of selection in this case *depends* on the ubiquity of variation available. Well, perhaps I am just repeating the question, or repeating my perplexity, in different terms. But now consider a charitable case for the importance of chance variation: variation is scarce and two initially identical lineages diverge considerably due to differences in the order of mutations, followed by cascading epistatic consequences. To be sure, there is an important sense in which the difference in outcomes cannot be due to selection in identical environments. And yet the difference in outcomes *depends* on different mutations having been accumulated by natural selection in a different order. Is selection just as "trivial" in this case as variation was in the previous case?

It will not be easy to disentangle the overall importance of chance variation from that of natural selection. But I believe Gould was right about "Darwinism" or, rather, how "Darwinism" played out in the Modern Synthesis, and has continued to play out, up to the recent past. Darwinians have been concerned not only to articulate and demonstrate the prevalence, the predominance, the ultimate control of natural selection. They have also been concerned to demonstrate that natural selection prevails and predominates over, and controls, chance variation. Natural selection is "anti-chance," as Dobzhansky put it. It is a "curb" on the "turmoil of mutation" (Dobzhansky 1974: 317, 331).

Chance variation was initially subordinated to natural selection in order to champion the importance of the latter relative to *directed* variation. I would even say (I argued above) that chance variation was initially *conceived* for the purpose of being thus subordinated. The trivialization of chance variation subsequently took on a life of its own even after the bogey of directed variation seemed well buried. But perhaps it is time to reassess the importance of chance variation. This issue will not be settled by counting cases and extrapolating. It is also a matter of conceptualizing or reconceptualizing. At the very least, it involves acknowledging that, in the case of evolution by natural selection of chance variation, the direction of evolutionary change may not be due to selection alone, but also to the order in which the variation arises. From identical starting points, in identical environments, selection may lead to different outcomes, depending on the order in which new mutations appear. The unitary direction of selection cannot in such cases be responsible for the divergent evolutionary outcomes. This shows that variation need not be predictably directed in order to impact the direction of evolutionary change. Variation is thus not only more important than the architects of the Modern Synthesis envisioned, but also more important than proponents of directed variation have realized.

Acknowledgments

I am especially grateful to Cosima Herter, Kostas Kampourakis, Dick Lewontin, Gerd Müller, Sally Otto, Massimo Pigliucci, Arlin Stoltzfus, Davide Vecchi, and Rasmus Winther for their insights and considerable patience in discussing these issues with me.

References

Beatty J (2009) Chance variation and evolutionary contingency: Darwin, Simpson, *The Simpsons*, and Gould. In Oxford Handbook of the Philosophy of Biology. M Ruse, ed.: 189–210. Oxford: Oxford University Press.

Blount ZD, Borland CZ, Lenski RE (2008) Historical contingency and the evolution of a key innovation in an experimental population of *Escherichia coli*. Proceedings of the National Academy of Sciences of the USA 105: 7899–7906.

Bull JJ, Molineux IJ (1992) Molecular genetics of adaptation in an experimental model of cooperation. Evolution 46: 882–895.

Coyne JA, Barton NH, Turelli M (1997) Perspective: A critique of Sewall Wright's shifting balance theory of evolution. Evolution 51: 643–671.

Darwin C (1859) On the Origin of Species by Means of Natural Selection, or the Preservation of Favoured Races in the Struggle for Life. 1st ed. London: John Murray.

Darwin C (1862) On the Various Contrivances by Which British and Foreign Orchids Are Fertilised by Insects. London: John Murray.

Darwin C (1868) The Variation of Animals and Plants Under Domestication. 2 vols. London: John Murray.

Darwin C (1859) On the Origin of Species by Means of Natural Selection, or the Preservation of Favoured Races in the Struggle for Life. 6th ed. London: John Murray.

Darwin C (1877) The Various Contrivances by Which Orchids Are Fertilised by Insects. 2nd ed. London: John Murray.

Darwin C (1993) The Correspondence of Charles Darwin, 1860. F Burckhardt et al., eds.: vol. 8. Cambridge: Cambridge University Press.

Darwin C (1999) The Correspondence of Charles Darwin, 1863. F Burckhardt et al., eds.: vol. 11. Cambridge: Cambridge University Press.

Darwin F, Seward AC, eds. (1903) More Letters of Charles Darwin. Vol. 2. London: John Murray.

Dawkins R (1997) Climbing Mount Improbable. New York: Norton.

Dobzhansky T (1951) Genetics and the Origin of Species. 3rd ed. New York: Columbia University Press.

Dobzhansky T (1970) Genetics of the Evolutionary Process. New York: Columbia University Press.

Dobzhansky T (1974) Chance and creativity in evolution. In Studies in the Philosophy of Biology: Reduction and Related Problems. FJ Ayala, T Dobzhansky, eds.: 307–338. Berkeley: University of California Press.

Fisher RA ([1930] 1958) The Genetical Theory of Natural Selection. 2nd ed. New York: Dover.

Gayon J (1998) Darwinism's Struggle for Survival: Heredity and the Hypothesis of Natural Selection. Cambridge: Cambridge University Press.

Gould SJ (1977) Ever Since Darwin. New York: Norton.

Gould SJ (1989) Wonderful Life: The Burgess Shale and the Nature of History. New York: Norton.

Gould SJ (2002) The Structure of Evolutionary Theory. Cambridge, MA: Harvard University Press.

Gould SJ, Lewontin RC (1979) The spandrels of San Marco and the panglossian paradigm: A critique of the adaptationist programme. Proceedings of the Royal Society of London B205: 581–598.

Gray A (1883) Letter to the editor. Nature 28: 78.

Gray A (1963 [1860]) Natural selection not inconsistent with natural theology. In Darwiniana: Essays and Reviews Pertaining to Darwinism. AH Dupree, ed.: 72–145. Cambridge, MA: Harvard University Press.

Gray A (1963 [1860]) [Review of] The Origin of Species by Means of Natural Selection. In Darwiniana: Essays and Reviews Pertaining to Darwinism. AH Dupree, ed.: 7–50. Cambridge, MA: Harvard University Press.

Griffiths AJF, Wessler SR, Lewontin RC, Gelbart WM (2004) An Introduction to Genetic Analysis. 8th ed. San Francisco: W.H. Freeman.

Hubby JL, Lewontin RC (1966) A molecular approach to the study of genic heterozygosity in natural populations. I. The number of different alleles at different loci in *Drosophila pseudoobscura*. Genetics 54: 577–594.

Jenkin F (1867) The Origin of Species. North British Review 46: 277–318.

Lennox J (1993) Darwin *was* a teleologist. Biology and Philosophy 8: 409–422.

Lenski RE, Rose MR, Simpson SC, Tadler SC (1991) Long-term experimental evolution in *Escherichia coli*. I. Adaptation and divergence during 2,000 generations. American Naturalist 138: 1315–1341.

Lenski RE, Travisano M (1994) Dynamics of adaptation and diversification: A 10,000-generation experiment with bacterial populations. Proceedings of the National Academy of Sciences of the USA 91: 6808–6814.

Lewontin RC (1967) The principle of historicity in evolution. In Mathematical Challenges to the Neo-Darwinian Interpretation of Evolution. PC Moorhead, MM Kaplan, eds.: 81–94. New York: Alan R. Liss.

Lewontin RC (1974) The Genetic Basis of Evolutionary Change. New York: Columbia University Press.

Lewontin RC, Hubby JL (1966) A molecular approach to the study of genic heterozygosity in natural populations. II. Amount of variation and degree of heterozygosity in natural populations of *Drosophila pseudoobscura*. Genetics 54: 595–609.

Provine WB (1986) Sewall Wright and Evolutionary Biology. Chicago: University of Chicago Press.

Romanes GJ (1883) Letter to the editor. Nature 27: 528–529.

Stoltzfus A (2006) Mutationism and the dual causation of evolutionary change. Evolution and Development 8: 304–317.

Travisano M, Mongold JA, Bennett AF, Lenski RE (1995) Experimental tests of the roles of adaptation, chance, and history in evolution. Science 267: 87–90.

Travisano M, Vasi F, Lenski RE (1995) Long-term experimental evolution in *Escherichia coli*. III. Variation among the replicate populations in correlated responses to novel environments. Evolution 49: 189–200.

Wade MJ, Goodnight CJ (1998) Perspective: The theories of Fisher and Wright in the context of metapopulations: When nature does many small experiments. Evolution 52: 1537–1553.

Weinrich DM, Delaney NF, DePristo MA, Hartl DL (2006) Darwinian evolution can follow only very few mutational paths to fitter proteins. Science 312: 111–114.

Wichman HA, Badgett MR, Scott LA, Boulianne CM, Bull JJ (1999) Different trajectories of parallel evolution during viral adaptation. Science 285: 422–424.

Wilson LG, ed. (1970) Sir Charles Lyell's Scientific Journals on the Species Question. Vol. 5. New Haven, CT: Yale University Press.

Wright S (1932) The roles of mutation, inbreeding, crossbreeding and selection in evolution. In Proceedings of the Sixth International Congress of Genetics 1: 356–366.

3 High-Dimensional Fitness Landscapes and Speciation

Sergey Gavrilets

The Modern Evolutionary Synthesis of the 1930s and 1940s remains the paradigm of evolutionary biology (Futuyma 1998; Gould 2002; Pigliucci 2007; Ridley 1993). The progress in understanding the process of evolution made during that period had been a direct result of the development of theoretical population genetics by Fisher, Wright, and Haldane, who built a series of mathematical models, approaches, and techniques showing how natural selection, mutation, drift, migration, and other evolutionary factors are expected to shape the genetic and phenotypic characteristics of biological populations. These theoretical advances provided "a great impetus to experimental work on the genetics of populations" (Sheppard 1954) and a "guiding light for rigorous quantitative experimentation and observation" (Dobzhansky 1955), and had many other far-reaching implications.

According to Provine (1978), the work of Fisher, Wright, and Haldane had significant influence on evolutionary thinking in at least four ways. First, their models showed that the processes of selection, mutation, drift, and migration were largely sufficient to account for microevolution. Second, they showed that some directions explored by biologists were not fruitful. Third, the models complemented and lent greater significance to particular results of field and laboratory research. Fourth, they stimulated and provided framework for later empirical research. Since the time of the Modern Synthesis, evolutionary biology has arguably remained one of the most mathematized branches of the life sciences, in which mathematical models and methods continuously guide empirical research, provide tools for testing hypotheses, explain complex interactions between multiple evolutionary factors, train biological intuition, identify crucial parameters and factors, evaluate relevant temporal and spatial scales, and point to the gaps in biological knowledge, as well as provide simple and intuitive tools and metaphors for thinking about complex phenomena.

In this chapter, I discuss two particular areas of theoretical evolutionary biology that have experienced significant progress since the late 1980s: a theory of fitness landscapes and a theory of speciation. I also outline two particular directions for theoretical studies on the origins of biodiversity which are especially important, in my opinion, for unification of different branches of the life sciences. One is the development of a theory of large-scale evolutionary diversification and adaptive radiation. The other is a quantitative theory of the origins of our own species.

Classical Fitness Landscapes

The theoretical notion of fitness landscapes (also known as "adaptive landscapes," "adaptive topographies," and "surfaces of selective value"), which emerged at the onset of the Modern Synthesis, has become a standard tool both for formal mathematical modeling and for the intuitive metaphorical visualizing of biological evolution, adaptation, and speciation. This notion was first introduced by Sewall Wright in a classic paper delivered at the 1932 International Congress of Genetics. Wright wanted to illustrate his ideas and mathematical results on the interaction of selection, random drift, mutation, and migration during adaptation in a nontechnical way accessible to biologists lacking quantitative skills (Wright 1932, 1988). Wright's metaphor of fitness landscapes is widely viewed as one of his most important contributions to evolutionary biology (Coyne et al. 1997; Pigliucci and Kaplan 2006; Provine 1986). Over the ensuing 70 years, the notion of fitness landscapes has been substantially expanded and has found numerous applications well outside of evolutionary biology (e.g., in computer science, engineering, economics, and biochemistry).

A key idea of evolutionary biology is that individuals in a population differ in fitness (due to the differences in genes and/or environments experienced). Differences in fitness that have genetic bases are the most important ones because it is the changes in genes that make innovations and adaptation permanent. The relationship between genes and fitness (direct or mediated via phenotype) is obviously of fundamental importance. In the most common modern interpretation, a fitness landscape specifies a particular fitness component (e.g., viability, that is, the probability to survive to the age of reproduction) as a function defined on a particular set of genotypes or phenotypes.

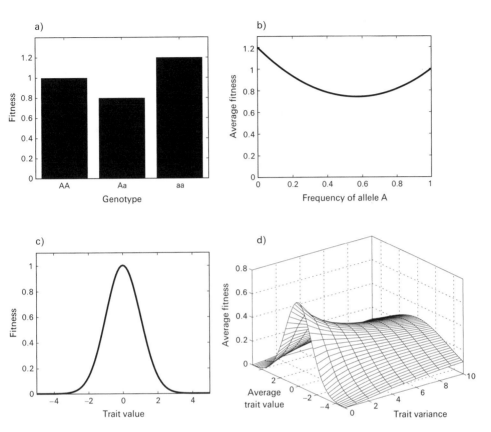

Figure 3.1
Examples of fitness landscapes. (a) Fitness landscape in a one-locus, two-allele model. (b) The average fitness landscape in a one-locus, two-allele model. (c) Fitness landscape for a quantitative trait model. (d) The average fitness landscape for a quantitative trait model.

For example, consider a very large, randomly mating diploid population under constant viability selection. Let us focus on a particular locus with two alleles, A and a, controlling fitness (viability). Then there are three different genotypes: two homozygotes, AA and aa, and a heterozygote, Aa. An example of a fitness landscape for this simple model is given in figure 3.1a. The fitness landscape illustrated in this figure corresponds to disruptive selection, that is, selection acting against intermediate genotypes (here, heterozygotes Aa). One may imagine an individual as a point on a fitness landscape and a population as a cloud of points which changes both its structure and its position as a result of action of different evolutionary factors (e.g, natural selection and sexual selection, mutation, recombination, drift, migration). The peaks and valleys of the landscape represent high-fitness and low-fitness combinations of genes (or phenotypic values), respectively; natural selection is imagined as a force pushing the population uphill, and adaptive evolution is visualized as hill-climbing.

In his original 1932 paper, Wright introduced two versions of fitness landscapes. The first corresponds to a relationship between a set of genes and individual fitness as illustrated in figure 3.1a. The second describes a relationship between variables characterizing the population's genetic state (e.g., allele frequencies) and the average fitness of the population. The fitness landscape for the average fitness can be derived from a fitness landscape for individual fitness in a straightforward way. For example, figure 3.1b illustrates the fitness landscape for the average fitness corresponding to the fitness landscape for individual fitness shown in figure 3.1a. In fitness landscapes for the average fitness, it is the population (rather than an individual) that is imagined as a point climbing the slope toward a nearby peak. The attractive feature of this interpretation of fitness landscapes is the fact that in some simple models, the change in allele frequencies induced by selection is directly proportional to the gradient of the average fitness (Wright 1931). In this case one can intuit the general features of the evolutionary change just from the shape of the corresponding fitness landscape, without the need to solve the underlying dynamic equations.

The generalization of the notion of fitness landscapes for the case of continuously varying traits (such as size, weight, or a concentration of a particular gene product) was introduced by Simpson (1953). An example of a fitness landscape for a single quantitative character is shown in figure 3.1c. The fitness landscape illustrated in this figure describes stabilizing selection, that is, selection favoring an intermediate optimum (here, at trait value 0). Figure 3.1d illustrates the average fitness landscape corresponding to the individual fitness landscape shown in figure 3.1c. In figure 3.1d the independent variables are the average and variance of the trait values in the population which jointly control the average fitness. Lande (1976, 1980) showed that the change in the average trait value induced by selection is proportional to the gradient of the average fitness. The theoretical work of Lande (1976, 1980), Barton (e.g., 1989a; Barton and Rouhani 1987; Barton and Turelli 1987), and others in the 1970s and 1980s made such landscapes an indispensable part of the theoretical toolbox of evolutionary biology.

In the original formulation and in most of the latter work, the fitness component under consideration was viability and, as such, it was a property of an individual. Later the notion of fitness landscapes was generalized to other fitness components, such as fertility (i.e., the number of offspring) or mating success (i.e., the probability of mating), which can be a property of a mating pair rather than of an individual (Gavrilets 2004). In most interpretations, fitness landscapes are static, that is, they do not change over time. However, models also exist in which landscapes change over time as a result of changes in the external environment. Frequency-dependent selection (e.g., Asmussen and Basnayake 1990; Cockerham et al. 1972; Dieckmann et al. 2004; Waxman and Gavrilets 2005), under which fitness continuously changes as the population evolves, can also be interpreted in terms of landscapes (or seascapes). Overall, fitness landscapes are an inherent and most crucial feature of all mathematical models dealing with natural or sexual selection. (Note that in many modeling papers, a technical term for specifying the relationship between genotype [or phenotype] and fitness is "fitness function" rather than "fitness landscape.")

In general, biological organisms have thousands of genes and/or gene products that can potentially affect fitness. This means that fitness landscapes are inherently multidimensional, as was already well realized by Wright himself. Unfortunately, the relationships between genotype (or phenotype) and fitness for real biological organisms are still poorly understood. Therefore, the dominant strategy for using fitness landscapes in theoretical evolutionary biology has been to make simplifying assumptions about their structure in an attempt to get a tractable model which, it is hoped, will capture some essential properties of the process under consideration. But are there any generic features of multidimensional fitness landscapes? Although we still miss precise and broad empirical evidence, some general features of fitness landscapes can be identified from available data, biological intuition, and mathematical reasoning.

To Sewall Wright, who used three-dimensional geographic landscapes as a metaphor for multidimensional relationships between genotype and fitness, the most prominent feature was the existence of many peaks of different height separated by many valleys of different depth. Different peaks can be viewed as alternative solutions to the problem of survival, which all biological organisms face. In Darwin's words, "the multifarious means for gaining the same end" (see Beatty 2008). Wright reasoned that nonlinear interactions of the effects of different loci and alleles on fitness coming from pleiotropy and epistasis will make the existence of multiple fitness peaks unavoidable. This picture of fitness landscapes (illustrated

(a)

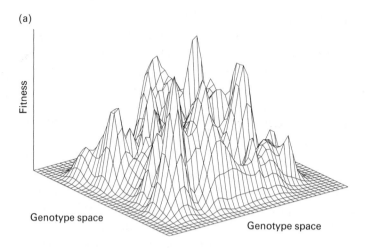

(b)

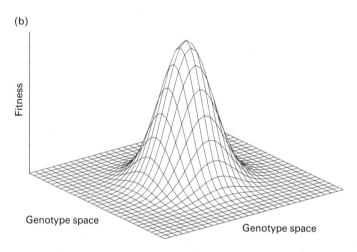

Figure 3.2
Two types of classical fitness landscapes. (a) A rugged landscape. (b) A single-peak landscape.

in figure 3.2a) is now known as that of rugged fitness landscapes (Kauffman 1993). Fitness peaks are important because of the expectation that natural selection will drive populations toward them. However, as soon as the population reaches a neighborhood of a local peak, any movement away from the peak will be prevented by selection.

It is important to realize that the peak reached by the population does not necessarily have the highest fitness. On the contrary, it is much more plausible that this peak has an intermediate height and that (much) higher fitness peaks exist nearby (Kauffman 1993; Kauffman and Levin 1987). Without some additional forces, a population evolving on a rugged landscape will stop changing after a relatively short, transient time. Selection is a force pushing the population uphill and thus preventing it from going downhill. Therefore, within the framework of rugged fitness landscapes, the problem of crossing fitness valleys, which is necessary for moving toward other peaks (and that would result in increased adaptation and/or evolutionary divergence), becomes of major importance.

There are two possible solutions to this problem. First, additional factors opposing selection and overcoming it, at least occasionally, can drive the population across a fitness valley. The factor that has received most attention in this regard is random genetic drift, which is particularly important in small populations (Kimura 1983; Lynch 2007). Second, temporal changes in the fitness landscape itself can result in temporary disappearance of fitness valleys. Sewall Wright's own solution was his shifting balance theory (Wright 1931, 1982), which relies on complex interactions of multiple evolutionary factors (selection, mutation, migration, and random drift).

The shifting balance theory focuses on a population spatially subdivided into a large number of small subpopulations (demes) exchanging migrants. Because demes are small and there are many of them, it is likely that one of them will make a transition by genetic drift across the fitness valley to an alternative (perhaps higher) peak. Wright separated the process of peak shift (i.e., evolution from one peak to another) in a deme into two steps: stochastic transition by random genetic drift from a neighborhood of an old fitness peak into the domain of attraction of a new peak (Wright's phase I), and deterministic movement toward the new peak once the deme is within its domain of attraction (Wright's phase II). Wright reasoned that once a new adaptive combination of genes realizing the new higher peak is established in a deme, the deme will have a higher population density. Then, as a result of higher emigration from such demes, the higher fitness peak will take over the whole system (Wright's phase III). Wright's argument was mainly verbal. However, the conclusions of later formal analyses did not support Wright's intuition. Recent formal modeling has shown that although the mechanisms underlying Wright's theory can, in principle, work, the conditions are rather strict (Coyne et al. 1997, 2000; Gavrilets 1996). Therefore, the mechanisms implied in the shifting balance theory can hardly provide a general route for adaptation and diversification (for a

the size of the expanding population is still small. An inherent feature of the shifting balance that severely constrains this process is the necessity to spread the new adaptive combination of genes from a local deme to the rest of the population. During this stage (i.e., phase III) new combinations of genes have to compete with the old ones, which outnumber them. Founder effect speciation avoids this difficulty by simply removing the need for the new combination of genes to compete with the old one: a local subpopulation grows to become a new species without interacting with the ancestral one. For several decades, buoyed by Mayr's authority, founder effect speciation (in various forms) was the dominant explanation of at least island speciation (Provine 1989).

The proponents of these theories offered only verbal schemes without trying to formalize them. Formal analyses of founder effect speciation using analytical models and numerical simulation were undertaken only in the 1980s and later (Barton 1989b; Barton and Charlesworth 1984; Charlesworth and Rouhani 1988; Charlesworth and Smith 1982; Gavrilets 2004; Lande 1980; Rouhani and Barton 1987). Contrary to prevailing wisdom at that time, the general conclusion of these analyses was that a founder event cannot result in a sufficiently high degree of reproductive isolation with a high enough probability to be a reasonable explanation for speciation. Convincing empirical evolutionary biologists that Mayr's theory cannot work was a very important contribution of theoreticians to our understanding of speciation.

As the preceding discussion shows, a number of beliefs and ideas held by the architects of the Modern Synthesis were later proven wrong or of limited biological significance and importance. Mathematical modeling played an important role in this continuous process of refining, extending, generalizing, and pruning evolutionary thought.

Properties of Multidimensional Landscapes Not Captured by Classical Theories

Both Wright and Fisher, along with other researchers utilizing the notion of fitness landscapes in their work, well realized that the dimensionality of biologically relevant fitness landscapes is extremely high (in thousands and millions). Still, they believed that the properties of three-dimensional "geographic" landscapes well captured those of multidimensional landscapes. However, the theoretical work of the past two decades (discussed below) has led to understanding that these expectations were not quite justified.

The extremely large dimensionality of fitness landscapes implies that the number of possible genotypes is astronomically high. (For example, with 1,000 genes, each of which can have only two alleles, the number of possible sequences is $2^{1000} \approx 10^{100}$.) Therefore, one should not expect them all to have different fitnesses—there should be a lot of redundancy in the genotype-to-fitness relationship, so that different genotypes must have similar fitnesses. The question is how these genotypes with similar fitnesses are distributed in the genotype space and whether high fitness genotypes may form connected networks expanding through the genotype space. As far as I am aware, it was Maynard Smith (1962, 1970) who was the first to suggest such a possibility. He explained such networks by an analogy with a word game where the goal is to transform one word into another by changing one letter at a time, with the requirement that all intermediate words are meaningful (as in the sequence WORD-WORE-GORE-GONE-GENE). More recent work has shown that connected networks of genotypes with similar fitnesses represent a generic property of multidimensional landscapes.

To illustrate these ideas, let us consider a two-dimensional lattice of square sites in which sites are independently painted black or white with probabilities P and $1 - P$, respectively (see figure 3.3). We will interpret black sites as viable genotypes and white sites as inviable genotypes. For each site, let its one-step neighbors be the four adjacent sites (directly above, below, on the left, and on the right). Let us say that two black sites are connected if there exists a sequence of black sites starting at one of them and going to another, such that subsequent sites in the sequence are neighbors. For any black site, let us define a *connected component* as the set of all black sites connected to the site under consideration. A simple numerical experiment shows that the number and the structure of connected components depend on the probability P. For small values of P there are many connected components of small size (see figure 3.3a). As P increases, the size of the largest connected component increases (see figure 3.3b). As P exceeds a certain threshold P_c known as the *percolation threshold*, the largest connected component (known as the *giant component*) emerges, which extends (percolates) through the whole system and includes a significant proportion of all black sites (see figure 3.3c). In this model, describing a so-called site percolation on an infinite two-dimensional lattice, the percolation threshold is $P_c \approx 0.593$ (e.g., Grimmett 1989).

Consider next a different model. Assume that there is a very large number L of diallelic loci. Now each genotype has L one-step neighbors

p = 0.20

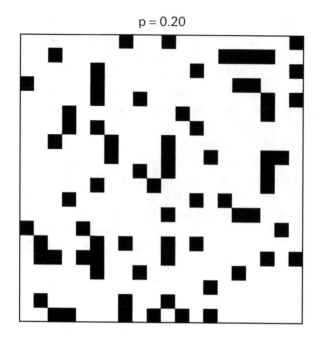

p = 0.40

Figure 3.3
Percolation in two dimensions for three different values of P.

p = 0.60

Figure 3.3
(Continued)

(single mutants). Let us assign fitnesses in exactly the same way as in the previous paragraph, that is, fitnesses are generated randomly and independently and are equal only to 1 (viable genotype) or 0 (inviable genotype), with probabilities P and $1 - P$, respectively. Similarly to the previous model (figure 3.3), viable genotypes will tend to form neutral networks. In this model, for small values of P, there are two qualitatively different regimes: subcritical, in which all connected components are relatively small (which takes place when $P < P_c$, where P_c is the percolation threshold), and supercritical, in which the majority of viable genotypes are connected in a single giant component, which takes place when $P > P_c$ (Gavrilets and Gravner 1997). A very important, though counterintuitive, feature of this model is that the percolation threshold is approximately the reciprocal of the dimensionality of the genotype space: $P_c \approx 1/L$, and thus P_c is very small if L is large (see Gavrilets 2004; Gavrilets and Gravner 1997). Therefore, increasing the dimensionality of the genotype space L, while keeping the probability of being viable P-constant, makes the formation of the giant component unavoidable.

The assumption that fitness can take only two values, 0 and 1, might be viewed as a serious limitation. To show that this is not so, let us consider the same genotype space as in the previous section (i.e., the set of L diallelic loci), but now assume that fitness, w, is a realization of a random variable having a uniform distribution between 0 and 1 (Gavrilets and Gravner 1997). Let us introduce threshold values w_1 and w_2, which differ by a small value, ε. Let us say that a genotype belongs to the (w_1,w_2)-fitness band if its fitness w satisfies the conditions $w_1 < w \le w_2$. Parameter ε can be viewed as the probability that a randomly chosen genotype belongs to the (w_1,w_2)-fitness band. One should be able to see that being a member of the (w_1,w_2)-fitness band is analogous to being viable in the previous model, with parameter ε playing the role of P in the previous model. Therefore, if the dimensionality of genotype space L is very large and $\varepsilon > 1/L$, there exists a giant component (i.e., a percolating nearly neutral network) of genotypes in the (w_1,w_2)-fitness band. Its members can be connected by a chain of single-gene substitutions resulting in genotypes that also belong to the network. If ε is small, the fitnesses of the genotypes in the (w_1,w_2)-fitness band will be very similar. Thus, with large L, extensive evolutionary changes can occur in a nearly neutral fashion via single substitutions along the corresponding nearly neutral network of genotypes belonging to a percolating cluster. Note that if one chooses $w_2 = 1$ and $w_1 = 1 - \varepsilon$, it follows that fitness landscapes have very high ridges (with genotype fitnesses between $1 - \varepsilon$ and 1) that continuously extend throughout the genotype space. In a similar way, if one chooses $w_2 = \varepsilon$ and $w_1 = 0$, it follows that the landscapes have very deep gorges (with genotype fitnesses between 0 and ε) that also continuously extend throughout the genotype space. I stress that the above conclusions apply not only for the uniform distribution of fitness values but also for any random distribution of fitnesses, provided the overall frequency of genotypes that belong to a (w_1,w_2)-fitness band is larger than $1/L$.

The above discussion illustrates two general points about scientific metaphors which one should keep in mind. The first is that specific metaphors (as well as mathematical models) are good for specific purposes only. The second is that accepting a specific metaphor necessarily influences and defines the questions that are considered to be important. The metaphor of "rugged adaptive landscapes" is very useful for thinking about local adaptation. However, its utility for understanding large-scale genetic and phenotypic diversification and speciation is questionable. The metaphor of rugged adaptive landscapes, with its emphasis on adaptive peaks and valleys, is to a large degree a reflection of the three-dimensional world we live in. However, genotypes and phenotypes of biological organisms differ in numerous characteristics, and thus the dimensionality of biologically realistic fitness landscapes is much larger than 3. Properties of multidimensional fitness landscapes are very different from those of low dimension. Consequently, it may be misleading to use three-dimensional analogies implicit in the metaphor of rugged adaptive landscapes in a multidimensional context. In particular, the problem of crossing fitness valleys may be nonexistent.

The networks of genotypes with similar fitnesses expanding throughout the genotype space can be graphically illustrated using a metaphor of holey fitness landscapes (Gavrilets 1997a, 2004; Gavrilets and Gravner 1997). A holey fitness landscape is a fitness landscape where relatively infrequent well-fit (or, as Wright put it, "harmonious") genotypes form a contiguous set that expands ("percolates") throughout the genotype space. An appropriate three-dimensional image of such a fitness landscape is an approximately flat surface with many holes representing genotypes that do not belong to the percolating network (see figure 3.4). Within the metaphor of holey landscapes, local adaptation and micro-evolution can be viewed as climbing from a hole toward a nearly neutral network of genotypes with fitnesses at a level determined by mutation-

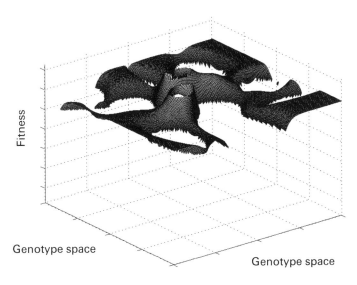

Figure 3.4
A holey fitness landscape.

selection-random drift balance. The process of climbing occurs on a shorter time scale than that necessary for speciation, clade diversification, and macroevolution. Once a corresponding fitness level is reached, the population will be prevented by selection from slipping off of this level to lower fitnesses, and by mutation, recombination, and gene flow from climbing to higher fitnesses. Speciation occurs when a population evolves to a genetic state separated from its initial state by a hole.

The earlier work on neutral and nearly neutral networks in multi-dimensional fitness landscapes concentrated exclusively on genotype spaces in which each individual is characterized by a discrete set of genes. However, many features of biological organisms that are actually observable and/or measurable are described by continuously varying variables such as size, weight, color, or concentration. A question of particular biological interest is whether (nearly) neutral networks are as prominent in a continuous phenotype space as they are in the discrete genotype space. Recent results provide an affirmative answer to this question. Specifically, Gravner et al. (2007) have shown that in a simple model of random fitness assignment, viable phenotypes are likely to form a large connected cluster even if their overall frequency is very low, provided the dimensionality of the phenotype space L (i.e., the number of phenotypic characters) is sufficiently large. In fact, the percolation threshold P_c for the probability of being viable scales with L as $1/2^L$ and thus decreases much faster than $1/L$, which is characteristic of the analogous discrete genotype space model.

Earlier work on nearly neutral networks was also limited to consideration of the direct relationship between genotype and fitness. Any phenotypic properties that usually mediate this relationship in real biological organisms were neglected. Gravner et al. (2007) studied a novel model in which phenotype is introduced explicitly. In their model, the relationships both between genotype and phenotype, and between phenotype and fitness, are of the many-to-one type, so that neutrality is present at both the phenotype and the fitness levels. Moreover, their model results in a correlated fitness landscape in which similar genotypes are more likely to have similar fitnesses. Gravner et al. (2007) showed that phenotypic neutrality and correlation between fitnesses can reduce the percolation threshold, making the formation of percolating networks easier.

Overall, the results of Gravner et al. reinforce the previous conclusion (Gavrilets 1997b, 2004; Gavrilets and Gravner 1997; Reidys et al. 1997; Reidys and Stadler 2001, 2002) that percolating networks of genotypes with approximately similar fitnesses (holey landscapes) is a general feature of multidimensional fitness landscapes (both uncorrelated and correlated, and in both genotype and phenotype spaces). To date, most empirical information on fitness landscapes in biological applications has come from studies of RNA (e.g., Fontana and Schuster 1998; Huynen et al. 1996; Schuster 1995), proteins (e.g., Lipman and Wilbur 1991; Martinez et al. 1996; Rost 1997), viruses (e.g., Burch and Chao 1999, 2004), bacteria (e.g., Elena and Lenski 2003; Woods et al. 2006), and artificial life (e.g., Lenski et al. 1999; Wilke et al. 2001). Although limited, these data provide support for the biological relevance of holey fitness landscapes.

The realization that biologically realistic fitness landscapes have properties fundamentally different from those implied during the Modern Synthesis represents a significant theoretical advance that took place relatively recently. The biological implications of this result concern a number of areas (Gavrilets 2004), including the dynamics of adaptation, maintenance of genetic variation, the role of genetic drift, genetic robustness, evolvability, the importance of chance and contingency in evolution, and speciation.

Modern Speciation Theory

As I have already mentioned, systematic attempts to lay foundations of a quantitative theory of speciation did not start until the 1960s and 1970s. The pioneering work of Balkau and Feldman (1973), Bazykin (1969), Crosby (1970), Dickinson and Antonovics (1973), and Maynard Smith (1966) laid foundations for future modeling efforts. Recent years have seen significant advances in speciation research (e.g., Coyne and Orr 2004; Dieckmann et al. 2004; Gavrilets 2003a, 2004; Howard and Berlocher 1998), and by now we have solid understanding of the factors promoting and restricting speciation, shaping its dynamics, as well as its characteristic time scales and patterns. As our understanding of the processes leading to the origin of new species increases, we appreciate more and more the importance of the insight of the founders of the Modern Synthesis that "speciation can occur in different ways" (Dobzhansky et al. 1977), and that "there are multiple answers to every aspect of speciation" (Mayr 1982).

Given a variety of speciation mechanisms, the question of their classification is of importance. Theoretical population genetics has identified a number of factors controlling evolutionary dynamics, such as mutation, random genetic drift, recombination, and natural and sexual selection. A straightforward approach for classifying different mechanisms and modes of speciation is according to the type and strength of the factors controlling or driving genetic divergence. In principle, any of the factors listed above can be used at any level of classification. However, traditionally the discussions of speciation in evolutionary biology are framed in terms of a classification in which the primary division is according to the level of migration between the diverging (sub)populations (Mayr 1942). In this classification the three basic (geographic) modes of speciation are allopatric, parapatric, and sympatric. The traditional stress on the spatial structure of (sub)populations as the primary factor of classification (rather than, say, on selection) reflects both the fact that it is most easily observed (relative to the difficulties in inferring the type and/or strength of selection acting in natural populations) and the growing realization that the spatial structure of populations is very important. Alternatively, it has been suggested to use a classification based on types of selection (Via 2001) or on a continuum of "geography/prezygotic isolating mechanisms" (Kirkpatrick and Ravigné 2002).

Sometimes very different biological mechanisms can be described by very similar mathematical models. Therefore, classifying mechanisms of speciation on the basis of similarity of the corresponding models may be of some use. Three general, partially overlapping sets of models can be identified. In the first set, which I will call "spontaneous clusterization" models, an initially random mating population accumulates a substantial amount of genetic variation by mutation, recombination, and random drift, and then splits into two or more partially or completely reproductively isolated clusters. Spontaneous clusterization models include those describing the accumulation of Dobzhansky-Muller genetic incompatibilities, speciation by hybridization, divergence in mating preferences, or allochronic speciation via divergence in the timing of mating (reviewed by Coyne and Orr 2004; Gavrilets 2004). Spontaneous clusterization can happen in any geographic context (i.e., allopatric, parapatric, or sympatric). This type of speciation can be imagined as population fragmentation on ridges in a holey fitness landscape with different clusters becoming reproductively isolated because they happen to be on opposite sides of a hole in the landscape. The fitness differences between genotypes which may be present are not of particular importance. This set of models is most advanced analytically.

In the second set of models, which can be called "adaptation with reproductive isolation as a by-product" models, the population is pulled genetically apart by some kind of selection (natural or sexual) for adaptation to a local abiotic or biotic environment. Reproductive isolation between diverging parts of the population emerges as a by-product of genetic or phenotypic divergence. This type of speciation can happen in any geographic context (i.e., allopatric, parapatric, or sympatric), but migration (gene flow) between the subpopulations actively opposes their genetic divergence and the evolution of reproductive isolation. Speciation again can be imagined as population fragmentation on ridges in a holey fitness landscape, but now fitness differences between genotypes along the ridge and at local fitness peaks are important. By now there is a variety of analytical models for this type of speciation.

The third set of models is similar to the second, but now there is an explicitly considered trait (or traits) that can evolve to directly decrease the probability of mating (and the level of gene flow) between diverging subpopulations. These are "reinforcement-type models" related to the classical idea of reinforcement (Dobzhansky 1940; Fisher 1930). These also are the most complex models, which are difficult to study analytically, and so far their analyses have been limited to numerical simulations. Imagining this kind of speciation in terms of fitness landscapes is very difficult (and not particularly useful), as there are several fitness components which are relevant simultaneously.

The most controversial scenario of speciation has traditionally been sympatric speciation. These controversies have attracted the attention of many theoreticians, and by now the great majority of theoretical work on speciation concerns speciation in the presence of gene flow between diverging populations driven by ecological selection (Gavrilets 2004; Kirkpatrick and Ravigné 2002). Most of this work is represented by numerical studies, but there now exist a number of simple analytical models of sympatric speciation (Gavrilets 2003b, 2004, 2006; Gavrilets and Waxman 2002). The theory of sympatric speciation is arguably the most developed part of theoretical speciation research. The general conditions for sympatric speciation as identified by recent theoretical research are (1) strong combined effects of disruptive selection and nonrandom mating, (2) strong association of the genes controlling traits subject to selection and those underlying nonrandom mating, (3) high levels of genetic variation, and (4) the absence of costs on being choosy in mate choice (Gavrilets 2004). Two most straightforward ways for sympatric speciation are provided by a "magic trait" mechanism and a habitat selection mechanism. The former describes situations in which there is a trait that is both subject to disruptive/divergent selection and simultaneously controls nonrandom mating (such as size in stickleback fish or color in *Heliconius* butterflies). The latter corresponds to situations in which organisms evolve stronger and stronger preferences for specific habitats where they form mating pairs and/or mate.

Mathematical models clearly show that, under certain biologically reasonable conditions, sympatric speciation is possible (Gavrilets 2004). However, in spite of the enormous interest in sympatric speciation and strong motivation to find examples, there are only a few cases (reviewed in Coyne and Orr 2004) where sympatric speciation is strongly implicated. One explanation for this discrepancy is that sympatric speciation is difficult to prove or it is difficult to rule out alternative scenarios. Another possibility is that conditions for sympatric speciation as identified by mathematical models are rarely satisfied in natural populations. Incorporating theoretical insights into empirical work and applying mathematical models to particular case studies (e.g., Gavrilets and Vose 2007; Gavrilets, Vose, et al. 2007) are crucial steps toward assessing the importance of sympatric speciation in nature.

Not surprisingly, there have been a number of theoretical developments that were not appreciated, predicted, or emphasized during the time of the Modern Synthesis. For example, from the theoretical point of view, the power of the phenomenon of spontaneous clusterization became apparent only recently. Although the recent theory of the reinforcement-type speciation provides some support for the verbal arguments made during the Modern Synthesis, it also identifies a number of limitations and weaknesses in these arguments (Servedio and Noor 2003). Mayr was very skeptical of the generality of ecological speciation and sympatric speciation, but recent work has shown that under certain conditions, both can be important. The potential role of sexual selection in speciation (Andersson 1994) is stressed by many modern theoretical studies, whereas it was almost completely neglected in the early discussions. Many models show that speciation can happen very rapidly after a long period of relative stability (stasis), while the earlier work emphasized continuity and small changes in evolution and speciation. The importance of conflicts (e.g., genomic or sexual) or coevolutionary interactions was not realized, while now models show that these factors can be a very powerful engine of speciation (Gavrilets and Waxman 2002). Overall, a diversity of new mechanisms for generating biodiversity are known now, but were unknown or underappreciated at the onset of the Modern Synthesis.

Two Focal Areas for Future Research on the Origins of Biodiversity

There are many exciting directions for empirical and theoretical research on the origins of biodiversity. Here I want to touch on two of them which are particularly important, in my opinion, for unification of different branches of life sciences. One is the development of a theory of large-scale evolutionary diversification. Ideally such a theory would link microevolutionary processes (e.g., selection, mutation, random drift, adaptation, coevolution, competition, etc.), studied by evolutionary biologists and ecologists, with macroevolutionary patterns (e.g., stasis, punctuation, dynamics of diversity and disparity, species selection), studied by paleontologists (Eldredge et al. 2005). The initial step in building such a theory would be a development of a theoretical framework for modeling adaptive radiation. The second question concerns the origins of our own species. Arguably, no area of evolutionary biology is more compelling to general audiences than those related to human origins; the topic underpins discussions of our place in the universe, of morality and cognition, and of our fate as a species. It is now recognized that many features of modern human behavior, psychology, and culture may be explainable to a certain extent in terms of selective factors that operated during the Pleistocene. Developing a modeling formalism for describing the action and effects of genetic, ecological, environmental, social, and cultural factors operating during the process of human origin would be a major breakthrough in (theoretical) evolutionary biology.

A Theory of Adaptive Radiation
Adaptive radiation is defined as the evolution of ecological and phenotypic diversity within a rapidly multiplying lineage (Schluter 2000; Simpson 1953). Classical examples include the diversification of Darwin's finches on the Galápagos islands, *Anolis* lizards on Caribbean islands, Hawaiian silverswords, and cichlids of the East African Great Lakes, among many others (Gillespie 2004; Givnish and Sytsma 1997; Losos 1998; Salzburger and Meyer 2004; Schluter 2000; Seehausen 2007; Simpson 1953). Adaptive radiation typically follows the colonization of a new environment or the establishment of a "key innovation" (e.g., nectar spurs in columbines, Hodges 1997) which opens new ecological niches and/or new paths for evolution.

Adaptive radiation is both a spectacular and a remarkably complex process, which is affected by many different factors (genetical, ecological, developmental, environmental, etc.) interweaving in nonlinear ways.

Different, sometimes contradictory, scenarios explaining adaptive radiation have been offered (Mayr 1963; Schluter 2000; Simpson 1953). Some authors emphasize random genetic drift in small founder populations (Mayr 1963), while others focus on strong directional selection in small founder populations (Eldredge 2003; Eldredge et al. 2005), strong diversifying selection (Schluter 2000), or relaxed selection (Mayr 1963). Which of these scenarios is more general is controversial. The large time scale involved and the lack of precise data on its initial and intermediate stages make identifying general patterns of adaptive radiation very difficult (Gillespie 2004; Losos 1998; Salzburger and Meyer 2004; Schluter 2000; Seehausen 2007; Simpson 1953). Further, it is generally unknown if the patterns identified in specific case studies apply to other systems.

The difficulties in empirical studies of general patterns of adaptive radiation, its time scales, driving forces, and consequences for the formation of biodiversity make theoretical approaches important. However, the phenomenon of adaptive radiation remains largely unexplored from a theoretical modeling perspective. Adaptive radiation can be viewed as an extension of the process of speciation (driven by ecological factors and subject to certain initial conditions) to larger temporal and spatial scales. As I already stated, a recent explosion in empirical speciation work (reviewed by Coyne and Orr 2004) was accompanied by the emergence of a quantitative theory of speciation (Gavrilets 2004). In contrast, there have been only few attempts to build genetically based models of large-scale evolutionary diversification.

Some recent work in my lab has begun to lay the foundations of a quantitative theory of adaptive radiation. Some of them are based on a model of adaptive radiation which is intended to be more abstract and general (Gavrilets and Vose 2005, 2009). Other attempts use models tailored for particular case studies such as cichlids in a crater lake (Barluenga et al. 2006; Gavrilets, Vose, et al. 2007), palms on an oceanic island (Gavrilets and Vose 2007; Savolainen et al. 2006), snails on seashores (Hollander et al. 2005, 2006; Sadedin et al. 2008), and butterflies in jungles (Duenez-Guzman et al. 2009; Mavárez et al. 2006). The general setup in all these models is similar. We typically start with a few individuals of a sexual diploid species colonizing a new environment (e.g., an island or a lake) in which a number of spatially structured empty ecological niches are available. Although the founders have low fitness, the abundant resources and the lack of competitors allow them to seed a population that is able to survive throughout the environment at low densities. The founders have no particular preference for the ecological niches available in the new environment. However, as selection acts on the new genetic variation supplied by mutation, different lineages can become adapted to and simultaneously develop genetic preferences for different ecological niches. The process of ecological and phenotypic diversification and speciation driven by selection for local adaptation is accompanied by the growth in the densities of emerging species. Eventually species utilizing different ecological niches evolve differences in mating preferences by a process analogous to reinforcement. In some models, rather than starting the simulations with a population of low-fitness individuals, we assumed that the initial population is represented by specialists perfectly adapted for one of the available niches. Our main interest is to develop a better understanding of the dynamics of invasion of empty ecological niches and subsequent diversification.

Although these efforts are still at very initial stages, some patterns do emerge across different models. The following summarizes these patterns:

• Traits controlling adaptation to ecological niches evolve faster, approach their optimum values closer, and maintain less genetic variation at (stochastic) equilibrium than traits controlling habitat preferences;

• Mating preference traits evolve at a slower pace than the ecological and habitat preference traits, maintain more genetic variation, and can fluctuate dramatically in time;

• Mating preferences can diverge both between and within species utilizing different ecological niches;

• Area effect: empty ecological niches get filled only on islands and in lakes of sufficiently large size;

• Effect of the number of loci: rapid and extensive diversification is most likely if the number of loci controlling ecological traits and habitat and mating preference traits is small;

• Timing of speciation: typically, there is a burst of speciation soon after colonization rather than a more or less continuous process of speciation;

• Overshooting effect: the diversity (i.e., the number of species) peaks early in the radiation;

• Hybridization and neutral gene flow: species can stably maintain their divergence in a large number of selected loci for very long periods of time in spite of substantial hybridization and gene flow that removes differentiation in neutral markers;

• Least action effect: speciation occurring after the initial burst usually involves a minimum phenotypic change;

• Differentiation in mating characters is often of a continuous nature without clearly defined, discrete morphs;

• Parallel diversification when new mating characters get shared across different ecological niches and/or when new ecological characters get shared across different "sexual morphs" is expected;

• While the characters controlling local adaptation and habitat preferences remain close to the optimum values, mating characters can change continuously in a neutral fashion;

• Given everything else the same, the typical stages of adaptive radiation are (1) divergence with respect to macrohabitat, (2) evolution of microhabitat choice and divergence with respect to microhabitat, (3) divergence with respect to "magic traits" (i.e., traits that simultaneously control the degree of local adaptation and nonrandom mating), and (4) divergence with respect to other traits controlling survival and reproduction.

Although some of these predictions are supported by empirical data (Gavrilets and Losos 2009), much more work (both theoretical and empirical) is necessary to really evaluate their biological and evolutionary significance. In evolutionary biology, comprehensive studies of a few model organisms have been very successful in identifying and understanding general evolutionary mechanisms and principles. In a similar way, comprehensive numerical studies of a few models of adaptive radiation will greatly benefit our understanding of large-scale diversification.

The Ultimate Speciation Event: The Origin of Our Own Species

Decades of intensive work by generations of evolutionary biologists have led to a dramatic increase in our understanding of how new species arise (Coyne and Orr 2004; Dieckmann et al. 2004; Gavrilets 2004; Howard and Berlocher 1998)—the central theme of Darwin's revolutionary book (Darwin 1859). I believe that the time is ripe for attacking the ultimate speciation event—the origin of our own species (Darwin 1871). Any general theory of the origin of humans will include a significant quantitative/mathematical component that will have to deal with a complex combination of ecological, genetic, cultural, and social factors, processes, and changes. Here, I want to illustrate one possible theoretical approach for modeling factors that likely were important during the earlier evolution of humans.

There are many features that make humans a "uniquely unique species," but the most crucial of them are related to the size and complexity of our brain (Geary 2004; Roth and Dicke 2005; Striedter 2005). Brain size in *Homo sapiens* increased in a runaway fashion over a period of a few hundred thousand years, but then stabilized or even slightly declined in the last 35,000–50,000 years (Geary 2004; Ruff et al. 1997; Striedter 2005). In humans, the brain is very expensive metabolically: it represents about 2% of the body's weight but utilizes about 20% of total body metabolism at rest (Holloway 1996). The burning question is what factors drove the evolution of human brain size and intelligence? A number of potential answers focusing on the effects of climatic (Vrba 1995), ecological factors (Russon and Begun 2004), and social factors have been hotly debated. One widely discussed set of ideas (Alexander 1990; Byrne and Whiten 1988; Dunbar 1998, 2003; Flinn et al. 2005; Geary 2004; Humphrey 1976; Roth and Dicke 2005; Striedter 2004; Whiten and Byrne 1997), coming under the rubric of the "social brain" hypothesis (sometimes also called the "Machiavellian intelligence" hypothesis), considers selective forces coming from social competitive interactions as the most important factor in the evolution of hominids, who at some point in the past became an ecologically dominant species (Alexander 1990; Flinn et al. 2005). These forces selected for more and more effective strategies of achieving social success (including deception, manipulation, alliance formation, exploitation of the expertise of others) and for the ability to learn and use them. In this scenario, the social success is translated into reproductive success (Betzig 1986, 1993; Zerjal et al. 2003) selecting for larger and more complex brains. Once a tool for inventing, learning, and using these strategies (i.e., a complex brain) was in place, it could be used for a variety of other purposes, including coping with environmental, ecological, technological, linguistic, and other challenges.

Modeling these processes requires one to build complex models that would include genes, memes (i.e., socially learned strategies), and competition for mating success. In an attempt to shed some light on the interaction of these processes, Gavrilets and Vose (2005) introduced an explicit genetic, individual-based, stochastic mathematical model of the coevolution of genes, memes, and mating behavior. In their model, genes control two properties of the brain: a learning ability characterizing the probability to learn a particular meme, and a cerebral capacity characterizing the number of memes a brain can learn. Both of these characteristics were treated as additive quantitative traits subject to stabilizing selection in order to capture the energetic costs of having a large brain. In turn, memes were characterized by their complexity (i.e., difficulty) and their "Machiavellian fitness" quantifying the advantage to an individual who has this meme. Meme complexity and fitness were negatively correlated, so that more efficient memes were more difficult to learn. The model also assumed that the effects of memes known to both competing individuals cancel each other. This assumption results in a need to continuously invent or learn new memes to be able to stay in the competition for mating success.

Due to its complexity, the model had to be studied numerically. Overall, the results of Gavrilets and Vose suggest that the mechanisms underlying this hypothesis can indeed result in a significant increase in the brain size and in the evolution of significant cognitive abilities on the time scale of 10,000–20,000 generations. Interestingly, Gavrilets and Vose show that in their model the dynamics of intelligence has three distinct phases. During the dormant phase only newly invented memes are present in the population. These memes are not learned by other individuals. During the cognitive explosion phase the population's meme count and the learning ability, cerebral capacity, and Machiavellian fitness of individuals rapidly increase in a runaway fashion. During the saturation phase natural selection resulting from the costs of having large brains checks further increases in cognitive abilities.

Both the learning ability and the cerebral capacity are selected against due to costs of having large brains, but having nonzero values of both traits is necessary for learning and using different memes. The process of transition from the dormant phase to the cognitive explosion phase is somewhat similar to that of a peak shift on a rugged landscape. As in the case of stochastic peak shifts on a rugged landscape, the transition from the dormant phase to the cognitive explosion phase is mostly limited by new genetic variation. The levels of cognitive abilities achieved during the cognitive explosion phase increase with the intensity of competition for mates among males and decrease with the number of loci controlling the brain size. The latter effect is explained by the fact that a larger number of loci implies weaker selection on each individual locus.

In the model, evolutionary processes occur at two different time scales: fast for memes and slow for genes. More complex memes provide more fitness benefits to individuals. However, during the cognitive explosion phase the complexity of memes present in the population does not increase but, on the contrary, decreases in time. This happens as a result of intense competition among memes: while complex memes give advantage to individuals on a slow (biological) time scale, they lose competition to simpler memes on a fast (social) time scale because they are more difficult to learn. Intriguingly, the model suggests that there may be a tendency toward a reduction in cognitive abilities (driven by the costs of having a large brain) as the reproductive advantage of having a large brain decreases and the exposure to memes increases in modern societies.

Much more effort is needed for building a comprehensive theory of the coevolution of genes, memes, groups, behaviors, and social networks that would be applicable to earlier human evolution. These efforts should include both the development of simple models that can be studied analytically and the performing of large-scale individual-based simulation studies of more complex and realistic models.

Conclusion

The emergence of a quantitative/mathematical theory of biological evolution was crucial for the success of the Modern Synthesis. The steady progress of empirical evolutionary research observed since the 1930s and 1940s was accompanied by many theoretical developments, including the theory of multidimensional fitness landscapes and the emergence of a dynamical theory of speciation and diversification on which I have focused in this chapter. We now have a much better understanding of evolutionary processes. Not unexpectedly, many of the new theoretical results show that certain expectations and intuitions prevalent earlier are wrong or have a limited scope. This is part of the scientific process. The development of adequate mathematical theories will remain crucial in the future for better understanding of other evolutionary processes. The common wisdom is that a picture is worth a thousand words. In the exact sciences, an equation is worth a thousand pictures. Two areas of theoretical research are, in my opinion, particularly important and poised for significant advances, as I attempted to illustrate above. The first is a theory of evolutionary diversification across multiple spatial and temporal scales that would link microevolutionary processes with macroevolutionary patterns. The second is a theory of human origins and factors shaping our behavior, social interactions, and history. In 2009 we celebrate the 150th anniversary of the publication of *The Origin of Species*. In 2021 we will celebrate the 150th anniversary of Darwin's other groundbreaking book, *The Descent of Man*. Significant theoretical progress can be achieved in the twelve years separating those two years.

Some Thoughts on an Extended Evolutionary Synthesis

The synthesis of several biological disciplines that occurred in the 1930s and 1940s and became known as the Modern Synthesis marked the beginning of the still ongoing process of unification of biological sciences. As our knowledge and understanding of particular areas of biology increases, the connections among them become clearer, resulting in a stronger and broader synthesis. Many developments in biology that have occurred since the 1970s and 1980s were not (and could not be) anticipated by Darwin or during the time of the Modern Synthesis. Many patterns and processes that were unknown or not viewed as particularly important and/or relevant earlier have become crucial for our understanding of the evolution of life on Earth in general, and of the place of our own species in this process in particular. All this is a normal process in the development of any scientific discipline. Do new developments and knowledge really challenge the ideas central at the time of the Modern Synthesis and require a dramatic reevaluation of the basics? Definitely not. Declaring the Modern Synthesis or the Darwinian theory dead, wrong, or in crisis because some of the beliefs or views held previously are not supported by newer data or theories, or because there are still gaps in our knowledge, means being ignorant of how the science develops. Do new developments and new knowledge in different areas of biology justify the need for something that can be called an Extended Evolutionary Synthesis? I think the answer to this question is a very subjective matter.

To me, many recent advances of evolutionary biology that are sometimes presented as focal points of a future Extended Evolutionary Synthesis (e.g., Pigliucci 2007) fit well in the grand scheme of variation, selection, and inheritance within the populational context laid down during the Modern Synthesis. Moreover, from the theoretical point of view, the general rules and patterns of evolutionary dynamics will not be dramatically different if, say, the contribution of large mutations to variation were more significant, or if inheritance via epigenetic effects were more common, or if group selection and autocatalytic selection mediated via niche construction were more powerful or widespread than currently thought. I expect the underlying dynamic equations to be similar to those describing more "mainstream" types of variation (via small mutation), selection (individual or pair selection), and inheritance (via "classical" genes). This implies that general evolutionary patterns as we already understand them will not be significantly altered. Finally, great advances having significant implications both for our understanding of evolution and for many practical questions concerning our lives have happened across many different areas of biology, so that singling out just a few for defining an extended synthesis does not seem justified. The unification of biological sciences will be achieved via continuous extension of evolutionary thinking into various branches of the life sciences and social sciences.

Acknowledgments

I thank A. Birand, C. Eaton, G. B. Müller, and M. Pigliucci for very useful comments on the manuscript. My research was supported by National Institutes of Health grant GM56693.

References

Alexander RD (1990) How Did Humans Evolve? Reflections on the Uniquely Unique Species. Special publication no. 1. Ann Arbor: University of Michigan, Museum of Zoology.

Andersson MB (1994) Sexual Selection. Princeton, NJ: Princeton University Press.

Asmussen MA, Basnayake E (1990) Frequency-dependent selection: The high potential for permanent genetic variation in the diallelic, pairwise interaction model. Genetics 125: 215–230.

Balkau BJ, Feldman MW (1973) Selection for migration modification. Genetics 74: 171–174.

Barluenga M, Stolting KN, Salzburger W, Muschick M, Meyer A (2006) Sympatric speciation in Nicaraguan crater lake cichlid fish. Nature 439: 719–723.

Barton, NH (1989a) The divergence of a polygenic system subject to stabilizing selection, mutation and drift. Genetical Research 54: 59–77.

Barton NH (1989b) Founder effect speciation. In Speciation and Its Consequences. D Otte, JA Endler, eds.: 229–256. Sunderland, MA: Sinauer.

Barton NH, Charlesworth B (1984) Genetic revolutions, founder effects, and speciation. Annual Review of Ecology and Systematics 15: 133–164.

Barton NH, Rouhani S (1987) The frequency of shifts between alternative states. Journal of Theoretical Biology 125: 397–418.

Barton NH, Turelli M (1987) Adaptive landscapes, genetic distance and the evolution of quantitative characters. Genetical Research 49: 157–173.

Bazykin AD (1969) Hypothetical mechanism of speciation. Evolution 23: 685–687.

Beatty J (2008) Chance variation and evolutionary contingency: Darwin, Simpson, (the Simpsons), and Gould. In The Oxford Handbook of Philosophy of Biology. M Ruse, ed.: 189–210. Oxford: Oxford University Press.

Betzig LL (1986) Despotism and Differential Reproduction: A Darwinian View of History. New York: Aldine.

Betzig LL (1993) Sex, succession, and stratification in the first six civilizations: How powerful men reproduced, passed their power onto their sons, and used power to defend their wealth, women, and children. In Social Stratification and Socioeconomic Inequality. L Ellis, ed.: vol. 1: 37–74. Westport, CT: Praeger.

Bolnick DI, Fitzpatrick BM (2007) Sympatric speciation: Models and empirical evidence. Annual Review of Ecology, Evolution and Systematics 38: 459–487.

Burch CL, Chao L (1999) Evolution by small steps and rugged landscapes in the RNA virus φ6. Genetics 151: 921–927.

Burch CL, Chao L (2004) Epistasis and its relationship to canalization in the RNA virus φ6. Genetics 167: 559–567.

Butlin R (1987) Speciation by reinforcement. Trends in Ecology and Evolution 2: 8–13.

Butlin RK (1995) Reinforcement: An idea evolving. Trends in Ecology and Evolution 10: 432–434.

Byrne RW, Whiten A (1988) Machiavellian Intelligence: Social Expertise and the Evolution of Intellect in Monkeys, Apes, and Humans. Oxford: Clarendon Press.

Carson HL (1968) The population flush and its genetic consequences. In Population Biology and Evolution. RC Lewontin, ed.: 123–137. Syracuse, NY: Syracuse University Press.

Carson HL, Templeton AR (1984) Genetic revolutions in relation to speciation phenomena: The founding of new populations. Annual Review of Ecology and Systematics 15: 97–131.

Charlesworth B, Rouhani S (1988) The probability of peak shifts in a founder population. II. An additive polygenic trait. Evolution 42: 1129–1145.

Charlesworth B, Smith DB (1982) A computer model for speciation by founder effect. Genetical Research 39: 227–236.

Cockerham CC, Burrows PM, Young SS, Prout T (1972) Frequency-dependent selection in randomly mating populations. American Naturalist 106: 493–515.

Coyne J, Orr HA (2004) Speciation. Sunderland, MA: Sinauer.

Coyne JA, Barton NH, Turelli M (1997) A critique of Sewall Wright's shifting balance theory of evolution. Evolution 51: 643–671.

Coyne JA, Barton NH, Turelli M (2000) Is Wright's shifting balance process important in evolution? Evolution 54: 306–317.

Crosby JL (1970) The evolution of genetic discontinuity: Computer models of the selection of barriers to interbreeding between subspecies. Heredity 25: 253–297.

Darwin C (1859) On the Origin of Species by Means of Natural Selection, or the Preservation of Favoured Races in the Struggle for Life. London: John Murray.

Darwin C (1871) The Descent of Man, and Selection in Relation to Sex. London: John Murray.

Dickinson H, Antonovics J (1973) Theoretical considerations of sympatric divergence. American Naturalist 107: 256–274.

Dieckmann U, Doebeli M, Metz JAJ, Tautz D, eds. (2004) Adaptive Speciation. Cambridge: Cambridge University Press.

Dobzhansky T (1955) A review of some fundamental concepts and problems of population genetics. Cold Spring Harbor Symposia on Quantitative Biology 20: 1–15.

Dobzhansky T, Ayala FJ, Stebbins GL, Valentine JW (1977) Evolution. San Francisco: W.H. Freeman.

Dobzhansky TG (1937) Genetics and the Origin of Species. New York: Columbia University Press.

Dobzhansky TG (1940) Speciation as a stage in evolutionary divergence. American Naturalist 74: 312–321.

Duenez-Guzman EA, Mavárez J, Vose MD, Gavrilets S (2009) Case studies and mathematical models of ecological speciation. 4. Butterflies in a jungle. Evolution (in press)

Dunbar RIM (1998) The social brain hypothesis. Evolutionary Anthropology 6: 178–190.

Dunbar RIM (2003) The social brain: Mind, language, and society in evolutionary perspective. Annual Review of Anthropology 32: 163–181.

Eldredge N (2003). The sloshing bucket: How the physical realm controls evolution. In Towards a Comprehensive Dynamics of Evolution: Exploring the Interplay of Selection, Neutrality, Accident, and Function. JP Crutchfield, P Schuster, eds.: 3–32. New York: Oxford University Press.

Eldredge N, Thompson JN, Brakefield PM, Gavrilets S, Jablonski D, Jackson JBC, Lenski RE, Lieberman BS, McPeek MA, Miller W (2005) Dynamics of evolutionary stasis. Paleobiology 31: 133–145.

Elena SF, Lenski RE (2003) Evolution experiments with microorganisms: The dynamics and genetic bases of adaptation. Nature Reviews Genetics 4: 457–469.

Fisher RA (1930) The Genetical Theory of Natural Selection. Oxford: Clarendon Press.

Flinn MV, Geary DC, Ward CV (2005) Ecological dominance, social competition, and coalitionary arms races: Why humans evolved extraordinary intelligence. Evolution and Human Behavior 26: 10–46.

Fontana W, Schuster P (1998) Continuity in evolution: On the nature of transitions. Science 280: 1451–1455.

Futuyma DJ (1998) Evolutionary Biology. 3rd ed. Sunderland, MA: Sinauer.

Gavrilets S (1996) On phase three of the shifting-balance theory. Evolution 50: 1034–1041.

Gavrilets S (1997a) Coevolutionary chase in exploiter-victim systems with polygenic characters. Journal of Theoretical Biology 186: 527–534.

Gavrilets S (1997b) Evolution and speciation on holey adaptive landscapes. Trends in Ecology and Evolution 12: 307–312.

Gavrilets S (2003a) Evolution and speciation in a hyperspace: The roles of neutrality, selection, mutation and random drift. In Towards a Comprehensive Dynamics of Evolution: Exploring the Interplay of Selection, Neutrality, Accident, and Function. In JP Crutchfield, P Schuster, eds.: 135–162. New York: Oxford University Press.

Gavrilets S (2003b) Models of speciation: What have we learned in 40 years? Evolution 57: 2197–2215.

Gavrilets S (2004) Fitness Landscapes and the Origin of Species. Princeton, NJ: Princeton University Press.

Gavrilets S (2005) "Adaptive speciation": It is not that simple. Evolution 59: 696–699.

Gavrilets S (2006) The Maynard Smith model of sympatric speciation. Journal of Theoretical Biology 239: 172–182.

Gavrilets S, Gravner J (1997) Percolation on the fitness hypercube and the evolution of reproductive isolation. Journal of Theoretical Biology 184: 51–64.

Gavrilets S, Losos J (2009) Adaptive radiation: Contrasting recent theory with data. Science 323: 732–737.

Gavrilets S, Vose A (2005) Dynamic patterns of adaptive radiation. Proceedings of the National Academy of Sciences of the USA 102: 18040–18045.

Gavrilets S, Vose A (2006) The dynamics of Machiavellian intelligence. Proceedings of the National Academy of Sciences of the USA 103: 16823–16828.

Gavrilets S, Vose A (2007) Case studies and mathematical models of ecological speciation. 2. Palms on an oceanic island. Molecular Ecology 16: 2910–2921.

Gavrilets S, Vose A (2009) Dynamic patterns of adaptive radiation: Evolution of mating preferences. In Speciation and Patterns of Diversity. R Butlin, J Bridle, D Schluter, eds.: 102–126. Cambridge: Cambridge University Press.

Gavrilets S, Vose A, Barluenga M, Salzburger W, Meyer A (2007) Case studies and mathematical models of ecological speciation. 1. Cichlids in a crater lake. Molecular Ecology 16: 2893–2909.

Gavrilets S, Waxman D (2002) Sympatric speciation by sexual conflict. Proceedings of the National Academy of Sciences of the USA 99: 10533–10538.

Geary DC (2004) The Origin of Mind: Evolution of Brain, Cognition, and General Intelligence. Washington, DC: American Psychological Association.

Gillespie R (2004) Community assembly through adaptive radiation in Hawaiian spiders. Science 303: 356–359.

Givnish T, Sytsma K, eds. (1997) Molecular Evolution and Adaptive Radiation. Cambridge: Cambridge University Press.

Gould SJ (2002) The Structure of Evolutionary Theory. Cambridge, MA: Belknap Press of Harvard University Press.

Gravner J, Pitman D, Gavrilets S (2007) Percolation on fitness landscapes: Effects of correlation, phenotype, and incompatibilities. Journal of Theoretical Biology 248: 627–645.

Grimmett G (1989) Percolation. Berlin: Springer-Verlag.

Haldane JBS (1964) Defense of beanbag genetics. Perspectives in Biology and Medicine 7: 343–359.

Hodges S (1997) A rapid adaptive radiation via a key innovation in aquilegia. In Molecular Evolution and Adaptive Radiations. T Givnish, K Sytsma, eds.: 391–405. Cambridge: Cambridge University Press.

Hollander J, Adams DC, Johannesson K (2006) Evolution of adaptation through allometric shifts in a marine snail. Evolution 60: 2490–2497.

Hollander J, Lindgarth M, Johannesson K (2005) Local adaptation but not geographical separation promotes assortative mating in a snail. Animal Behaviour 70: 1209–1219.

Holloway R (1996) Evolution of the human brain. In Handbook of Human Symbolic Evolution. A Lock, CR Peters, eds.: 74–125. Oxford: Clarendon Press.

Howard DJ (1993) Reinforcement: Origin, dynamics, and fate of an evolutionary hypothesis. In Hybrid Zones and the Evolutionary Process. RG Harrison, ed.: 46–69. New York: Oxford University Press.

Howard DJ, Berlocher SH (1998) Endless Forms: Species and Speciation. New York: Oxford University Press.

Humphrey NK (1976) The social function of intellect. In Growing Points in Ethology. PPG Bateson, RA Hinde, eds.: 303–317. Cambridge: Cambridge University Press.

Huynen MA, Stadler PF, Fontana W (1996) Smoothness within ruggedness: The role of neutrality in adaptation. Proceedings of the National Academy of Sciences of the USA 93: 397–401.

Kaneshiro KY (1980) Sexual isolation, speciation and the direction of evolution. Evolution 34: 437–444.

Kauffman SA (1993) The Origins of Order: Self-Organization and Selection in Evolution. Oxford: Oxford University Press.

Kauffman SA, Levin S (1987) Towards a general theory of adaptive walks on rugged landscapes. Journal of Theoretical Biology 128: 11–45.

Kimura M (1983) The Neutral Theory of Molecular Evolution. New York: Cambridge University Press.

Kirkpatrick M, Ravigné V (2002) Speciation by natural and sexual selection: Models and experiments. American Naturalist 159: S22–S35.

Lande R (1976) Natural selection and random genetic drift in phenotypic evolution. Evolution 30: 314–334.

Lande R (1980) Genetic variation and phenotypic evolution during allopatric speciation. American Naturalist 116: 463–479.

Lenski RE, Ofria C, Collier TC, Adami C (1999) Genome complexity, robustness and genetic interactions in digital organisms. Nature 400: 661–664.

Levin DA (1993) Local speciation in plants: The rule not exception. Systematic Biology 18: 197–208.

Lipman DJ, Wilbur WJ (1991) Modelling neutral and selective evolution of protein folding. Proceedings of the Royal Society of London B245: 7–11.

Losos JB (1998) Ecological and evolutionary determinants of the species-area relationship in Caribbean anoline lizards. In Evolution on Islands. PR Grant, ed.: 210–224. Oxford: Oxford University Press.

Lynch M (2007) The Origins of Genome Architectures. Sunderland, MA: Sinauer.

Mallet J, Joron M (1999) Evolution of diversity in warning color and mimicry: Polymorphisms, shifting balance, and speciation. Annual Review of Ecology and Systematics 30: 201–233.

Martinez MA, Pezo V, Marlière P, Wain-Hobson S (1996) Exploring the functional robustness of an enzyme by in vitro evolution. EMBO Journal 15: 1203–1210.

Mavárez J, Salazar CA, Bermingham E, Salcedo C, Jiggins CD, Linares, M (2006) Speciation by hybridization in *Heliconius* butterflies. Nature 441: 868–871.

Maynard Smith J (1962) Disruptive selection, polymorphism and sympatric speciation. Nature 195: 60–62.

Maynard Smith J (1966) Sympatric speciation. American Naturalist 100: 637–650.

Maynard Smith J (1970) Natural selection and the concept of a protein space. Nature 225: 563–564.

Mayr E (1942) Systematics and the Origin of Species. New York: Columbia University Press.

Mayr E (1954) Change of genetic environment and evolution. In Evolution as a Process. C Barigozzi, ed.: 1–19. New York: Alan R. Liss.

Mayr E (1959) Where are we? Cold Spring Harbor Symposia on Quantitative Biology 24: 1–14.

Mayr E (1963) Animal Species and Evolution. Cambridge, MA: Belknap Press of Harvard University Press.

Mayr E (1982) The Growth of Biological Thought. Cambridge, MA: Belknap Press of Harvard University Press.

Noor MAF (1999) Reinforcement and other consequences of sympatry. Heredity 83: 503–508.

Orr HA (2002) The population genetics of adaptation: The adaptation of DNA sequences. Evolution 56: 1317–1330.

Orr HA (2006a) The distribution of fitness effects among beneficial mutations in Fisher's geometric model of adaptation. Journal of Theoretical Biology 238: 279–285.

Orr HA (2006b) The population genetics of adaptation on correlated fitness landscapes: The block model. Evolution 60: 1113–1124.

Pigliucci M (2007) Do we need an extended evolutionary synthesis? Evolution 61: 2743–2749.

Pigliucci M, Kaplan J (2006) Making Sense of Evolution: The Conceptual Foundations of Evolutionary Biology. Chicago: University of Chicago Press.

Provine WB (1978) The role of mathematical population geneticists in the evolutionary synthesis of the 1930s and 1940s. Studies in the History of Biology 2: 167–192.

Provine WB (1986) Sewall Wright and Evolutionary Biology. Chicago: University of Chicago Press.

Provine WB (1989) Founder effects and genetic revolutions in microevolution and speciation: A historical perspective. In Genetics, Speciation and the Founder Principle. LV Giddings, KY Kaneshiro, WW Anderson, eds.: 43–76. Oxford: Oxford University Press.

Reidys CM, Forst CV, Schuster P (2001) Replication and mutation on neutral networks. Bulletin of Mathematical Biology 63: 57–94.

Reidys CM, Stadler PF (2001). Neutrality in fitness landscapes. Applied Mathematics and Computation 117: 321–350.

Reidys CM, Stadler PF (2002) Combinatorial landscapes. SIAM Review 44: 3–54.

Reidys CM, Stadler PF, Schuster P (1997) Generic properties of combinatory maps: Neutral networks of RNA secondary structures. Bulletin of Mathematical Biology 59: 339–397.

Ridley M (1993) Evolution. Boston: Blackwell Scientific.

Rost B (1997) Protein structures sustain evolutionary drift. Folding and Design 2: S19–S24.

Roth G, Dicke U (2005) Evolution of the brain and intelligence. Trends in Cognitive Sciences 9: 250–257.

Rouhani S, Barton NH (1987) The probability of peak shifts in a founder population. Journal of Theoretical Biology 126: 51–62.

Ruff CB, Trinkaus E, Holliday TW (1997) Body mass and encephalization in Pleistocene *Homo*. Nature 387: 173–176.

Russon AE, Begun DR, eds. (2004) The Evolution of Thought: Evolutionary Origins of Great Ape Intelligence. Cambridge: Cambridge University Press.

Sadedin S, Hollander J, Panova M, Johannesson K, Gavrilets S (2008) Case studies and mathematical models of ecological speciation. 3: Ecotype formation in a Swedish snail. Molecular Ecology (submitted)

Salzburger W, Meyer A (2004) The species flocks of East African cichlid fishes: Recent advances in molecular phylogenetics and population genetics. Naturwissenschaften 91: 277–290.

Savolainen V, Anstett M-C, Lexer C, Hutton I, Clarkson JJ, Norup MV, Powell MP, Springate D, Salamin N, Baker WJ (2006) Sympatric speciation in palms on an oceanic island. Nature 441: 210–213.

Schluter D (2000) The Ecology of Adaptive Radiation. Oxford: Oxford University Press.

Schuster P (1995) How to search for RNA structures: Theoretical concepts in evolutionary biotechnology. Journal of Biotechnology 41: 239–257.

Seehausen O (2007) African cichlid fish: A model system in adaptive radiation research. Proceedings of the Royal Society of London B 273: 1987–1998.

Servedio MR, Noor MAF (2003) The role of reinforcement in speciation: Theory and data. Annual Review of Ecology and Systematics 34: 339–364.

Sheppard PM (1954) Evolution in bisexually reproducing organisms. In Evolution as a Process. J Huxley, AC Hardy, EB Ford, eds.: 201–218. London: Allen and Unwin.

Simpson GG (1953) The Major Features of Evolution. New York: Columbia University Press.

Striedter GF (2004) Principles of Brain Evolution. Sunderland, MA: Sinauer.

Templeton AR (1980) The theory of speciation via the founder principle. Genetics 94: 1011–1038.

Via S (2001) Sympatric speciation in animals: The ugly duckling grows up. Trends in Ecology and Evolution 16: 381–390.

Vrba ES (1995) The fossil record of African antelopes (Mammalia, Bovidae) in relation to human evolution and paleoclimate. In Paleoclimate and Evolution, with Emphasis on Human Origins. ES Vrba, GH Denton, TC Partridge, LH Burckle, eds.: 385–424. New Haven, CT: Yale University Press.

Wade MJ, Goodnight CJ (1998) Perspective. The theories of Fisher and Wright in the context of metapopulations: When nature does many small experiments. Evolution 52: 1537–1553.

Goodnight CJ, Wade MJ (2000) The on-going synthesis: A reply to Coyne et al. Evolution 54: 317–324.

Waxman D (2006) Fisher's geometrical model of evolutionary adaptation: Beyond spherical geometry. Journal of Theoretical Biology 241: 887–895.

Waxman D, Gavrilets S (2005) Issues of terminology, gradient dynamics and the ease of sympatric speciation in Adaptive Dynamics. Journal of Evolutionary Biology 18: 1214–1219.

Waxman D, Welch JJ (2005) Fisher's microscope and Haldane's ellipse. American Naturalist 166: 447–457.

Whiten A, Byrne RW, eds. (1997) Machiavellian Intelligence II. Extensions and Evaluations. Cambridge: Cambridge University Press.

Wilke CO, Wang JL, Ofria C, Lenski RE, Adami C (2001) Evolution of digital organisms at high mutation rates leads to survival of the flattest. Nature 412: 331–333.

Woods R, Schneider D, Winkworth CL, Riley MA, Lenski RE (2006) Tests of parallel molecular evolution in a long-term experiment with *Escherichia coli*. Proceedings of the National Academy of Sciences of the USA 103: 9107–9112.

Wright S (1931) Evolution in Mendelian populations. Genetics 16: 97–159.

Wright S (1932) The roles of mutation, inbreeding, crossbreeding and selection in evolution. In Proceedings of the Sixth International Congress on Genetics. DF Jones, ed.: vol. 1: 356–366.

Wright S (1982) The shifting balance theory and macroevolution. Annual Review of Genetics 16: 1–19.

Wright S (1988) Surfaces of selective value revisited. American Naturalist 131: 115–123.

Zerjal T, Xue Y, Bertorelle G, Wells RS, Bao W, Zhu S, et al. (2003) The genetic legacy of the Mongols. American Journal of Human Genetics 72: 717–721.

4 Multilevel Selection and Major Transitions

David Sloan Wilson

Multilevel selection (MLS) theory addresses a fundamental issue in evolutionary biology that was not featured strongly in the Modern Synthesis. The concept of major evolutionary transitions and human evolution as a major transition has made MLS theory more relevant than ever before. This chapter will provide a brief overview of MLS theory, major evolutionary transitions, and human evolution as a major transition, so that these subjects can become part of an extended evolutionary synthesis.

MLS Theory and Its Relation to the Modern Synthesis

Darwin's theory of natural selection is framed in terms of individual organisms surviving and reproducing better than other organisms, such as a more drought-resistant plant, a better concealed insect, a faster-running deer, and so on. These traits are *locally advantageous*; individuals possessing them are more fit than individuals in their immediate vicinity that do not.

In contrast, traits that help other organisms or that cause whole groups to function adaptively are usually not locally advantageous. Examples include helping to raise the offspring of others, watching out for predators in a way that protects everyone in the vicinity, and conserving shared resources when they are scarce. These traits are clearly "for the good of the group," but they do not give individuals possessing the trait a fitness advantage, compared with other individuals in their immediate vicinity. They are *locally disadvantageous*.

The evolution of traits that are "for the good of the group" but locally disadvantageous is not a trivial problem. Most traits associated with human morality have this "for the good of the group" quality, in addition to self-sacrificial traits in nonhuman species. Darwin proposed a straightforward solution in *Descent of Man* (1871: ch. 4) and elsewhere, as represented by this canonical passage in *Descent* (p. 166):

It must not be forgotten that although a high standard of morality gives but a slight or no advantage to each individual man and his children over other men of the same tribe, yet then an increase in the number of well-endowed men and advancement in the standard of morality will certainly give an immense advantage to one tribe over another. There can be no doubt that a tribe including many members who, from possessing in a high degree the spirit of patriotism, fidelity, obedience, courage, and sympathy, were always ready to aid one another, and to sacrifice themselves for the common good, would be victorious over most other tribes, and this would be natural selection. At all times throughout the world tribes have supplanted other tribes; and as morality is one important element in their success, the standard of morality and the number of well-endowed men will thus everywhere tend to rise and increase.

The human traits listed by Darwin are manifestly adaptive at the group level, despite their local disadvantage. Groups of individuals who aid each other will outcompete other groups, even if such individuals are selectively disadvantageous within groups. Natural selection can operate at more than one level of the biological hierarchy, each level favoring a different set of traits. This was the birth of what later became known as MLS theory. Darwin did not comment on the irony that morality, by this account, is primarily a within-group phenomenon and can lead to the evolution of behaviors, such as between-group conflict, that can qualify as immoral from a third-person perspective. Also, competition between groups need not take the form of direct conflict. Groups that function better as collective units for any reason will differentially contribute to the total gene pool, just as drought-resistant plants "outcompete" drought-susceptible plants in desert environments without any direct interactions.

The three fathers of population genetics theory—Ronald Fisher, J. B. S. Haldane, and Sewall Wright—all considered the problem of group-level selection, but only briefly (see Sober and Wilson 1998: ch. 1 for a review). Creating a mathematical framework for evolution in general pushed this particular problem into the shadows. Even Sewall Wright's shifting balance theory, which bears a superficial resemblance to group-level selection, addressed the question of how individual-level traits with a complex genetic basis can evolve. When genetic interactions are epistatic, multiple local equibria exist that are not equally adaptive at the individual level, leading to Wright's famous metaphor of a multi-peak adaptive landscape. A multigroup population structure is required for a population to inhabit more than one peak of an adaptive landscape, and there must be a way for the populations occupying the highest peaks to replace the populations occupying the lowest peaks. As strange as it might seem, Wright's first consideration of the evolution of altruistic social behaviors was a brief discussion and sketch of a model in his 1945 book review of George Gaylord Simpson's *Tempo and Mode of Evolution*. This illustrates the degree to which the issue at the heart of MLS theory was eclipsed by other issues at the heart of the Modern Synthesis.

Nevertheless, Fisher, Haldane, and Wright all confirmed Darwin's original insight, however briefly. Traits that are selectively disadvantageous within groups can evolve by causing groups to outcompete other groups. Between-group competition can take a variety of forms, such as direct conflict, fissioning at different rates, or contributing more dispersers to the total gene pool. In all cases, the local disadvantage of "for the good of the group" traits must be counterbalanced by an advantage at a larger scale for the traits to evolve in the total population.

One achievement of the Modern Synthesis of the 1940s was to make population genetics theory part of mainstream evolutionary biology. However, the problem of group-level selection remained in the shadows during the next 20 years. Moreover, many biologists did not share Darwin's original insight and naively assumed that adaptations can evolve at any level of the biological hierarchy—for the good of the individual, group, species, or even ecosystem—without requiring special conditions. When the need for higher-level selection was acknowledged, it was often assumed that it could easily prevail against lower-level selection. This position, which in retrospect is called "naive group selection," is illustrated by the final paragraph of the textbook *Principles of Animal Ecology* (Allee et al. 1949):

The probability of survival of individual living things, or of populations, increases with the degree to which they harmoniously adjust themselves to each other and their environment. This principle is basic to the concept of the balance of nature, orders the subject matter of ecology and evolution, underlies organismic and developmental biology, and is the foundation for all sociology.

To some extent, the architects of the Modern Synthesis shared this confusion. Because evolution is a population-level process, it is easy to assume that parameters such as mutation rate, sexual reproduction, and reproductive isolation have evolved to make evolution an *efficient* population-level process. Yet, many traits, such as mutation rate and sexual reproduction, are selectively disadvantageous within populations, creating a conflict between levels of selection. These issues are still being debated under headings such as "the evolution of evolvability" (e.g., Wagner and Altenberg 1996; Pepper 2003) and "lineage selection" (e.g., Nunney 1999; Jablonski 2008), so the architects of the Modern Synthesis can be forgiven for not having fully articulated or resolved them at the time.

The problem of higher-level adaptations did not begin to occupy center stage until the 1960s, when the Modern Synthesis was firmly established. A number of authors, including most famously John Maynard Smith in England and George C. Williams in America, began to question the veracity of naive group selectionist claims. In *Adaptation and Natural Selection*, Williams (1966) interpreted population genetics theory for a broad audience of biologists, including Darwin's original insight that group-level adaptations require a process of group-level selection and tend to be undermined by lower-level selection. Then he evaluated the evidence and made a strong claim: even though group-level adaptations can evolve in *principle*, they seldom evolve in *practice*. Higher-level selection is almost invariably weak compared with lower-level selection, and most interpretations of adaptations as "for the good of the group" are just plain wrong. As he put it (p. 93), "group-level adaptations do not, in fact, exist."

Williams's assessment was based less on empirical evidence than on theoretical arguments and the principle of parsimony, which dictates that simpler explanations (individual selection) be preferred over more complex explanations (group selection) whenever possible. In this fashion, broad topics such as territoriality and dominance were interpreted as individual-level adaptations based on plausibility arguments without anything close to a rigorous empirical test of selection within and among groups. The closest that Williams came to a rigorous empirical test concerned the evolution of sex ratios, in which within-group selection favors an even sex ratio and between-group selection favors either a male- or female-biased sex ratio, depending upon whether population growth or reguation is favored at the group level. Williams thought it was "abundantly clear" that most species have an even sex ratio, declaring that "I would regard the problem of sex ratio as solved" (p. 272).

Williams's categorical rejection of group selection was widely accepted, and *Adaptation and Natural Selection* became as influential as the books associated with the Modern Synthesis published in the 1940s and 1950s.

The rejection of group selection required an explanation of seemingly other-oriented behaviors in individualistic terms. Alternatives were provided by a number of theoretical frameworks, such as kin selection theory (benefiting one's own genes in the bodies of others), reciprocity (benefiting others in expectation of return benefits), and selfish gene theory (the gene as the fundamental unit of selection for the evolution of all traits). During this period, it became almost mandatory for authors to assure their readers that group selection was not being invoked.

Ironically, group selection *was* being invoked. Almost immediately, it began to emerge that all evolutionary theories of social behavior assume the existence of multiple groups, that the traits labeled "altruistic" and "cooperative" are selectively disadvantageous within groups and require group-level selection to evolve in the total population. The various theoretical frameworks all obeyed the central logic of MLS theory and differed primarily in perspective. Several examples will be provided to illustrate the striking fact that the rejection of group selection persisted much longer than the theoretical and empirical basis for rejecting group selection.

First, one year after the publication of *Adaptation and Natural Selection*, William D. Hamilton (1967) published an influential article titled "Extraordinary Sex Ratios," which documented many examples of extreme female-biased sex ratios, especially in small species of arthropods that live in highly subdivided populations. In his mathematical model to explain the evolution of female-biased sex ratios, Hamilton assumed that a large number of groups ("hosts") are colonized at random by N individuals. In other words, the multigroup population structure could not have been more explicit. Female-biased sex ratios are selectively disadvantageous within groups but cause the group to contribute differentially to the total gene pool, exactly as Williams postulated in his use of sex ratio as an empirical test of within- versus between-group selection. Yet, aside from a footnote in which Hamilton noted the "pleasing" influence of both within- and between-group selection, he primarily described his model in terms of females maximizing their number of grand-offspring under conditions of local mate competition. This difference in perspective enabled sex ratio theory to become a hot topic in evolutionary biology without anyone noticing that it refuted Williams's best evidence against group selection, a fact that he finally acknowledged in the 1990s (Williams 1992: 49).

As a second example, Hamilton (1963, 1964) originally formulated inclusive fitness theory (termed "kin selection" by Maynard Smith 1964) as the net effect of an altruistic allele on all copies of itself identical by descent. Thus, a single altruisic act would result in a fitness decrement of c for the actor and a fitness increment of b for the recipient, which must be weighted by the probability r that the recipient possesses the same gene identical by descent. When br–c > 0, there are more copies of the altruistic allele than there were before. However, natural selection is based on relative fitness, not absolute fitness. What if the altruistic act increases the number of nonaltruistic genes in the vicinity even more? In the early 1970s, Hamilton reformulated his theory on the basis of an equation derived by George Price (1970, 1972) that is more explicit about the fact that social interactions among relatives inherently imply a multigroup population structure. When it was viewed through the lens of the Price equation, Hamilton immediately saw that altruism expressed among relatives is locally disadvantageous, just as any other form of altruism, and evolves only by virtue of groups of altruists differentially contributing to the total gene pool. High degrees of genetic relatedness signify high genetic variation among groups, increasing the importance of between-group selection compared to within-group selection. Hamilton's theory remained a powerful explanation for the evolution of altruism, but emerged as a *kind of* group selection rather than *an alternative* to group selection, as he clearly acknowleged in 1975. Yet, Hamilton's new formulation did not cause the field as a whole to reconsider the rejection of group selection that was now in full swing.

As a third example, game theory became a popular theoretical framework for studying the evolution of cooperation and altruism in the 1980s (e.g., Axelrod and Hamilton 1981; Maynard Smith 1982). As with inclusive fitness theory, game theorists did not think that they were invoking group selection. The core game theory models assume pairwise interactions, but two-person game theory easily generalizes to n-person game theory, where n refers to the groups of individuals that actually interact and determine each other's fitness. In other words, evolutionary game theory models assume a multigroup population structure, just like all other models of social behavior. The groups can form at random, as groups of relatives, or by other processes such as assortative interactions. Traits labeled "cooperative" and "altruistic" are selectively disadvantageous or at best neutral within groups, and evolve only by virtue of the differential contributions of groups to the total gene pool. One reason that the central logic of MLS theory is not more obvious in game theory models is because the fitness of individual strategies is averaged across groups rather than partitioned into within- and between-group components. Thus, a conditional strategy such as tit-for-tat can evolve in the total population, even though individuals employing the tit-for-tat strategy never beat their social partners in within-group social interactions. It is very easy to label the strategy that evolves in the total population "individually advantageous" without checking to see the scale (within-group versus between-group) at which the advantage resides.

As a fourth example, Williams (1966) discussed the concept of average effects as part of his exposition of population genetics theory. The average effect of a gene is its fitness averaged across all genotypic and social contexts, and provides the "bottom line" of what evolves in the total population. Williams accurately called it a "bookkeeping method." As such, it includes group selection as an evolutionary force, no matter how weak or strong. A gene coding for altruism that evolves by group selection, despite its selective disadvantage within groups, has a higher average effect than the alternative gene for selfishness, which is just another way of saying that it evolves. Nevertheless, the concept of average effects soon began to be interpreted as an argument against group selection. Genes were called "the fundamental unit of selection," and all genes that evolve were termed "selfish genes" (Dawkins 1976). Only in retrospect has it become obvious that the "replicator" concept is no argument at all against group selection. Similarly, the concept of extended phenotypes (Dawkins 1982) notes that genes can have effects that extend beyond the body of the individual containing the gene. True enough, but this fact by itself says nothing about the fundamental issue that MLS theory was designed to solve—the fact that some genes are locally disadvantageous and can evolve only by being advantageous at a larger scale.

To summarize, what was originally a strong empirical claim—lower-level selection invariably trumping higher-level selection—has been authoritatively rejected. In its place, we have a collection of theoretical frameworks that all assume the existence of multiple groups and the central logic of MLS, but that differ in how they calculate the bottom line of what evolves in the total population. The coexistence of multiple equivalent theoretical frameworks is called pluralism (D. S. Wilson 2008; D. S. Wilson and E. O. Wilson 2007). Today, a pluralistic scientist might prefer inclusive fitness theory for its heuristic value, but would cheerfully acknowledge that altruism is selectively disadvantageous within groups and requires group selection, as these terms are used in MLS theory. Pluralism has its merits, but it sometimes obscures how much has changed since the rejection of group selection in the 1960s. Darwin was right:

traits can evolve by virtue of benefiting whole groups, despite being selectively disadvantageous within groups. Within-group selection does not invariably trump between-group selection, and their relative importance must be determined on a case-by-case basis. (See Sober and Wilson 1998 for a review of the turbulent history of group selection, and Okasha 2007, and D. S. Wilson and E. O. Wilson 2007, 2008 for more detailed reviews of contemporary MLS theory.)

How does MLS theory relate to the modern synthesis? It falls squarely within the paradigm of microevolution, population genetics models, and an emphasis on adaptation and natural selection established by the Modern Synthesis. However, the central focus of MLS theory—whether "for the good of the group" traits can evolve—didn't occupy center stage until the 1960s, 20 years after the Modern Synthesis had become established. MLS theory can therefore be truly regarded as an *extension* of the Modern Synthesis. The fact that *Adaptation and Natural Selection* became as influential as the books associated with the Modern Synthesis illustrates that MLS theory is an *important* extension. Finally, the revival of MLS theory shows that it *continues* to be an important extension in contemporary research.

Major Evolutionary Transitions

The balance between levels of selection is not static, but can itself evolve. When between-group selection sufficiently dominates within-group selection, the group becomes so functionally organized that it becomes a higher-level organism in its own right.

This scenario was proposed by the cell biologist Lynn Margulis (1970) to explain the evolution of the eukaryotic cell not by small mutational steps from bacterial cells, but as symbiotic associates of bacteria. Few biologists noticed that Margulis's theory, which involves between-group selection trumping within-group selection, was diametrically opposed to the dogma that within-group selection invariably trumps between-group selection.

Later, John Maynard Smith and Eörs Szathmáry (1995, 1999) expanded Margulis's theory to include other major transitions in the history of life, including the origin of life itself as groups of cooperating molecular reactions, the first cells, multicellular organisms, social insect colonies, and human evolution. In each case, mechanisms evolve that suppress selection within groups, causing between-group selection to become the dominant evolutionary force.

A major transition includes a number of hallmarks: (1) it is a rare event; (2) it results in momentous consequences once it occurs—lower-level organisms are no match for the new superorganism, which becomes ecologically dominant and radiates through evolutionary time to become thousands of species; (3) the transition is never complete—within-group selection is only suppressed, not entirely eliminated. Even multicellular organisms have a disturbing number of genetic elements that spread by intragenomic conflict rather than "for the good of the group" (Burt and Trivers 2006).

The concept of major evolutionary transitions is one of the most important developments in evolutionary biology. The architects of the Modern Synthesis imagined all evolution to be the result of small mutational steps—individuals from individuals. The concept of major transitions identifies an entirely new pathway—individuals from groups. Although major transitions fall squarely within MLS theory, they were not predicted by MLS theorists prior to Margulis's bold symbiotic cell theory. These foundational developments have surely gone beyond the Modern Synthesis.

Human Evolution as a Major Transition

Although Maynard Smith and Szathmáry (1995, 1999) were bold about expanding the concept of major transitions, they were timid about applying it to human evolution, restricting themselves to a discussion of the genetic basis of language, which by itself does not obviously relate to major transitions. Now it appears likely that human evolution was a full-fledged major transition, from groups *of* organisms to groups *as* organisms. The reason that we are so unique among primates is that our ancestors became the primate equivalent of a single organism or a social insect colony (Boehm 1999; D. S. Wilson 2007; Wilson, Van Vugt, and O'Gorman 2008).

Recall that the key ingredient of a major transition is the suppression of fitness differences within groups, causing between-group selection to become the primary evolutionary force. In most primate species, including our closest ancestors, intense within-group competition limits the opportunities for cooperation among members of the group. This is in contrast to extant human hunter-gatherer societies, which are fiercely egalitarian. What accounts for this shift, and when did it occur in human evolution?

Humans are incomparably better at throwing projectiles than other primates, an ability that required whole-body anatomical changes and evolved early in the hominid lineage. Although the original purpose of throwing was presumably to deter predators and competing scavengers, it could also be used to suppress bullying and other domineering behavior within groups (Bingham 1999). This is a specific version of a more general hypothesis of guarded egalitarianism, originally advanced by Boehm (1993, 1999) on the basis of the egalitarian nature of most extant hunter-gatherer societies. However it was accomplished, guarded egalitarianism provides the key ingredient of an evolutionary transition.

It has been common in the past to regard advanced human cognitive abilities, such as a theory of mind, as the first step of human evolution that made widespread cooperation possible (e.g., Tomasello 1999). Now it appears that the sequence needs to be reversed (e.g., Tomasello et al. 2005). The first event was the suppression of fitness differences within groups, based upon adaptations such as the ability to throw stones, which did not require a change in social cognition. Then, between-group selection favored forms of mental cooperation in addition to physical cooperation. After all, symbolic thought and the social transmission of behaviors are fundamentally cooperative activities that are unlikely to take place among uncooperative individuals. Even human capacities that we take for granted, such as the communicative nature of our eyes, our ability to point, and awareness of others that emerges early in infancy, are forms of cooperation that appear to be uniquely human (reviewed by Tomasello et al. 2005).

In retrospect, human evolution has all the hallmarks of a major transition. It was a rare event, occurring only once among primates. It had momentous consequences; cooperation enabled our ancestors to spread over the planet, eliminating other hominids and many other species along the way. We also diversified to occupy all climatic zones and hundreds of ecological niches, although by cultural evolution rather than genetic evolution. The advent of agriculture enabled us to increase the scale of society by many orders of magnitude through a process of cultural multilevel selection (e.g., Turchin 2005). Finally, the transition was not complete. Within-group selection still takes place and is merely suppressed compared to between-group selection.

Thinking of human evolution as a major evolutionary transition is so new that most of the implications remain to be discovered, providing yet another area of study that was not anticipated by the Modern Synthesis.

A Postscript on the Study of Human Behavior and Culture from an Evolutionary Perspective

This chapter is mostly about MLS, but I would like to end with a comment on all aspects of human behavior and culture from an evolutionary perspective. It was obvious to everyone in Darwin's time that, if true, his theory would revolutionize our understanding of ourselves. Yet, by the early twentieth century, studying human behavior and culture from an evolutionary perspective was largely off-limits. The Modern Synthesis respected this boundary for the most part, saying much about biology but remaining cautious about human behavior and culture. In his book *Mankind Evolving*, for example, Dobzhanksy (1962: 345) stated that "I know no better criterion of wisdom and values" than the following passage from an ancient Chinese sage:

Every system of moral laws must be based upon the man's own consciousness, verified by the common experience of mankind, tested by due sanction of historical experience and found without error, applied to the operations and processes of nature in the physical universe and found to be without contradiction, laid before the gods without question or fear, and able to wait a hundred generations and have it confirmed without a doubt by a Sage of posterity.

This passage might be edifying, but it is difficult to know what it has to do with evolution in any concrete sense.

In 1973, Dobzhansky made his oft-quoted statement "Nothing in biology makes sense except in the light of evolution." Two years later, E. O. Wilson (1975) provided an example in the form of *Sociobiology: The New Synthesis*, which claimed that the social behavior of all species, from ants to primates, could be understood on the basis of the same evolutionary principles. Sociobiology was celebrated as a major advance—except for the last chapter on humans, which ignited a storm of controversy. Wilson was attempting to relate evolutionary theory to human behavior in a much more concrete sense than Dobzhansky in *Mankind Evolving*, but this was not admissible in 1975.

It wasn't until the 1990s that terms such as "evolutionary psychology" and "evolutionary anthropology" began to gain currency, and even then they had the air of scandal about them. In short, although evolutionary biology developed into an enormously sophisticated science during the twentieth century, its widespread application to human behavior and culture has only taken place only since the mid-1990s. If the theme of the present volume is how evolutionary theory has gone beyond the

Modern Synthesis, then the inclusion of human behavior and culture counts as one of the most important recent extensions, including but not restricted to the concept of human evolution as a major evolutionary transition.

References

Allee WC, Emerson AE, Park O, Park T, Schmidt KP (1949) Principles of Animal Ecology. Philadelphia: Saunders.

Axelrod R, Hamilton WD (1981) The evolution of cooperation. Science 211: 1390–1396.

Bingham PM (1999) Human uniqueness: A general theory. Quarterly Review of Biology 74: 133–169.

Boehm, C. (1993) Egalitarian behavior, society and reverse dominance hierarchy (and replies). Current Anthropology 34: 227–254.

Boehm, C (1999) Hierarchy in the Forest: The Evolution of Egalitarian Behavior. Cambridge, MA: Harvard University Press.

Burt A, Trivers R (2006) Genes in Conflict: The Biology of Selfish Genetic Elements. Cambridge, MA: Belknap Press of Harvard University Press.

Darwin C (1871) The Descent of Man and Selection in Relation to Sex. 2 vols. New York: Appleton.

Dawkins R (1976) The Selfish Gene. Oxford: Oxford University Press.

Dawkins R (1982) The Extended Phenotype. Oxford: Oxford University Press.

Dobzhansky T (1962). Mankind Evolving: The Evolution of the Human Species. New Haven, CT: Yale University Press.

Dobzhansky T (1973) Nothing in biology makes sense except in the light of evolution. American Biology Teacher 35: 125–129.

Hamilton WD (1963) The evolution of altruistic behavior. American Naturalist 97: 354–356.

Hamilton WD (1964) The genetical evolution of social behavior: I and II. Journal of Theoretical Biology 7: 1–52.

Hamilton WD (1967) Extraordinary sex ratios. Science 156: 477–488.

Hamilton WD (1975) Innate social aptitudes in man: An approach from evolutionary genetics. In Biosocial Anthropology. R. Fox, ed.: 133–153. London: Malaby Press.

Jablonski D (2008) Species selection: Theory and evidence. Annual Review of Ecology, Evolution and Systematics 39: 501–524.

Margulis L (1970) Origin of Eukaryotic Cells: New Haven, CT: Yale University Press.

Maynard Smith J (1964) Group selection and kin selection. Nature 201: 1145–1147.

Maynard Smith J (1982) Evolution and the Theory of Games. New York: Cambridge University Press.

Maynard Smith J, E Szathmáry (1995) The Major Transitions in Evolution. New York: Oxford University Press.

Maynard Smith J, Szathmáry E (1999) The Origins of Life: From the Birth of Life to the Origin of Language. Oxford: Oxford University Press.

Nunney L (1999) Lineage selection: Natural selection for long-term benefit. In Levels of Selection in Evolution. L Keller, ed.: 238–252. Princeton, NJ: Princeton University Press.

Okasha S (2007) Evolution and the Levels of Selection. Oxford: Oxford University Press.

Pepper JW (2003) The evolution of evolvability in genetic linkage patterns. Biosystems 69: 115–126.

Price GR (1970) Selection and covariance. Nature 277: 520–521.

Price GR (1972) Extension of covariance selection mathematics. Annals of Human Genetics 35: 485–490.

Sober E, Wilson DS (1998) Unto Others: The Evolution and Psychology of Unselfish Behavior. Cambridge, MA: Harvard University Press.

Tomasello M (1999) The Cultural Origins of Human Cognition. Cambridge, MA: Harvard University Press.

Tomasello M, Carpenter M, Call J, Behne T, Moll H (2005) Understanding and sharing intentions: The origins of cultural cognition. Behavioral and Brain Sciences 28: 675–691.

Turchin P (2005) War and Peace and War: Life Cycles of Imperial Nations. New York: Pi Press.

Wagner GP, Altenberg L (1996) Complex adaptations and the evolution of evolvability. Evolution 50: 967–976.

Williams GC (1966) Adaptation and Natural Selection: A Critique of Some Current Evolutionary Thought. Princeton, NJ: Princeton University Press.

Williams GC (1992) Natural Selection: Domains, Levels and Challenges. Oxford: Oxford University Press.

Wilson DS (2007) Evolution for Everyone: How Darwin's Theory Can Change the Way We Think About Our Lives. New York: Delacorte.

Wilson DS (2008) Social semantics: Toward a genuine pluralism in the study of social behavior. Journal of Evolutionary Biology 21: 368–373.

Wilson DS, Van Vugt M, O'Gorman R (2008) Multilevel selection and major evolutionary transitions: Implications for psychological science. Current Directions in Psychological Science 17: 6–9.

WilsonDS, Wilson, EO (2007) Rethinking the theoretical foundation of sociobiology. Quarterly Review of Biology 82: 327–348.

Wilson DS, Wilson EO (2008) Evolution "for the good of the group." American Scientist 96: 380–389.

Wilson EO (1975) Sociobiology: The New Synthesis. Cambridge, MA: Harvard University Press.

Wright S (1945) Tempo and Mode in Evolution: A critical review. Ecology 26: 415–419.

III EVOLVING GENOMES

5 Integrating Genomics into Evolutionary Theory

Gregory A. Wray

Many of the major advances in evolutionary biology have grown out of synthesis between disparate disciplines. Indeed, synthesis was present right from the beginning. Darwin was a consummate integrator of information: in formulating his theory of natural selection he drew key insights not only from the scientific literature and his own extensive observations of natural history, but also from geology and sociology. He also drew on medicine, plant and animal breeding, and the nascent fields of embryology and paleontology to provide material evidence in support of his ideas. Many decades later, the Modern Synthesis integrated fundamental advances from Mendelian and quantitative genetics into evolutionary thinking, while a more contemporary view of paleontology played a smaller but significant part of the integration. The Modern Synthesis is often considered the most pivotal era in the history of post-Darwinian thinking about evolution. Witness the title and topic of the present volume: the Modern Synthesis is the benchmark against which all other advances in evolutionary theory are measured (Pigliucci 2007).

A compelling case could be made, however, that information about the material basis for heredity has been just as transformative to evolutionary biology as the Modern Synthesis. Understanding the structure of DNA, the physical nature and variety of mutations, the molecular consequences of different kinds of mutations, and the mechanisms by which genes produce traits have all led to profound insights into evolutionary processes and mechanisms (Lynch 2007). Although the impact of molecular biology on evolutionary thinking was spread over several decades rather than concentrated into just a few years, the insights that have emerged from it are as profound as any that emerged from the Modern Synthesis. A prominent example is Kimura's Neutral Theory, which was motivated by empirical observations of genetic variation. The development of the Neutral Theory was not a natural extension of the Modern

Synthesis, and could not have happened in a premolecular era. Yet Kimura's ideas have utterly transformed how evolutionary biologists model and analyze evolutionary processes at a genetic level.

A century and a half after the publication of the *Origin of Species*, evolutionary biology is once again in a period of extraordinary integration and synthesis (Feder and Mitchell-Olds 2003; Rose and Oakley 2007; Pagel and Pomiankowski 2007). A quick perusal of the chapters in this book reveals that the impetus for this excitement is coming from several sources. Unquestionably, however, one of the most important of these is the availability of genome-scale data sets from many species and from many individuals within some species. As methods for gathering genomic data become more robust and as prices for doing so drop, consideration of these very large and rich data sets will become routine in every facet of evolutionary biology. The impact will be profound. In this chapter, I discuss some of the opportunities and challenges that the genomic era brings to evolutionary biology, and some of the ways current research into genome evolution is extending the Modern Synthesis.

Extending the Modern Synthesis to the Genome

Although genomics is one of the youngest branches of biology, two distinct phases in its history are already over. The first is the era when information was limited to genome sequences from just a handful of widely divergent species. The number of genome sequences is rising exponentially: prokaryotic genomes are being sequenced on a daily basis, and eukaryotic genome sequences appear almost weekly. Alignments of entire mammalian genomes and reconstructed ancestral genome sequences for internal nodes (Ma et al. 2006; Blanchette et al. 2004) signal the beginning of a new era in understanding how genomes evolve. Sequenced genomes from closely related species provide particularly appealing subjects for evolutionary analyses, and this information is now available for several clades (e.g., Kellis et al. 2003; Stark et al. 2007; Rhesus Macaque Genome Sequencing and Analysis Consortium 2007). It is possible to apply comparative methods in a serious way to sequences at the scale of tens of kb up to entire genomes (e.g., A. G. Clark et al. 2003; Doniger and Fay 2007; Hahn 2007), based on information that is accessible through the Web. As ultrahigh throughput sequencing technologies become more robust and affordable, it is becoming possible to generate whole-genome sequences and very large population samples of targeted regions at costs that a single lab group can contemplate.

The second phase in the history of genomics that is already over is the era when data consisted only of DNA sequences. Today, much of the excitement centers around functional data at a genome-wide scale. Microarrays, which measure mRNA levels for thousands of genes at once, were the first technology to provide data of this kind (Eisen et al. 1998). Evolutionary biologists began using microarrays as costs dropped, providing the first glimpses of evolutionary differences in gene expression throughout the genome (e.g., Oleksiak et al. 2002; Khaitovitch et al. 2005). Ultrahigh-throughput DNA sequencing will largely supplant microarrays within the next few years, circumventing the need to design custom arrays for each species (a major impediment to comparative analyses) and providing the first comprehensive and unbiased sampling of gene expression. Several other kinds of genome-scale functional assays are making their way into evolutionary studies, including assays for alternate splicing (Calarco et al. 2007), binding sites for transcription factors (Moses et al. 2006; Odom et al. 2007), DNA methylation (Zhang et al. 2008), chromatin configuration (Babbitt and Kim 2008), and microRNA binding sites (Wang et al. 2008). Although these remain expensive technologies to apply in a comparative context, prices continue to drop, and they will undoubtedly see increasing application by evolutionary biologists.

Do these genome-scale data sets offer anything to evolutionary biologists that analyses of single genes do not? After all, a genecentric focus has proven hugely successful for more than half a century. The answer is in the affirmative for a couple of basic reasons. For one, there are interesting evolutionary phenomena that are apparent only at the scale of hundreds or thousands of genes. For instance, we would not know about the properties of very weak negative selection, including bias against certain codons (Akashi 1996; Nielsen et al. 2007) and spurious transcription start sites (Hahn et al. 2003; Froula and Francino 2007), were it not possible to examine sequences from whole genomes. Second, genome-scale data sets provide much more accurate and less biased information than any single gene, or even dozens of genes, can provide. Each gene is unique with respect to levels of variation and fixed differences, reflecting a distinct combination of the influences of negative selection to maintain the function of its product, any recent adaptive changes, the genomic region where it resides, and the presence of duplicates nearby or far away (Gillespie 1992; Li 1997). Surveying thousands of genes provides a much clearer understanding of general trends by averaging across these distinct individual histories. Third, full genome

sequences make it much easier to study rare events, such as gene duplications and losses, transpositions and other large-scale rearrangements,
changes in centromere positioning, and so forth. The vast majority of
what we know about the origin and fate of mutations is based on the
most abundant kind of mutation (single base substitutions) because they
can be quantified within segments as tiny as a few kb or even less; in
contrast, we know much less about the frequency with which rarer mutations arise and persist, the conditions that influence these processes, and
their phenotypic and fitness consequences. At the scale of entire genomes,
even rare events can be studied quantitatively. Fourth, comprehensive
surveys of all genes allow one to screen for functional differences. For
instance, microarrays measure which genes show the biggest differences
in expression between species or populations or environments, a powerful complement to genetic association studies. Like quantitative genetics,
this approach is unbiased and comprehensive, in the sense that it can
query most or all genes rather than just a hand-picked set of genes;
however, it is sometimes far cheaper than quantitative genetics and
can be applied in many species for which breeding designs are not
practical.

In studying multiple genome-scale data sets, evolutionary biologists
are venturing into largely uncharted territory for both theory and data
analysis. The opportunities are exciting, but the challenges are not trivial.
To take just one example (but an important one): nearly all of population genetic theory is predicated on the assumption of minimal epistatic
interaction among genes; in other words, the basic assumption is that
one can ignore the rest of the genome when considering segregating
variation in any given gene. Yet there is growing evidence that epistasis
is pervasive (Gibson and Dworkin 2004). Genetic background often
has a strong effect on the expressivity of a mutation, whether this is
measured as an organismal trait or an intermediate phenotype such as
gene expression (e.g., Brem and Kruglyak 2005). Effects of linkage are
also generally ignored in studies of population genetics and molecular
evolution. Another kind of interaction that is often ignored is gene
duplication. One of the lessons from whole-genome sequencing has
been the discovery that many genes are tandemly duplicated. The presence of a nearby paralog may affect patterns of nucleotide substitution
by relieving functional constraint (e.g., Lynch et al. 2001; Hittinger and
Carroll 2007).

The reality is that genes do not evolve in isolation, but rather in the
context of the rest of the genome. Statistical methods for estimating the

magnitude of epistatic interactions at genomic scales are beginning to appear (Jannink and Jansen 2001; Yang et al. 2007; Pattin et al. 2008), but have not been widely applied. Multilocus models of selection that incorporate nonadditive contributions to fitness have been developed (Gavrilets and de Jong 1993; Beerenwinkel et al. 2007), but empirical applications again remain limited. Now that it is possible to obtain genotypes from hundreds or thousands of markers throughout the genome, it begins to be possible to apply these tests and models more widely.

The following sections discuss three extensions to the Modern Synthesis that are emerging out of the tumult and excitement of evolutionary genomics.

Updating Models of Genetic Information

The first extension from a genomics perspective is to update working models of genetic information. The dominant gene models that evolutionary biologists have used for decades are beginning to show their age. This is true of the gene models used in population genetics, molecular evolution, and evolutionary genetics—although for somewhat different reasons in each case. Progress in molecular biology has rendered the gene models used in all three of these areas of evolutionary biology outdated, and genomic data sets are pushing them past the breaking point. Each model is considered in turn below, along with some of the ways in which current research is providing fruitful extensions.

Population Genetics
The traditional gene model of population genetic theory is highly abstract. A gene is considered in isolation from the rest of the genome, on the assumption that the majority of genetic variation is additive. Gene-by-environment interactions are assumed to be minimal, and the effects of a mutation are assumed to be static even if new mutations arise or the environment changes. The abstract model also ignores information that is often available about specific genes, such as the function of its product and whether it belongs to a gene family. This is not to say that gene-by-environment interactions have never been studied, of course, nor that gene function is never taken into account (see, for example, Via and Lande 1985); the point is that such studies are the exception.

Simple models are convenient from a mathematical perspective, and abstraction was not only justifiable but necessary during the time of the Modern Synthesis when the physical nature of genes and mutations were a complete mystery. But a simple, abstract gene model is becoming increasingly limiting in an era where we know an enormous amount about how genes function, how different kinds of mutations alter that function, how mutations can interact to influence traits, and how genes differ from one another in terms of function and trait associations.

Unfortunately, it is difficult to gauge how well a simple, abstract gene model performs relative to one that incorporates more information. Part of the problem is that we have only a dim idea of the general extent of nonadditive interactions among segregating variants and of gene-by-environment interactions. What is clear, however, is that these kinds of assumption-violating interactions are not rare. Gene-by-environment interactions are pervasive (reviewed in West-Eberhard 2003). Although less well documented, epistasis apparently is also widespread (Gibson and Dworkin 2004; Hermisson and Wagner 2004; Azevedo et al. 2006). Population geneticists have modeled epistasis (e.g., Wagner and Mezey 2000; Schlosser and Wagner 2008), and quantitative genetics provides powerful tools for measuring its effects (e.g., H. Li et al. 2007; Aylor and Zeng 2008), but most empirical studies focus on main effects and relatively few have attempted to uncover the general extent of epistasis. The largest relevant data sets come from studies of gene expression at a genome-wide scale (Rockman and Kruglyak 2006). Recent studies have demonstrated the evolutionary impact of epistasis across genomes (e.g., Brem and Kruglyak 2005; Cooper et al. 2008) and that ecologically relevant environmental variation can influence the transcription of a large proportion of genes (e.g., Idaghdour et al. 2008; Sambandan et al. 2008). The fitness consequences of these effects are rarely known, but many are hypothesized to be adaptive, including stress responses, immune system function, induced defenses, and various forms of phenotypic plasticity (West-Eberhard 2003; Pigliucci 2005).

Limitations in the abstract gene model become clear when information about a specific gene is available. Most of these concern the added predictive power that this information can bring. When modeling the likely fate of a mutation, it can be helpful to know, for instance, that a gene is the product of a recent duplication and might be under relaxed selective constraint, or that it lies within an inversion, and the resulting lack of recombination might influence its fate independent of fitness consequences. Likewise, it is useful to know something about the function of the gene's product: immune system components and testis-expressed genes, for example, generally evolve faster than most other kinds of genes in animals (W.-H. Li 1997). Simply knowing that a mutation has occurred in a histone gene is much more useful for predicting its fate than the most sophisticated population genetic model that ignores gene function. Similarly, mutations in different positions within a gene can also have very different fates, because they are more or less likely to alter the function of the protein and thereby affect fitness. This effect is very clear, for instance, when comparing the fate of mutations within the active site of an enzyme or the DNA binding domain of a transcription factor with the rest of the gene.

Relevant information about specific genes and mutations can be incorporated into population genetic models. One area where population genetic theory has already developed explicit models to take advantage of ancillary information is to distinguish between genes that reside on autosomes from those that reside on sex chromosomes. These models generally predict subtle effects, but analyses of genome-scale data sets allow one to test their predictions (e.g., Nachman and Crowell 2000; Lu and Wu 2005). Whether a gene resides on an autosome or a sex chromosome is a qualitative datum, but many other relevant kinds of information are quantitative. Extending population genetic models to incorporate these other kinds of information will require parameterization.

Increasingly, the necessary information is available. Genomic sequences are at hand for many of the species commonly studied by evolutionary biologists, so that it is often possible to obtain information about gene copy number, gene product function, and chromosomal position (and much more information than this for the major model organisms). In clades where the genomes of multiple species have been sequenced or where variation has been surveyed genome-wide, it is also possible to derive a quantitative expectation of the likely fate of a mutation over longer time scales. Genome-wide analyses are beginning to reveal distinct patterns among functional classes of genes in terms of linkage disequilibrium, mutational spectrum, and population structure within species (e.g., Voigt et al. 2006; R. M. Clark et al. 2007), as well as in terms of sequence substitution rates between species (e.g., A. G. Clark et al. 2003; Haygood et al. 2007). Although these studies are primarily exploratory, they provide the basis for building an expectation about the fate of alleles in populations that are parameterized for individual genes, functional classes of genes, and nucleotide positions within genes.

Molecular Evolution

In contrast to population genetics, the gene models used in studies of molecular evolution have recognized differences among mutations from the outset. The traditional molecular evolution gene model consists of a sequence of DNA that begins with an ATG codon, ends with one of the stop codons, and contains several more codons in between. The distinction between synonymous and nonsynonymous nucleotide substitutions provides a crude model for interpreting the fitness consequences of mutations, and the ratio of these two classes of mutations has been a staple of analyses for decades (Gillespie 1992; W.-H. Li 1997; Hartl and Clark 2006). But this is a rough and imprecise model: many, perhaps the majority, of nonsynonymous substitutions have no fitness consequence because they don't affect protein function, while some synonymous substitutions have fitness consequences because they alter splicing or codon usage (e.g., Kimchi-Sarfaty et al. 2007).

Other kinds of mutations are generally ignored. Insertions and deletions in multiples of three bases need not alter the reading frame, although they will if they fall across an intron-exon junction. In-frame indels are almost always eliminated from alignments prior to quantitative analysis, even though they are at least as likely to affect fitness as nonsynonymous substitutions. Three other kinds of mutations—indels that shift the reading frame, premature stop codons, and changes in the position of the start codon—can all have large functional consequences because they alter protein length. These kinds of mutations are usually assumed to result in loss of protein function, and therefore to carry a large negative fitness component. Comparisons within and across genomes reveal that this is clearly not always the case (e.g., Ng et al. 2008).

An even larger deficiency in the codon-based gene model is that it ignores a large fraction of mutations that affect gene function. Transcriptional initiation is regulated by sequences that lie almost entirely outside coding sequences; transcripts are spliced to remove introns in a sequence-dependent manner; many genes utilize alternative transcription start sites; 3' untranslated regions often contain sites that regulate message stability and trafficking; and additional noncoding sequences regulate chromatin configuration or encode microRNA molecules that regulate transcript turnover or translation (Lewin 2007; Latchman 2007).

Most of these various kinds of functional noncoding sequences have been studied by molecular biologists for decades. Regulatory sequences have received relatively little attention in studies of molecular evolution, but this is becoming increasingly difficult to justify (Chen and Rajewsky 2007; Wray 2007; Carroll 2008). Genome sequence comparisons between species have estimated that a roughly comparable number of functionally constrained sites lie in noncoding and coding regions of eukaryotic genomes (Shabalina and Kondrashov 1999; Shabalina et al. 2001; Andolfatto 2005). In humans there may be more segregating mutations that alter transcriptional regulation than mutations affecting protein sequence (Rockman and Wray 2002), and positive selection on 5' noncoding regions (just one part of the regulatory landscape) was apparently at least as extensive as positive selection on all coding sequences during human origins (Haygood et al. 2007). Evidence is accumulating that the traditional focus on coding sequences misses out on a large fraction of mutations of adaptive significance within a genome. And this may be as much a qualitative as a quantitative blind spot: coding and noncoding mutations may contribute differentially to particular kinds of traits, such as morphology, reproduction, or immune function (Carroll 2008; Haygood et al. submitted).

Quantitative Genetics

The gene model of quantitative genetics is based simply on physical locations within the genome. The subjects of quantitative genetic study are appropriately called loci (positions) rather than genes, because they may or may not reside within a gene. Until quite recently, identifying the causal variants that define quantitative trait loci (QTL for short) has been quite difficult except in unusual cases. As a result, quantitative genetics has largely ignored the molecular consequences of causal mutations (biochemical consequence if coding, regulatory consequence if not), whether certain kinds of mutations are more likely to produce trait or fitness consequences than others, and how causal mutations actually alter organismal traits of interest.

Recently, however, the focus of quantitative genetics is shifting away from simply identifying QTL that are devoid of any functional context and toward a model of identifying genes with testable functional involvement and precise mutational bases that can be experimentally investigated. This has been made possible by the ability to carry out high-resolution mapping from large numbers of genetic markers (10^4–10^6) distributed across the genome, which in turn makes feasible the identification of causal mutations (also known as QTN, or quantitative trait nucleotides) in a growing number of cases. Knowing what kinds of genes and mutations contribute to adaptation in organismal traits adds a whole new dimension to evolutionary genetics, for instance, providing insights into how mutations produce trait differences and whether parallel traits have parallel genetic bases (e.g., Shapiro et al. 2004; Prud'homme et al. 2006; Tishkoff et al. 2007).

Another extension is the use of scans for positive selection (described earlier), which provide a valuable complement to traditional quantitative genetic approaches by identifying genes that may be involved in adaptation throughout the genome. There are many cases where quantitative genetic approaches cannot be applied, for instance, when hybrids can't be made between species or generation times are too long. In such cases, genome-wide scans for positive selection provide one of the few comprehensive and unbiased approaches to identifying the genetic basis for trait differences among species.

The conventional gene models of population genetics, quantitative genetics, and molecular evolution all need to be, and are being, extended in order to accommodate new information. Much of the impetus comes from outside of evolutionary biology, primarily from advances in molecular biology but increasingly from genome-scale data sets. The inescapable fact is that we now live in a world where data sets are immensely larger than they were just a few years ago. Larger scales of data provide not only a more accurate, but also a more comprehensive, view of evolutionary processes than ever before. Importantly, this brings both new information and new challenges. On the one hand, larger data sets allow quantification of rare processes and the identification of spatial organization with the genome. On the other hand, leveraging these much larger data sets will require further extensions to both the theory and the analytical tools of evolutionary biology (Singh 2003; Lynch 2007; Pagel and Pomiankowski 2007).

Moving Beyond the Gene-in-a-Bubble Approach

The second extension to the Modern Synthesis that is emerging from genomics involves integrating approaches from different branches of evolutionary biology. It seems intuitively obvious that answering complex questions in evolutionary biology will sometimes require drawing on a combination of methods from phylogenetics, population genetics, evolutionary genetics, molecular evolution, evolutionary ecology, and evolutionary developmental biology. In practice, however, most publications in evolutionary biology draw on the methods of just one of these areas.

Why is this so? Following the Modern Synthesis, evolutionary biology became increasingly genecentric, with attention focused on the gene as the primary unit of selection (Hamilton 1963; Williams 1966; Dawkins 1976). About the same time, it became possible to sequence first proteins, and later DNA, technologies which utterly transformed evolutionary biology. And in so doing, this strongly reinforced the gene-centric perspective. The single gene became the prime unit of analysis for population genetics, and molecular evolution in particular. These two notably active and fertile areas of evolutionary biology during the last third of the twentieth century developed rather different gene models (see previous section).

But they shared one thing in common: the gene in their genecentric models resided within a conceptual bubble, hermetically sealed away from environmental influences, regulatory processes, trait associations, and interactions with the rest of the genome. Empiricists worked with inbred lines, reared organisms under uniform environmental conditions, sequenced a single gene of interest, and focused on DNA to the exclusion of traits; theorists built models and developed statistical tests aligned with this gene-in-a-bubble approach. (In contrast, research in quantitative genetics often incorporated gene interactions and environmental influences.)

Isolating genes from the environment, the rest of the genome, and trait associations is a powerful approach for addressing certain kinds of questions, because it factors out "messiness" of various kinds. Treating genes in isolation also made sense in an era when relevant data about influences external to a single gene were sparse. But the gene-in-a-bubble approach is ultimately limiting. Real organisms live in environments that vary in space and time; alleles function in a diversity of genetic backgrounds; and genes affect traits that have real ecological consequences.

An important extension at this point is popping the bubble surrounding the single gene and placing it into a broader and more biologically realistic context. This requires examining a gene from the perspectives of population genetics, evolutionary ecology, quantitative genetics, evolutionary developmental biology, and more. Genomic data sets provide an impetus to break down the boundaries between these traditionally distinct disciplines by offering a common currency for analysis and modeling. Genome sequences, for example, facilitate the development of genetic markers for quantitative genetics, provide data for developing accurate background models of sequence evolution testing, produce data that allow testing for selection throughout the genome, furnish the information necessary for carrying out functional analyses such as microarrays, allow inference of changes in regulatory sequences such as enhancers and microRNAs, and identify candidate genetic differences that may explain trait differences.

Integrating approaches among traditionally distinct disciplines within evolutionary biology can yield unique insights. Studies are beginning to appear that identify which genes are involved in the evolution of a particular trait, discern whether these genes have been under positive or balancing selection, reveal how changes in the function of these genes alter the trait of interest, and measure the impact of the trait consequences in the natural environment. Examples of studies that integrate some or all of these perspectives include the evolution of reduced armor in freshwater populations of three-spined sticklebacks (Shapiro et al. 2004, 2006), the evolution of wing and abdominal color pattern within the genus Drosophila (Gompel et al. 2005; Prud'homme et al. 2006; Jeong et al. 2008), and the evolution of lactose tolerance and malaria resistance during very recent human evolution (Hamblin and Di Rienzo 2000; Enattah et al. 2002; Tishkoff et al. 2007).

Extending Analyses across Scales of Genetic Organization

The third kind of extension of the Modern Synthesis inspired by genomics is applying both theoretical and analytical approaches to studying evolutionary processes across the full scale of genetic organization. Although the focus has been at the scale of a single gene for several decades, the lower and upper bounds of this scale are increasingly accessible to study by evolutionary biologists. At the smallest end of the scale are single mutations, while at the largest lie whole genomes. In between fall other less commonly discussed but important scales of genetic organization: haplotypes and genic partitions (exons, introns, 5' and 3' untranslated regions, and regulatory regions) lie between mutations and genes in scale, while gene networks and chromosomes occupy distinct organizational dimensions at a scale between genes and whole genomes.

The challenge is to extend both the theoretical and the analytical approaches of population genetics, quantitative genetics, and molecular evolution across this entire range of scales. Increasingly, data sets are appearing that either put a strain on traditional methods of evolutionary analysis or simply can't be analyzed within existing frameworks. This problem is particularly acute at the whole-genome scale. Very large data sets are often subjected to far more statistical comparisons than traditional evolutionary studies (often by several orders of magnitude).

Correcting for multiple comparisons is clearly important, but off-the-shelf methods designed for other purposes may be inappropriate because their assumptions are violated by the nature of the data. In addition, there is rarely a clear understanding of the distribution of trait measures, and the appropriate statistical test is not always obvious. Finally, missing information and variation in data quality are often a much larger problem with genomic data sets than in traditional evolutionary analyses. Visual checks on data quality are simply not possible because of scale, and statistical tests that assume complete data sets are often inappropriate.

Gene networks are a second important scale of genetic organization where it will be necessary to extend traditional methods of evolutionary analysis. Understanding how genetic variation affects network function, for instance, will require modeling networks to provide testable predictions, as well as a way to incorporate the structure of interactions among genes when analyzing results. Some existing modeling approaches (based, for instance, on Boolean, Bayesian, or algebraic topology methods) and statistical methods (e.g., path analysis) provide promising possibilities. A special challenge for association studies is how to incorporate information about known interactions within gene networks, as this violates assumptions of independence that are part of traditional approaches (Lynch and Walsh 1998); a few studies have developed and applied methods to do this (e.g., Walsh et al. 2008).

Not all the challenges in extending evolutionary theory and analysis to other scales of genetic organization are statistical: data representation and visualization present their own problems. No one wants to read a table that is 1,000 lines long (E. coli-sized), much less 20,000 lines long (human genome-sized). A common visualization tool is the "heat map," which represents a value for each gene, typically a test statistic or functional measure, as a color value. These displays can pack thousands of results into a small space, albeit at the cost of severe data reduction and the ability to read the names of individual genes. There are also challenges in building databases that provide ready access to results so that analyses can be conducted by other investigators.

Considerable progress has been made toward addressing these challenges. The analysis of genomic data sets is the primary area of research for increasing numbers of statisticians, mathematicians, and computer scientists. Some of their research has direct applications for evolutionary analyses, including false discovery rate to correct for multiple comparisons (Storey and Tibshirani 2003), hidden Markov models for modeling sequence evolution (Siepel and Haussler 2004), and appropriate statistical frameworks for association studies (Pritchard et al. 2000). The transfer of expertise is not all in one direction: phylogenetic methods are routinely (but unfortunately not universally) applied in mainstream genomic studies to distinguish orthology from paralogy. Although phylogenetic inference is more computationally intensive than pattern-matching methods such as reciprocal best BLAST hits, it generally produces a lower error rate. On the data representation front, visual displays are gradually evolving beyond the heat maps that became almost ubiquitous following the invention of microarrays. Finally, databases have begun to move beyond simply serving as data repositories, and are setting standards for organizing and archiving information.

Evolutionary biology is transitioning from an era of data limitation to one of data abundance, and even superabundance, in a limited but growing number of areas. Put simply, the challenge is shifting away from how to gather data and toward how to analyze, integrate, and make sense of very large data sets (Singh 2003).

Summary and Prospects

A century and a half after the publication of the *Origin of Species*, evolutionary biology is entering a time of extraordinary expansion. The foundation that Darwin built was tested and vastly strengthened, initially with the elucidation of transmission genetics and later with the nature of the genetic material. Today we are in the midst of another exciting, and potentially equally transformative, period in the history of evolutionary biology. Information from molecular biology, developmental biology, and, most recently, genomics is prompting substantial changes to the genecentric view that emerged during and shortly after the Modern Synthesis. I have argued that three specific extensions are under way already. First, enormous advances in understanding how genes function and how they work together to produce developmental and physiological processes are prompting substantial and highly informative updates to the gene models used in evolutionary studies. Second, technological advances now allow us to study genes in the context of the rest of the genome and the environment rather than as isolated entities, revealing much about gene interactions, phenotypic plasticity, and developmental roles. And third, expanding analyses beyond genes as the focal unit is allowing researchers to study the evolution of the hereditary material at a wide range of scales, from single mutations through gene networks to entire genomes.

The advent of genome-scale data sets is prompting a wide range of exciting new studies. The results are revealing evolutionary phenomena of which we were previously unaware, such as codon bias and biases in the distribution of positive selection on coding and noncoding sequences depending on gene function. Genomic data are also being used to evaluate predictions that were formerly untestable, such as genome-wide estimates of epistasis and the long-term fate of gene duplications. Technological advances are providing entirely new ways to identify the genetic basis for trait evolution through whole-genome association studies and genome-scale functional assays such as microrrays. Applying the traditional approaches of population genetics, evolutionary genetics, and molecular evolution to genomic data sets poses nontrivial challenges. Balanced against these challenges, however, are extraordinary opportunities to better understand evolutionary processes and mechanisms. Theorists and quantitative biologists, in particular, are entering a period of exceptional opportunity as data sets expand in scale, scope, and kind.

It is too soon to know how extensively and in what ways genomic data will influence our understanding of evolution. But one point is already clear: these data are building upon and extending the robust framework of the Modern Synthesis in ways that we could not have imagined even a decade ago.

Acknowledgments

David Garfield and Jenny Tung provided many helpful comments. My research is supported by grants from the National Science Foundation and the National Institutes of Health.

References

Akashi H (1996) Molecular evolution between Drosophila melanogaster and D-simulans: Reduced codon bias, faster rates of amino acid substitution, and larger proteins in D-melanogaster. Genetics 144: 1297–1307.

Andolfatto P (2005) Adaptive evolution of non-coding DNA in Drosophila. Nature 437: 1149–1152.

Aylor DL, Zeng, ZB (2008) From classical genetics to quantitative genetics to systems biology: Modeling epistasis. PLoS Genetics 4: 1000029.

Azevedo L, Suriano G, van Asch B, Harding RM, Amorim A (2006) Epistatic interactions: How strong in disease and evolution? Trends in Genetics 22: 581–585.

Babbitt GA, Kim, Y (2008) Inferring natural selection on fine-scale chromatin organization in yeast. Molecular Biology and Evolution 25: 1714–1727.

Beerenwinkel N, Pachter L, Sturmfels B, Elena SF, Lenski RE (2007) Analysis of epistatic interactions and fitness landscapes using a new geometric approach. BMC Evolutionary Biology 7: 60.

Blanchette M, Kent WJ, Riemer C, Elnitski L, Smit AFA, Roskin KM, Baertsch R, Rosenbloom K, Clawson H, Green ED, Haussler D, Miller W (2004) Aligning multiple genomic sequences with the threaded blockset aligner. Genome Research 14: 708–715.

Brem RB, Kruglyak L (2005) The landscape of genetic complexity across 5,700 gene expression traits in yeast. Proceedings of the National Academy of Sciences of the USA 102: 1572–1577.

Calarco JA, Saltzman AL, Ip JY, Blencowe B (2007) Technologies for the global discovery and analysis of alternative splicing. Advances in Experimental Medicine and Biology 623: 64–84.

Carroll SB (2008) Evo-devo and an expanding evolutionary synthesis: A genetic theory of morphological evolution. Cell 134: 25–36.

Chen K, Rajewsky N (2007) The evolution of gene regulation by transcription factors and microRNAs. Nature Reviews Genetics 8: 93–103.

Clark AG, Glanowski S, Nielsen R, Thomas PD, Kejariwal A, Todd MA, Tanenbaum DM, Civello D, Lu F, Murphy B, Ferriera S, Wang G, Zheng XG, White TJ, Sninsky JJ, Adams MD, Cargill M (2003) Inferring nonneutral evolution from human-chimp-mouse orthologous gene trios. Science 302: 1960–1963.

Clark RM, Schweikert G, Toomajian C, Ossowski S, Zeller G, Shinn P, Warthmann N, Hu TT, Fu G, Hinds DA, Chen HM, Frazer KA, Huson DH, Schölkopf B, Nordborg M, Raetsch G, Ecker JR, Weigel D (2007) Common sequence polymorphisms shaping genetic diversity in Arabidopsis thaliana. Science 317: 338–342.

Cooper TF, Remold SK, Lenski RE, Schneider D (2008) Expression profiles reveal parallel evolution of epistatic interactions involving the CRP regulon in Escherichia coli. PLoS Genetics 4: e35.

Dawkins R (1976) The Selfish Gene. New York: Oxford University Press.

Doniger SW, Fay JC (2007) Frequent gain and loss of functional transcription factor binding sites. PLoS Computational Biology 3: 932–942.

Dworkin I, Gibson G (2006) Epidermal growth factor receptor and transforming growth factor-β signaling contributes to variation for wing shape in Drosophila melanogaster. Genetics 173: 1417–1431.

Eisen MB, Spellman PT, Brown PO, Botstein D (1998) Cluster analysis and display of genome-wide expression patterns. Proceedings of the National Academy of Sciences of the USA 95: 14863–14868.

Enattah NS, Sahi T, Savilahti E, Terwilliger JD, Peltonen L, Järvelä I (2002) Identification of a variant associated with adult-type hypolactasia. Nature Genetics 30: 233–237.

Feder ME, Mitchell-Olds T (2003) Evolutionary and ecological functional genomics. Nature Reviews Genetics 4: 649–655.

Froula JL, Francino MP (2007) Selection against spurious promoter motifs correlates with translational efficiency across bacteria. PLoS ONE 2: e745.

Gavrilets S, de Jong G (1993) Pleiotropic models of polygenic variation, stabilizing selection, and epistasis. Genetics 134: 609–625.

Gibson G, Dworkin I (2004) Uncovering cryptic genetic variation. Nature Reviews Genetics 5: 681–691.

Gillespie JH (1992) The Causes of Molecular Evolution. New York: Oxford University Press.

Gompel N, Prud'homme B, Wittkopp PJ, Kassner VA, Carroll, SB (2005) Chance caught on the wing: Cis-regulatory evolution and the origin of pigment patterns in Drosophila. Nature 433: 481–487.

Hahn MW (2007) Detecting natural selection on cis-regulatory DNA. Genetica 129: 7–18.

Hahn MW, Stajich JE, Wray GA (2003) The effects of selection against spurious transcription factor binding sites. Molecular Biology and Evolution 20: 901–906.

Hamblin MT, Di Rienzo A (2000) Detection of the signature of natural selection in humans: Evidence from the Duffy blood group locus. American Journal of Human Genetics 66: 1669–1679.

Hamilton WD (1963) Evolution of altruistic behavior. American Naturalist 97: 354–356.

Hartl DL, Clark AG (2006) Principles of Population Genetics. 4th rev. ed. Sunderland, MA: Sinauer.

Haygood R, Babbitt CC, Fedrigo O, Wray GA (submitted) Positive selection in the human genome exhibits strong contrasts between coding and noncoding sequences.

Haygood RH, Fedrigo O, Hanson B, Yokoyama K-D, Wray GA (2007) Promoter regions of many neural- and nutrition-related genes have experienced positive selection during human evolution. Nature Genetics 39: 1140–1144.

Hermisson J, Wagner GP (2004) The population genetic theory of hidden variation and genetic robustness. Genetics 168: 2271–2284.

Hittinger CT, Carroll SB (2007) Gene duplication and the adaptive evolution of a classic genetic switch. Nature 449: 677–681.

Idaghdour Y, Storey JD, Jadallah SJ, Gibson G (2008) A genome-wide gene expression signature of environmental geography in leukocytes of Moroccan amazighs. PLoS Genetics 4: e1000052.

Jannink J-L, Jansen R (2001) Mapping epistatic quantitative trait loci with one-dimensional genome searches. Genetics 157: 445–454.

Jeong S, Rebeiz M, Andolfatto P, Werner T, True J, Carroll SB (2008) The evolution of gene regulation underlies a morphological difference between two Drosophila sister species. Cell 132: 783–793.

Kellis M, Patterson N, Endrizzi M, Birren B, Lander ES (2003) Sequencing and comparison of yeast species to identify genes and regulatory elements. Nature 423: 241–254.

Khaitovich P, Hellmann I, Enard W, Nowick K, Leinweber M, Franz H, Weiss G, Lachmann M, Pääbo S (2005) Parallel patterns of evolution in the genomes and transcriptomes of humans and chimpanzees. Science 309: 1850–1854.

Kimchi-Sarfaty C, Oh JM, Kim IW, Sauna ZE, Calcagno AM, Ambudkar SV, Gottesman MM (2007) A "silent" polymorphism in the MDR1 gene changes substrate specificity. Science 315: 525–528.

Latchman DS (2007) Eukaryotic Transcription Factors. 5th ed. Burlington, MA: Academic Press.

Lewin B (2007) Genes IX. Boston: Jones and Bartlett.

Li H, Gao G, Li J, Page GP, Zhang K (2007) Detecting epistatic interactions contributing to human gene expression using the CEPH family data. BMC Proceedings suppl. 1: S67.

Li W-H (1997) Molecular Evolution. Sunderland, MA: Sinauer.

Liu Y, Duan W, Paschall J, Saccone NL (2007) Artificial neural networks for linkage analysis of quantitative gene expression phenotypes and evaluation of gene x gene interactions. BMC Proceedings suppl. 1: S47.

Lu J, Wu CI (2005) Weak selection revealed by the whole-genome comparison of the X chromosome and autosomes of human and chimpanzee. Proceedings of the National Academy of Sciences of the USA 102: 4063–4067.

Lynch M (2007) Origins of Genome Architecture. Sunderland, MA: Sinauer.

Lynch M, O'Hely M, Walsh B, Force A (2001) The probability of preservation of a newly arisen gene duplicate. Genetics 159: 1789–1804.

Lynch M, Walsh B (1998) Genetics and Analysis of Quantitative Traits. Sunderland, MA: Sinauer.

Ma J, Zhang LX, Suh BB, Raney BJ, Burhans RC, Kent WJ, Blanchette M, Haussler D, Miller W (2006) Reconstructing contiguous regions of an ancestral genome. Genome Research 16: 1557–1565.

Moses AM, Pollard DA, Nix DA, Iyer VN, Li XY, Biggin MD, Eisen MB (2006) Large-scale turnover of functional transcription factor binding sites in Drosophila. PLoS Computational Biology 2: 1219–1231.

Nachman MW, Bauer VL, Crowell SL, Aquadro CF (1998) DNA variability and recombination rates at X-linked loci in humans. Genetics 150: 1133–1141.

Nachman MW, Crowell SL (2000) Estimate of the mutation rate per nucleotide in humans. Genetics 156: 297–304.

Ng PC, Levy S, Huang J, Stockwell TB, Walenz BP, Li K, Axelrod N, Busam DA, Strausberg RL, Venter JC (2008) Genetic variation in an individual human exome. PLoS Genetics 4: e1000160.

Nielsen R, DuMont VLB, Hubisz MJ, Aquadro CF (2007) Maximum likelihood estimation of ancestral codon usage bias parameters in Drosophila. Molecular Biology and Evolution 24: 228–235.

Odom DT, Dowell RD, Jacobsen ES, Gordon W, Danford TW, MacIsaac KD, Rolfe PA, Conboy CM, Gifford DK, Fraenkel, E (2007) Tissue-specific transcriptional regulation has diverged significantly between human and mouse. Nature Genetics 39: 730–732.

Oleksiak MF, Churchill GA, Crawford DL (2002) Variation in gene expression within and among natural populations. Nature Genetics 32: 261–266.

Pagel M, Pomiankowski A, eds. (2007) Evolutionary Genomics and Proteomics. Sunderland, MA: Sinauer.

Pattin KA, White BC, Barney N, Gui J, Nelson HH, Kelsey KT, Andrew AS, Karagas MR, Moore JH (2008) A computationally efficient hypothesis testing method for epistasis using multifactor dimensionality reduction. Genetic Epidemiology 33: 87–94.

Pigliucci M (2005) Evolution of phenotypic plasticity: Where are we going now? Trends in Ecology and Evolution 20: 481–486.

Pigliucci M (2007) Do we need an extended evolutionary synthesis? Evolution 61: 2743–2749.

Pritchard JK, Stephens M, Donnelly P (2000) Inference of population structure using multilocus genotype data. Genetics 155: 945–959.

Prud'homme B, Carroll SB (2006) Monkey see, monkey do. Nature Genetics 38: 740–741.

Prud'homme B, Gompel N, Rokas A, Kassner VA, Williams TM, Yeh SD, True JR, Carroll SB (2006) Repeated morphological evolution through cis-regulatory changes in a pleiotropic gene. Nature 440: 1050–1053.

Rhesus Macaque Genome Sequencing and Analysis Consortium (2007) Evolutionary and biomedical insights from the rhesus macaque genome. Science 316: 222–234.

Rockman MV, Kruglyak L (2006) Genetics of global gene expression. Nature Reviews Genetics 7: 862–872.

Rockman MV, Wray GA (2002) Abundant raw material for cis-regulatory evolution in humans. Molecular Biology and Evolution 19: 1991–2004.

Ronald J, Brem RB, Whittle J, Kruglyak L (2005) Local regulatory variation in Saccharomyces cerevisiae. PLoS Genetics 1: 213–222.

Rose MR, Oakley TH (2007) The new biology: Beyond the Modern Synthesis. Biology Direct 2: 30.

Sambandan D, Carbone MA, Anholt RRH, Mackay TEC (2008) Phenotypic plasticity and genotype by environment interaction for olfactory behavior in Drosophila melanogaster. Genetics 179: 1079–1088.

Schlosser G, Wagner GP (2008) A simple model of co-evolutionary dynamics caused by epistatic selection. Journal of Theoretical Biology 250: 48–65.

Shabalina SA, Kondrashov, AS (1999) Pattern of selective constraint in C-elegans and C-briggsae genomes. Genetical Research 74: 23–30.

Shabalina SA, Ogurtsov AY, Kondrashov VA, Kondrashov AS (2001) Selective constraint in intergenic regions of human and mouse genomes. Trends in Genetics 17: 373–376.

Shapiro MD, Bell MA, Kingsley, DM (2006) Parallel genetic origins of pelvic reduction in vertebrates. Proceedings of the National Academy of Sciences of the USA 103: 13753–13758.

Shapiro MD, Marks ME, Peichel CL, Blackman BK, Nereng KS, Jonsson B, Schluter D, Kingsley DM (2004) Genetic and developmental basis of evolutionary pelvic reduction in threespine sticklebacks. Nature 428: 717–723.

Siepel A, Haussler D (2004) Combining phylogenetic and hidden Markov models in biosequence analysis. Journal of Computational Biology 11: 413–428.

Singh RS (2003) Darwin to DNA, molecules to morphology: The end of classical population genetics and the road ahead. Genome 46: 938–942.

Stark A, Lin MF, Kheradpour P, Pedersen JS, Parts L, Carlson JW, Crosby MA, Rasmussen MD, Roy S, Deoras AN, Ruby JG, Brennecke J, Hodges E, Hinrichs AS, Caspi A, Park SW, Han MV, Maeder ML, Polansky BJ, Robson BE, Aerts S, van Helden J, Hassan B, Gilbert DG, Eastman DA, Rice M, Weir M, Hahn MW, Park Y, Dewey CN, Pachter L, Kent WJ, Haussler D, Lai EC, Bartel DP, Hannon GJ, Kaufman TC, Eisen MB, Clark AG, Smith D, Celniker SE, Gelbart WM, Kellis M (2007) Discovery of functional elements in 12 Drosophila genomes using evolutionary signatures. Nature 450: 219–232.

Storey JD, Tibshirani R (2003) Statistical significance for genomewide studies. Proceedings of the National Academy of Sciences of the USA 100: 9440–9445.

Tishkoff SA, Reed FA, Ranciaro A, Voight BF, Babbitt CC, Silverman JS, Powell K, Mortensen HM, Hirbo JB, Osman M, Ibrahim M, Omar SA, Lema G, Nyambo TB, Ghori J, Bumpstead S, Pritchard JK, Wray GA, Deloukas P (2007) Convergent adaptation of human lactase persistence in Africa and Europe. Nature Genetics 39: 31–40.

Via S, Lande R (1985) Genotype-environment interaction and the evolution of phenotypic plasticity. Evolution 39: 505–522.

Voight BF, Kudaravalli S, Wen XQ, Pritchard JK (2006) A map of recent positive selection in the human genome. PLoS Biology 4: 659–659.

Wagner GP, Mezey J (2000) Modeling the evolution of genetic architecture: A continuum of alleles model with pairwise A x A epistasis. Journal of Theoretical Biology 203: 163–175.

Walsh T, McClellan JM, McCarthy SE, Addington AM, Pierce SB, Cooper GM, Nord AS, Kusenda M, Malhotra D, Bhandari A, Stray SM, Rippey CF, Roccanova P, Makarov V, Lakshmi B, Findling RL, Sikich L, Stromberg T, Merriman B, Gogtay N, Butler P, Eckstrand K, Noory L, Gochman P, Long R, Chen Z, Davis S, Baker C, Eichler EE, Meltzer PS, Nelson SF, Singleton AB, Lee MK, Rapoport JL, King MC, Sebat J (2008) Rare structural variants disrupt multiple genes in neurodevelopmental pathways in schizophrenia. Science 320: 539–543.

Wang XW, Gu J, Zhang MQ, Li YD (2008) Identification of phylogenetically conserved microRNA cis-regulatory elements across 12 Drosophila species. Bioinformatics 24: 165–171.

West-Eberhard MJ (2003) Developmental Plasticity and Evolution. New York: Oxford University Press.

Williams GC (1966) Natural selection, the costs of reproduction, and a refinement of Lacks' Principle. American Naturalist 100: 687–690.

Wray GA (2007) The evolutionary significance of cis-regulatory mutations. Nature Reviews Genetics 8: 206–216.

Yang J, Zhu J, Williams RW (2007) Mapping the genetic architecture of complex traits in experimental populations. Bioinformatics 23: 1527–1536.

Zhang X, Shiu SH, Cal A, Borevitz JO (2008) Global analysis of genetic, epigenetic and transcriptional polymorphisms in Arabidopsis thaliana using whole genome tiling arrays. PLoS Genetics 4: e1000032.

6 Complexities in Genome Structure and Evolution

Michael Purugganan

In 1977, the first genome was sequenced—the viral species φX174, a mere 5.4 kb in length—and this sequencing was a landmark in biology (Sanger, Nicklen, and Coulson 1977). It was 18 years later, in 1995, when the genome of a living organism was first announced, the 1.8 Mb *Haemophilus influenzae* bacterial genome (Flesichmann et al. 1995), and in 2001, we crossed another major milestone when it was announced that the human genome sequence—3 Gb in length—had been completed by two groups (International Human Genome Sequencing Consortium 2001; Venter et al. 2001). Today, whole-genome sequencing projects proliferate ever faster; as of early 2008, more than 180 genome sequences had been completed across all major kingdoms, and genome sequencing technology has advanced to the point that (depending on the size of the organism's genome) single investigator laboratories can contemplate sequencing the genome of their favorite organism. And the frenzy is not confined to model species; in early 2008, the 2.2 Gb platypus genome was completed (Warren et al. 2008), as was the 372 Mb papaya genome (Ming et al. 2008).

The advent of genomics also made possible the study of comparative genomics, possible, which, taken in an evolutionary framework, allows us to examine the diversification of genome structure and the function of their component genes. Genomics has provided enabling technologies for the evolutionary sciences, producing large-scale information about genome structure and function across multiple species. This has resulted in expanding knowledge of the complete inventory of genes found in organismal genomes, and in the process has begun to bring to light several issues that continue to excite the interests of molecular evolutionists. In this chapter, I will discuss four issues that modern evolutionary biology has either learned or needs to grapple with in the age of genomics. These topics are only a handful of the myriad opportunities that genomic science presents to the advancement of evolutionary thought, but they illustrate both the possibilities and the challenges that have surfaced in the genomics era, and may help extend the scope of the Modern Synthesis.

Transposable Elements and the Genomic Ecosystem

The genome is structurally dynamic, an observation that was first advanced by Barbara McClintock in her groundbreaking studies on transposable elements (McClintock 1984). Early conceptions of the genome as a stable unit began to give way to a view of genomes as malleable structures that could evolve quite rapidly, composed of various entities that colonize the unique environment of the genomic "ecosystem."

Since about 1990, the genome has been increasingly seen as a vibrant structure subject to the movement of genetic elements within genomes and between species. Lateral gene transfer, for example, is ubiquitous between the nuclear and organellar genomes—both the eukaryotic mitochondria and the plant-specific chloroplast (Blanchard and Schmidt 1996; Huang et al. 2005). Moreover, transfers of genome segments between species are now known to be widespread (Richardson and Palmer 2007), and are believed to be a central feature of bacterial biology (Hao and Golding 2008). These have led to the notion of the genome as an evolutionary mosaic, built up of pieces cobbled together from various taxa during the long evolutionary history of life.

Transposable elements, however, have remained the major players in contributing to genome dynamism. The first, molecularly characterized in the early 1980s, were the Tn elements in bacteria, the P elements of *Drosophila,* and the Ac/Ds system of maize (Berg and Howe 1989). Since their discovery, these mobile elements are known to fall into several major types. Of these elements, the DNA transposable elements operate by excision and insertion into various places in the genome, while retrotransposons propagate by making RNA copies of themselves that are reverse transcribed and inserted into other locations in the genome (Wessler 2006). Other, smaller elements are known—for example, the MITE transposons of plants—whose mode of transposition still remains to be completely elucidated (Jiang et al. 2003).

Transposable elements comprise a large fraction of genomes, making up as high as 44% of the human genome. Numerous transposon families are observed, with over 4 million transposon copies in the human genome belonging to at least 848 families and subfamilies (Feschotte and Pritham 2007). It is the large number of transposons that have colonized and inhabit genomes, their life cycles and impact on genome structure, that bear thought on their place in molecular evolutionary processes (Wessler 2006; Feschotte and Pritham 2007).

One major question in evolutionary genetics has been the evolutionary impact of transposable elements, and whether their possible roles in facilitating genome structural evolution may explain their persistence despite their ability to insert in gene coding regions with consequent deleterious mutational impacts. The contribution of transposable elements to within-species genome variation can be substantial: a study in maize, for example, shows that intergenic regions of the genome can be dramatically divergent even between individuals in a population as a result of multiple transposon insertions and deletions (Q. H. Wang and Dooner 2006). There is also evidence to indicate that element movement can be induced by unusual environmental conditions, as has been observed under the stress of tissue culture (Komatsu, Simamoto, and Kyozuka 2003), although it is unclear what other environmental conditions may trigger element activity.

Genetic studies have shown that most transposon insertions are deleterious, producing hypomorphic loss-of-function alleles. There are certainly indications, however, that transposons can be a force in reshaping gene structure and regulation. Studies in transposable elements have shown that they can harbor regulatory elements that are co-opted by genes into which these elements are inserted, providing a means of rewiring genetic regulatory networks (White, Habera, and Wessler 1994; Wessler, Bureau, and White 1995). Clear examples are observed in plant genomes, as in the case of a ~500-bp regulator sequence in the tomato *LAT59* gene, which is derived from a retrotransposon (White, Habera, and Wessler 1994). Another example is a retrotransposon insertion in the upstream promoter regions of members of the pea rbcsE gene family, which may drive regulatory diversification of different gene family members (White, Habera, and Wessler 1994).

Transposon insertions into coding sequences are also known to create new introns or induce alternative splicing patterns that can create new transcripts and, possibly, protein products (Purugganan and Wessler 1992; Purugganan 1993). In a study of retrotransposon insertions in the maize *Wx* gene, it was shown that insertions of retroelements in the *Wx-Stonor, wx-B5,* and *wx-G* alleles lead to alternatively spliced products (Varagona, Purugganan, and Wessler 1992). The alternative splicing

need of molecular systematists and population geneticists for neutral molecular markers, to infer taxon or population relationships and demographic processes, both of which could be done only in an unbiased fashion with the use of neutral loci.

The primacy of the neutral model began to be seriously questioned in the 1990s, when molecular evolutionists rediscovered their lingering interest in selection and adaptation (Kreitman 1996; Orr 2005). This led to critiques of the neutral model and accumulating information on examples of selection at the molecular level (Kreitman 1996). There had always been clear molecular examples of selection—including diversifying selection at the mammalian MHC locus and plant self-incompatibility genes (Ohta 1991; Klein et al. 1998; Kamau and Charlesworth 2005)—but other instances of selection at the genetic level began to be systematically uncovered. Interest in the nature of selection is strengthened by increasingly sophisticated methods of detecting the action of selection on genes. The historical information imprinted in DNA sequences provides insights into the relative roles of drift, selection, and recombination in generating and maintaining genetic variation at the molecular level, and theoretical methods exist that can identify which of these mechanisms have acted on specific loci (Kreitman 2000; Nielsen 2001). This is aided by better means of using coalescence methods to model the evolutionary process at the molecular level, and to disentangle demographic effects from the signature left by selection (Lawton-Rauh 2008). Combining these methods with genomics data, it is now apparent that the genomes of many organisms are shaped in large part by selection (but see Lynch 2007), and that one can identify the genes that have experienced positive selection in the past.

There are several footprints of selection on genes that can be identified in molecular evolutionary analyses (Nielsen 2001). One is an increase in the rate of fixation of amino acid substitutions between species when compared with the relative levels of replacement polymorphisms within species or populations, which can be detected by the McDonald-Kreitman test (McDonald and Kreitman 1991), or an extension of this method that uses a Poisson random field formulation (Bustmante et al. 2002), which provides a means to estimate selection on a large number of genes across genomes. One of the first applications of the latter method was a comparison between protein-coding genes in *Drosophila melanogaster* and *Arabidopsis thaliana*, two model systems that have been used extensively for molecular evolutionary analyses. This method shows that there is rampant negative selection across the inbred *A. thaliana* genome in comparison to the outbred fruit fly (Bustamante et al. 2002). Moreover, using an EST approach, this method can be used to identify positively selected genes between *A. thaliana* and its sister species *A. lyrata* (Barrier et al. 2003). By screening 304 genes between these two species, 14 were identified with high ratios of nonsynoynymous (Ka) to synonymous (Ks) nucleotide substitutions (Ka/Ks > 1), and using data from polymorphisms at these genes within *A. thaliana*, a Poisson random field analysis was able to confirm that these loci had higher selective coefficients than other *Arabidopsis* genes (Barrier et al. 2003).

The availability of whole-genome sequences between closely related species, as well as within-species molecular population genomics data, now allows a whole-genome analysis of selection on genes. The recent completion of the chimpanzee sequence, for example, allowed investigators to compare amino acid divergence for 3,377 genes, of which 9% show an excess of amino acid divergence suggestive of positive selection, indicating that adaptive selection has shaped the protein sequences of a significant proportion of protein-coding genes in these primate species (Bustamante et al. 2005). These positively selected genes include ones involved in defense/immunity, gametogenesis, apoptosis, and sensory perception (Bustamante et al. 2005).

Another unambiguous signature of positive selection is a "selective sweep," which is recognized in part as significantly reduced nucleotide variation across a genomic region in proximity to a selected gene. This reduction in polymorphism levels at the selected gene occurs because an allele under positive selection increases in frequency in the population at a much faster rate than the rate at which new neutral mutations accumulate. The reduction in nucleotide variation, however, extends outside of the selected allele to linked neutral sites, as these sites hitchhike with the target of selection (Maynard-Smith and Haigh 1974). For partial selective sweeps, one can also observe a region of elevated linkage disequilibrium surrounding a selected gene. The physical extent of a sweep (whether a few hundred bps or several hundred kb) is governed by the strength of selection, effective population size, and effective recombination rate in the selected region. Population bottlenecks also reduce nucleotide variation levels, but this is manifested genome-wide, rather than the more localized decrease in polymorphisms associated with selective sweeps (Wright et al. 2005).

Selective sweeps have been demonstrated in a variety of contexts, spurred by genomic information that provides sequence data for extended tracts of organismal genomes. Validation of the notion of selective sweeps comes from domesticated taxa, where these regions of reduced nucleotide variation have been observed in the *tb1* (Clark et al. 2004) and *Y1* genes of maize (Palaisa et al. 2004) and the *Wx* gene of rice (Olsen et al. 2006). Using genome-wide data with approximately 1.2 million single nucleotide polymorphisms in African-American, Caucasian and Chinese groups, 101 selective sweeps have also been observed in humans (Voight et al. 2006).

Since selective sweeps are a clear signature of positive selection, they can be used to identify genes associated with adaptive evolution. This novel mapping approach, which scans the genome for a selection signature of low variation across a localized genomic region (Nielsen 2005), is known as adaptive trait locus (Luikart et al. 2003), hitchhiking (Harr, Kauer, and Schlötterer 2002), or selective sweep mapping (Pollinger et al. 2005). These methods have been successfully used in identifying the warfarin resistance locus in rats (Kohn, Pelz, and Wayne 2000), and several selected loci in *Drosophila* (Harr, Kauer and Schlötterer 2002) and humans (Sabeti et al. 2006). A recent study in dogs used selective sweep mapping to identify the *FGFR3* gene, mutations of which lead to foreshortened limbs in dachsunds, and *TRYP1*, which is associated with coat color in Large Munsterlander breeds (Pollinger et al. 2005). These initial successes suggest that this approach may permit rapid fine-mapping of evolutionarily selected genes and is a potentially powerful addition to the genomics tool kit.

Selection was the hallmark of the Darwinian conception of evolutionary change (Darwin 1859). The Modern Synthesis affirmed the nature of selection, by providing the mathematical and genetic framework for its analysis, and subsequent workers have, within the framework of the Synthesis, showed the contributions of neutral drift in the dynamics of the evolutionary process. Genome-wide data have now demonstrated that selection leaves its imprint on genome variation, privileging this evolutionary force once again in molecular evolutionary studies, paving the way for a new stage in the study of the genetic basis of evolutionary adaptation.

Genome Networks and Epistasis

The evolutionary synthesis developed in the middle of the twentieth century in the absence of knowledge of the molecular basis of genes, a state of knowledge that changed a few decades later as molecular genetic research began to identify and isolate genes within genomes that underlie various phenotypic characteristics. In concert with the development of molecular evolution and population genetics, the advances in molecular genetics have implicitly expanded the Modern Synthesis. Today, a major thrust of evolutionary genetics is the study of gene polymorphisms at the molecular level, and how these may contribute to population or species divergence. The study of the genetic architecture of evolutionary traits has progressed from consideration of genes as statistical entities in breeding populations to the point where it is possible to address the molecular basis of adaptive phenotypes at a concrete, molecular level.

A major issue in the study of the genetic architecture of adaptive evolution is the role of epistasis, or gene interactions in the specification of phenotypic variants in populations and adaptive trajectories (Wolf, Brodie, and Wade 2000). The relative contributions of epistatic gene interactions in adaptive evolution was the subject of intense debates between R. A. Fisher and Sewall Wright, two of the architects of the Modern Synthesis, and this debate continues today as investigators seek to quantify the extent of epistasis and its adaptive significance in populations (Wolf, Brodie, and Wade 2000).

The discussion on epistasis since the evolutionary synthesis proceeded in the absence of any understanding of the molecular basis for gene interactions, and in many cases epistasis was simply attributed to amorphous, ill-defined genetic background effects. The advent of molecular genetics has provided concrete manifestations of the nature of epistatic interactions, in the form of genetic pathways and networks that explicitly specify the interactions of the genes within the context of the development of organismal phenotypes (Cork and Purugganan 2004). In the last few years, genomic science and systems biology have taken the analysis of gene networks to a whole new level, by allowing investigators to examine genome-wide networks and identify the regulatory interactions of a large fraction of genes within genomes (Lee et al. 2002), specifying physical interaction maps of gene products within an organism (Ito et al. 2001; Kelley and Ideker 2005), and even experimental rewiring of genetic networks (Bennett and Hasty 2008). This growing conception of gene function embedded in a web of network interactions now provides the physical basis of epistasis, in the same manner that the first molecular identification of genes at the DNA level in the 1980s provided a molecular basis for the study of the genetic basis of phenotypic evolution.

The study of epistasis is poised to expand beyond its generic conception as a statistical departure from nonadditivity of genes to a molecular characterization of evolutionarily relevant gene interactions investigated at the molecular level. A recent study in flowering time variation in the model plant *A. thaliana* illustrates this new direction (Caicedo et al. 2004). Flowering in *Arabidopsis* is repressed by the *FLC* gene, which encodes a MADS-box transcriptional regulator that appears to down-regulate the *FT* flowering time gene (Michaels and Amasino 1999). Molecular population genetic analysis reveals multiple haplotypes at *FLC* in a species-wide survey, including several putatively hypomorphic transposable element insertion alleles found at low (<5%) frequency. There also appears to be a large number of single nucleotide polymorphisms (SNPs), primarily located at the first intron, which differentiates two major *FLC* haplotype groups (Caicedo et al. 2004).

Association studies of these two major *FLC* haplotype groups indicate that they are responsible for some of the natural variation in flowering time observed in *A. thaliana* (Caicedo et al. 2004). However, this phenotypic effect is dependent on the allele state of an upstream regulator of *FLC:* the *FRI* locus (Johanson et al. 2000). *FRI* is present naturally as either active alleles or, at high frequency, may have large deletions that can inactivate *FRI* (Gazzani et al. 2003). The observed effect of *FLC* is seen only in the presence of an active *FRI* allele; in genotypes that have an inactive *FRI* deletion allele, there is no effect of the two *FLC* genotypes. A recent independent mapping study (Scarcelli et al. 2007) has confirmed this *FLC/FRI* interaction in wild accessions of *A. thaliana*.

This is a classic example of an epistatic interaction that affects a crucial life history trait, and there is also evidence that it may be maintained in part by selection. The interaction appears to contribute to an observed latitudinal cline in flowering time among field-grown plants. Moreover, there is a significant level of linkage disequilibrium between these two genes within *A. thaliana*. Together, the role of this interaction in clinal variation as well as the significant level of disequilibrium is consistent with the possibility of epistatic selection on these loci (Caicedo et al. 2004).

The ability to carry out large-scale, genome-wide experimentation has also begun to shed light on the evolutionary importance of epistasis. Synergistic epistasis has been shown to play a role in the functional behavior of gene duplicates in the yeast genome (Jasnos and Korona 2007). Epistatic buffering appears to be important in maintaining fitness in yeast deletion strains, mitigating to some degree the phenotypic expression of mutants (Jasnos and Korona 2007). These experimental results are confirmed, in part, by artificial gene network modeling, which has demonstrated that recombination between networks leads to selection for genetic robustness, but that this selection also results in negative epistasis between mutations (Azevedo et al. 2006).

These and other studies suggest that one of the key debates in the evolutionary synthesis—the role of epistasis—may now be fruitfully studied in the context of genetic networks in the age of genomics. This will provide a sound genomic basis for our understanding of the role of epistasis in the evolutionary process. Indeed, other studies of the evolution of genetic networks have shown how they evolve at the macroevolutionary level, and the time is now right to examine their microevolutionary dynamics, again expanding the scope of our understanding of the evolutionary dynamics that the Modern Synthesis formulated in the last century and contributing to an Extended Evolutionary Synthesis.

The Evolutionary Potential of the Epigenome

A key tenet of the Modern Synthesis is that heritable phenotypic variation within populations provides the basis for adaptive evolution, and that genetic variation underlying phenotypic traits determines the adaptive response to natural selection. It is now widely accepted that variation among individuals in the degree of methylation of genes has also been found to produce heritable, altered states of gene expression (Kakutani 2002; see also chapter in this volume) and novel phenotypes, and provides an additional layer of epigenetic variation in populations (Kalisz and Purugganan 2004; Bossdorf, Richards, and Pigliucci 2008). These epigenetic variants can affect ecologically important traits, including floral symmetry (Cubas, Vincent, and Coen 1999), plant and seed pigmentation levels (Chandler, Eggleston, and Dorweiler 2000), pathogen resistance (Stokes, Kunkel, and Richards 2002), and several other developmental and phenological traits (Kakutani 2002).

Epigenetic variants can have differing molecular underpinnings, but changes in the degree of nucleotide (particularly cytosine) methylation are a major component of epigenetic modification (Bird 2002). Studies have shown that epialleles are stable across generations and can thus be inherited, and that methylation differences can translate into changes in gene expression levels that lead to phenotypic variants (Bird 2002). Stable epialleles have been observed in different species, particularly in plants, including the A. thaliana SUPERMAN (Jacobsen and Meyerowitz 1997) and FWA (Soppe et al. 2000) genes. Examples of epigenetic silencing via methylation are also associated with gene duplications and repeated sequences in the genome, including the *PAI* tryptophan biosynthetic gene family (Bender and Fink 1995), which in many *Arabidopsis* ecotypes includes three unlinked genes (*PAI1–PAI3*). Natural inactive variants of *PAI2* are hypermethylated throughout the locus (Bender and Fink 1995), and are associated with a stable duplication polymorphism in the genome. Only Arabidopsis ecotypes that have an inverted duplication of the gene at *PAI1* (referred to as the *PAI1-PAI4* locus) display the hypermethylation and transcriptional repression of *PAI2* (Bender and Fink 1995).

Other naturally occurring epialleles have been described, one of the most remarkable being in *Linaria vulgaris*, where radially symmetric floral mutants of the wild-type bilaterally symmetric flowers were described by Linnaeus, and are still known to exist in natural populations (Cubas, Vincent, and Coen 1999). Molecular genetic studies revealed that the radial forms are caused by naturally occurring epialleles of the *CYCLOIDEA* gene, which encodes a transcriptional activator involved in the developmental process generating flower asymmetry (Cubas, Vincent, and Coen 1999).

Variation in the degree of methylation is also observed between individuals, which is a key prerequisite if this heritable mechanism is to contribute to evolutionary diversification (Kalisz and Purugganan 2004). A genetic mapping study of *A. thaliana* demonstrated natural variation in methylation levels at rDNA loci found in nucleolar organizing regions (NOR) (Riddle and Richards 2002). Additionally, *A. thaliana* shows among-accession methylation-sensitive polymorphism, with ~ 34% differences in methylation-sensitive variation of amplified fragment-length polymorphism (AFLP) markers (Cervera et al. 2002). A recent study using a tiling microarray has also demonstrated extensive variation in methylation signals on a genomic scale between *A. thaliana* accessions (Vaughn et al. 2007).

If epialleles can directly contribute to variation within populations and be stably inherited across generations, then they should behave in a way similar to sequence-based allelic variation with respect to phenotypes and fitness effects (Kalisz and Purugganan 2004). To determine the possible significance of epialleles in adaptive evolution, their frequency and stability in natural populations must be determined. Although it is clear that these changes can be inherited over several generations in the laboratory, it is unclear whether they are stable over large numbers of generations over evolutionary time (Kalisz and Purugganan 2004). There are two ways in which methylation-associated variation could differ from sequence-based polymorphisms (Kalisz and Purugganan 2004). First, if the rate of formation of new epialleles differs from the rate of stable nucleotide mutations and/or the average effect of an epiallele differs from those of stable mutations, then rates of evolutionary change could be affected. Second, if epialleles persist for multiple generations, but are not permanent, then they could play a transitory role in adaptation if the rate of formation of new methylated epialleles is greater than the nucleotide mutation rate. Another possible result of epiallele formation is that even in plant populations that are monomorphic at the nucleotide level, one could observe phenotypic variation if epiallelic variants are present (Kalisz and Purugganan 2004). All this makes clear that although epigenetic changes remain compatible with the Modern Synthesis, by simply representing an alternative heritable mechanism, dissecting the details could potentially lead to new insights into the dynamics of the evolutionary process that are not currently considered.

Conclusion

Genome science has provided us with an unprecedented glimpse into the structure of the genetic components of organisms and how these are expressed in complex interacting networks. The challenge is to use the large amount of data generated by genomic sciences both to test aspects of the evolutionary synthesis and to take into account novel phenomena that were unknown at the time the Modern Synthesis was formulated. In this chapter, I have discussed several features in which genomic science provides either new features to consider (such as transposable elements) or new insights into long-standing but still ill-understood concepts, such as selection and epistasis.

All these new phenomena need to be accommodated within the framework of evolutionary genetics, and this is already in process. Certainly, the extent of selection that has left imprints on genome sequences lies squarely within our conception of the Modern Synthesis, and indeed is the result of the predictive power of modern evolutionary genetics. The nature of epistasis, while a controversial feature of the early formulations of the Modern Synthesis, is nevertheless increasingly gaining prominence, and provides new avenues of research into the genetic architecture of evolutionary change. The dynamics of transposable elements and epigenetic variation within genomes still remains poorly studied, and more work in this area is clearly needed.

This is just a small glimpse of the possibilities. The number of areas that can still be explored remains large, including the nature of regulatory evolution, the contribution of epigenetic variation, and evolutionary constraints at the level of genome networks. The impact of genomics on the study of the evolutionary process has been revolutionary, and has impinged on all aspects of evolutionary thought, some of which we have not even considered in this chapter: from areas as diverse as phylogenetic analysis of large data sets to the development of new theoretical methodologies. It is clear that data and approaches from genome science provide the basis for extending our understanding of evolutionary genetics as set forth in the Modern Synthesis and for possible elaborations of its conceptual framework.

References

Azevedo RBR, Lohaus R, Srinivasan S, Dang KK, Burch CL (2006) Sexual reproduction selects for robustness and negative epistasis in artificial gene networks. Nature 440: 87–90.

Bender J, Fink G (1995) Epigenetic control of an endogenous gene family is revealed by a novel blue fluorescent mutant of *Arabidopsis*. Cell 83: 725–734.

Berg D, Howe M (1989) Mobile DNA. MM Howe, ed. Washington, DC: ASM Press.

Bird A (2002) DNA methylation patterns and epigenetic memory. Genes and Development 16: 6–21.

Blanchard J, Schmidt GW (1996). Mitochondrial DNA migration events in yeast and humans: Integration by a common end-joining mechanism and alternative perspectives on nucleotide substitution patterns. Molecular and Biological Evolution 13: 537–548.

Bustamante CD, Nielsen R, Sawyer SA, Olsen KM, Purugganan MD, Hartl DL (2002) The cost of inbreeding in *Arabidopsis*. Nature 416: 531–534.

Barrier M, Bustamante CD, Yu JY, Purugganan MD (2003). Selection on rapidly evolving proteins in the *Arabidopsis* genome. Genetics 163: 723–733.

Bennett MR, Hasty J (2008). Genome rewired. Nature 452: 824–825.

Bossdorf O, Richards CL, Pigliucci M (2008) Epigenetics for ecologists. Ecology Letters 11: 106–115.

Bustamante CD, Fiedel-Alon A, Williamson S, Nielsen R, Hubisz MT, Glanowski S, Tanenbaum DM, White TJ, Sninsky JJ, Hernandez RD, Civello D, Adams MD, Cargill M, Clark, AG (2005) Natural selection on protein-coding genes in the human genome. Nature 437: 1153–1157.

Caicedo AL, Stinchcombe JR, Olsen KM, Schmitt J, Purugganan MD (2004) Epistatic interaction between *Arabidopsis FRI* and *FLC* flowering time genes generates a latitudinal cline in a life history trait. Proceedings of the National Academy of Sciences of the USA 101: 15670–15675.

Cam HP, Noma KI, Ebina H, Levin HL, Grewal IS (2008) Host genome surveillance for retrotransposons by transposon-derived proteins. Nature 451: 431–436.

Cervera MT, Ruiz-Garcia L, Martinez-Zapater JM (2002) Analysis of DNA methylation in *Arabidopsis thaliana* based on methylation-sensitive AFLP markers. Molecular Genetics and Genomics 268: 543–552.

Chandler VL, Eggleston WB, Dorweiler JE (2000) Paramutation in maize. Plant Molecular Biology 43: 121–145.

Clark RM, Linton E, Messing J, Doebley JF (2004) Pattern of diversity in the genomic region near the maize domestication gene *tb1*. Proceedings of the National Academy of Sciences of the USA 101: 700–707.

Cordaux R, Udit S, Batzer MA, Feschotte C (2006). Birth of a chimeric primate gene by capture of the transposase gene from a mobile element. Proceedings of the National Academy of Sciences of the USA 103: 8101–8106.

Cork JM, Purugganan MD (2004) The evolution of molecular genetic pathways and networks. Bioessays 26: 479–484.

Cubas P, Vincent C, Coen E (1999) An epigenetic mutation responsible for natural variation in floral symmetry. Nature 401: 157–161.

Darwin C (1859). The Origin of Species. London: John Murray.

Feschotte C, Pritham EJ (2007) DNA transposons and the evolution of eukaryotic genomes. Annual Review of Genetics 41: 331–368.

Fleischmann RD et al. (1995) Whole genome random sequencing and assembly of *Haemophilus influenzae* Rd genome. Science 269: 496–512.

Fridell RA, Prett A-M, Searles LL (1990). A retrotransposon 412 insertion within an exon of the *Drosophila melanogaster vermilion* gene is spliced from the precursor mRNA. Genes and Development 4: 559–566.

Gazzani S, Gendall AR, Lister C, Dean C (2003) Analysis of the molecular basis of flowering time variation in *Arabidopsis* accessions. Plant Physiology 132: 1107–1114.

Hao W, Golding GB (2008) Uncovering rate variation of lateral gene transfer during bacterial genome evolution. BMC Genomics 9: 235.

Harr B, Kauer M, Schlötterer C (2002) Hitchhiking mapping: A population-based fine-mapping strategy for adaptive mutations in *Drosophila melanogaster*. Proceedings of the National Academy of Sciences of the USA 99: 12949–12954.

Huang CY, Grünheit N, Ahmadinejad N, Timmis JN, Martin W (2005) Mutational decay and age of chloroplast and mitochondrial genomes transferred recently to angiosperm nuclear chromosomes. Plant Physiology 138: 1723–1733.

International Human Genome Sequencing Consortium (2001) Initial sequencing and analysis of the human genome. Nature 409: 860–921.

Ito T, Chiba T, Ozawa R, Yoshida M, Hattori M, Sakaki Y (2001) A comprehensive two-hybrid analysis to explore the yeast protein interactome. Proceedings of the National Academy of Sciences of the USA 98: 4569–4574.

Jacobsen SE, Meyerowitz EM (1997) Hypermethylated *SUPERMAN* epigenetic alleles in *Arabidopsis*. Science 277: 1100–1103.

Jasnos L, Korona R (2007) Epistatic buffering of fitness loss in yeast double deletion strains. Nature Genetics 39: 550–554.

Jiang N, Bao ZR, Zhang XY, Hirochika H, Eddy SR, McCouch SR, Wessler SR (2003) An active DNA transposon family in rice. Nature 421: 163–167.

Johanson U, West J, Lister C, Michaels S, Amasino R, Dean C (2000) Molecular analysis of *FRIGIDA*, a major determinant of natural variation in *Arabidopsis* flowering time. Science 290: 344–347.

Kakutani T (2002) Epialleles in plants: Inheritance of epigenetic information over generations. Plant and Cell Physiology 43: 1106–1111.

Kalisz S, Purugganan MD (2004) Epialleles via DNA methylation: Consequences for plant evolution. Trends in Ecology and Evolution 19: 309–314.

Kamau E, Charlesworth D (2005) Balancing selection and low recombination affect diversity near the self-incompatibility loci of the plant *Arabidopsis lyrata*. Current Biology 15: 1773–1778.

Kelley R, Ideker T (2005) Systematic interpretation of genetic interactions using protein networks. Nature Biotechnology 23: 561–566.

Kimura M (1977) Preponderance of synonymous changes as evidence for the neutral theory of molecular evolution. Nature 267: 275–276.

Kimura M (1979) Neutral theory of molecular evolution. Scientific American 241: 98–105.

Klein J, Sato A, Nagl S, O'hUigin C (1998) Molecular trans-species polymorphism. Annual Review of Ecology and Systematics 29: 1–21.

Kohn MH, Pelz H-J, Wayne RK (2000) Natural selection mapping of the warfarin-resistance gene. Proceedings of the National Academy of Sciences of the USA 97: 7911–7915.

Komatsu M, Shimamoto K, Kyozuka J (2003) Two-step regulation and continuous retrotransposition of the rice LINE-type retrotransposon Karma. Plant Cell 15: 1934–1944.

Kreitman M (1996) The neutral theory is dead. Long live the neutral theory. Bioessays 18: 678–683.

Kreitman M (2000) Methods to detect selection in populations with applications to the human. Annual Review of Genomics and Human Genetics 1: 539–559.

Lawton-Rauh A (2008) Demographic factors shaping genetic variation. Current Opinion in Plant Biology 11: 103–109.

Lee TI, Rinaldi NJ, Robert F, Odom DT, Bar-Joseph Z, Gerber GK, Hannett NM, Harbison CT, Thompson CM, Simon I, Zeitlinger J, Jennings EG, Murray HL, Gordon DB, Ren B, Wyrick JJ, Tagne J-B, Volkert TL, Fraenkel E, Gifford DK, Young RA (2002) Transcriptional regulatory networks in Saccharomyces cerevisiae. Science 298: 799–804.

Lerat E, Capy P (1999) Retrotransposons and retroviruses: Analysis of the envelope gene. Molecular Biology and Evolution 16: 1198–1207.

Long M (2001). Evolution of novel genes. Current Opinion in Genetics and Development 11: 673–680.

Long M, Langley CH (1993). Natural selection and the origin of jingwei, a chimeric processed functional gene in Drosophila. Science 260: 91–95.

Luikart G, England PR, Tallmon D, Jordan S, Taberlet P (2003) The power and promise of population genomics: From genotyping to genome typing. Nature Reviews Genetics 4: 981–994.

Lynch M (2007) The Origins of Genome Architecture. Sunderland, MA: Sinauer.

Marillonnet S, Wessler SR (1997) Retrotransposon insertion into the maize waxy gene results in tissue-specific RNA processing. Plant Cell 9: 967–978.

Maynard-Smith J, Haigh J (1974) The hitch-hiking effect of a favorable gene. Genetic Research 23: 23–35.

McClintock B (1984). The signficance of the responses of the genome to challenge. Science 226: 792–801.

McDonald JH, Kreitman M (1991). Adaptive protein evolution at the Adh locus in Drosophila. Nature 351: 652–654.

Michaels SD, Amasino RM (1999) FLOWERING LOCUS C encodes a novel MADS domain protein that acts as a repressor of flowering. Plant Cell 11: 949–956.

Ming R et al. (2008) The draft genome of the transgenic tropical fruit tree papaya (Carica papaya Linnaeus). Nature 452: 991–996.

Nielsen R (2001) Statistical tests of selective neutrality in the age of genomics. Heredity 86: 641–647.

Nielsen R (2005) Molecular signatures of natural selection. Annual Review of Genetics 39: 197–218.

Ohta T (1991) Role of diversifying selection and gene conversion in evolution of MHC. Proceedings of the National Academy of Sciences of the USA 88: 6716–6720.

Olsen KM, Caicedo AL, Polato N, McClung A, McCouch S, Purugganan MD (2006) Selection under domestication: Evidence for a sweep in the rice *Waxy* genomic region. Genetics 173: 975–983.

Orr HA (2005) The genetic theory of adaptation: A brief history. Nature Reviews Genetics 6: 119–127.

Palaisa K, Morgante M, Tingey S, Rafalski A (2004) Long-range patterns of diversity and linkage disequilibrium surrounding the maize *Y1* gene are indicative of an asymmetric selective sweep. Proceedings of the National Academy of Sciences of the USA 101: 9885–9890.

Pollinger JP, Bustamante CD, Fledel-Alon A, Schmutz S, Gray MM, Wayne RK (2005) Selective sweep mapping of genes with large phenotypic effects. Genome Research 15: 1809–1819.

Purugganan MD (1993) Transposable elements as introns: Evolutionary connections. Trends in Ecology and Evolution 8: 239–243.

Purugganan MD, Wessler SR (1992) The splicing of transposable elements and its role in intron evolution. Genetica 86: 295–303.

Richardson AO, Palmer JD (2007) Horizontal gene transfer in plants. Journal of Experimental Botany 58: 1–9.

Riddle NC, Richards EJ (2002) The control of natural variation in cytosine methylation in *Arabidopsis*. Genetics 162: 355–363.

Sabeti PC, Schaffner SF, Fry B, Lohmueller J, Varilly P, et al. (2006) Positive natural selection in the human lineage. Science 312: 1614–1620.

Sanger F, Nicklen S, Coulson AR (1977). DNA sequencing with chain-terminating inhibitors. Proceedings of the National Academy of Sciences of the USA 74: 5463–5467.

Scarcelli N, Cheverud JM, Schaal BA, Kover PX (2007) Antagonistic pleiotropic effects reduce the potential adaptive value of the *FRIGIDA* locus. Proceedings of the National Academy of Sciences of the USA 104: 16986–16991.

Soppe WJJ, Jacobsen SE, Alonso-Blanco C, Jackson JP, Kakutani T, Koornneef M, Peeters AJM (2000) The late flowering phenotype of *fwa* mutants is caused by gain-of-function epigenetic alleles of a homeodomain gene. Molecular Cell 6: 791–802.

Stokes TL, Kunkel BN, Richards EJ (2002) Epigenetic variation in *Arabidopsis* disease resistance. Genes and Development 16: 171–182.

Varagona MJ, Purugganan MD, Wessler SR (1992). Alternative splicing induced by insertion of retrotransposons into the maize *waxy* gene. Plant Cell 4: 811–820.

Vaughn MW, Tanurdžic M, Lippman Z, et al. (2007) Epigenetic natural variation in *Arabidopsis thaliana*. PLoS Biology 5: 1617–1629.

Venter JC et al. (2001). The sequence of the human genome. Science 291: 1304–1351.

Voight BF, Kudaravalli S, Wen XQ, Pritchard JK (2006) A map of recent positive selection in the human genome. PLoS Biology 4: 446–458.

Wallace MR, Andersen LB, Saulino AM, Gregory PE, Glover TW, Collins FS (1991). A de novo Alu insertion results in neurofibromatosis type 1. Nature 353: 864–866.

Wang W, Brunet FG, Nevo E, et al. (2002) Origin of *sphinx*, a young chimeric RNA gene in *Drosophila melanogaster*. Proceedings of the National Academy of Sciences of the USA 99: 4448–4453.

Wang QH, Dooner HK (2006) Remarkable variation in maize genome structure inferred from haplotype diversity at the *bz* locus. Proceedings of the National Academy of Sciences of the USA 103: 17644–17649.

Warren WC, et al. (2008) Genome analysis of the platypus reveals unique signatures of evolution. Nature 453: 175–183.

Wessler SR (2006) Eukaryotic transposable elements and genome evolution special feature: Transposable elements and the evolution of eukaryotic genomes: Proceedings of the National Academy of Sciences of the USA 103: 17600–17601.

Wessler SR, Bureau TE, White SE (1995) LTR-retrotransposons and MITEs: Important players in the evolution of plant genomes. Current Opinion in Genetics and Development 5: 814–821.

White SE, Habera LF, Wessler SR (1994) Retrotransposons in the flanking regions of normal plant genes: A role for copia-like elements in the evolution of gene structure and expression. Proceedings of the National Academy of Sciences of the USA 91: 11792–11796.

Wolf JB, Brodie ED, Wade MJ (2000) Epistasis and the Evolutionary Process. New York: Oxford University Press.

Wright SI, Bi IV, Schroeder SG, Yamasaki M , Doebley JF, et al. (2005) The effects of artificial selection of the maize genome. Science 308: 1310–1314.

Zeyl C, Bell G (1996) Symbiotic DNA in eukaryotic genomes. Trends in Ecology and Evolution 11: 10–15.

Zilberman D, Henikoff S (2004) Silencing of transposons in plant genomes: Kick them when they're down. Genome Biology 5: 249.

IV   INHERITANCE AND REPLICATION

7 Transgenerational Epigenetic Inheritance

Eva Jablonka and Marion J. Lamb

Since the 1990s, a growing number of evolution-oriented biologists have expressed the view that the foundations of the Modern Synthesis—the evolutionary paradigm that was constructed during the 1930s and 1940s and has dominated views of evolution for the past 60 years—need rethinking. They believe that the construction of a new, extended, evolutionary synthesis is under way. Challenges to the Modern Synthesis view have been coming from many directions, most notably from developmental biology, microbiology, ecology, animal behavior, and cultural studies. In this chapter we focus mainly on developmental biology, in particular on molecular studies of epigenetics, and on one specific challenge: the challenge of "soft inheritance." Soft inheritance occurs when new variations that are the result of environmental effects are transmitted to the next generation (Mayr 1982). In order to understand the nature of this challenge, we first summarize the assumptions about heredity and development that were built into the late twentieth-century version of the Modern Synthesis. This synthesis defined itself not only by what it included, but also by what it explicitly excluded or marginalized. We emphasize these excluded and marginalized issues because they are among those that are being most strongly contested today. The late twentieth-century version of the Modern Synthesis assumed:

1. Heredity occurs through the transmission of germ-line genes. Genes are discrete units of DNA that are located in chromosomes. Hereditary variations are the result of differences in DNA base sequence. *There are no inherited variations that cannot be expressed in terms of inherited genetic differences.*

2. Hereditary variation is the consequence of (1) the many random combinations of preexisting alleles that are generated by the sexual processes; and (2) new variations (mutations) resulting from accidental

changes in DNA. *Hereditary variation is not affected by the developmental history of the individual. There is no "soft inheritance."*

3. Heritable variations usually have small effects, and evolution is typically gradual. Through the selection of individuals with phenotypes that make them slightly more adapted to their environment than are other individuals in the population, some alleles increase in frequency. *Mutation pressure is not an important factor in evolution. With a few exceptions, macroevolution is continuous with microevolution, and does not require any additional processes.*

4. The ultimate unit of selection is the gene. Although genes interact and the interactions are often nonlinear, the additive fitness effects of single genes (which can be extracted from the fitness effects of the developmental networks in which they participate) drive evolution by natural selection. *The genetic-developmental network and the phenotype it generates are not heritable and cannot be a unit of evolution.*

5. Morphological innovations, like all innovations, are the results of gene mutations that, when beneficial, accumulate over time and lead to a qualitatively new form. *Generic, physical-chemical properties of biological matter, which underlie plasticity, have no role in morphological and physiological innovations other than specifying the boundaries of the forms that are possible.*

6. The targets of selection are individuals, which are well-defined entities. Although conspecifics in groups interact and may co-evolve with each other as well as with their symbionts and parasites, group selection and community selection are rare. Species selection may exist, but is of marginal significance. *The community is only rarely a target of selection, and species selection cannot explain the main patterns of macroevolution.*

7. Evolution occurs through modifications from a common ancestor, and is based on vertical descent. *Horizontal transfer of genes or other types of information has only minor significance, and does not alter the basic branching structure of phylogenies. The main pattern of evolutionary divergence is, at all times and for all taxa, treelike, not weblike.*

Biologists are now questioning each of these assumptions, arguing that:

1. Heredity involves more than DNA. There are heritable variations that are independent of variations in DNA sequence, and they have a

degree of autonomy from DNA variations. These non-DNA variations can form an additional substrate for evolutionary change, and also guide genetic evolution (Jablonka and Lamb 1995, 2005; Jablonka and Raz 2009).

2. Soft inheritance, the inheritance of developmentally induced and regulated variations, exists, and is likely to be important. It involves both non-DNA variations and developmentally induced variations in DNA sequences (Jablonka and Lamb 2005, 2008).

3. The rate at which heritable variations appear is sometimes higher in stressful conditions, and the spectrum of variations may be different, involving amplification, transposition, and massive, heritable gene activation and inactivation (see, for example, Levy and Feldman 2004; Cullis 2005). Such changes can lead to saltational evolution (Jablonka and Lamb 2008; Lamm and Jablonka 2008). Furthermore, variations in the expression and organization of a small set of genes that seems to be common to development in all animal phyla can have dramatic phenotypic effects (Carroll 2005). Macroevolution may be a consequence of changes in these core genes, as well as of the operation of stress-induced mechanisms that result in systemic mutations and genome repatterning.

4. It is the network of developmental interactions, rather than the gene, that is the focus of selection. A gene's expression and the scope of its effects depend not only on its own intrinsic nature, but also—and often much more—on the regulatory structure of the developmental network in which it is integrated (Wilkins 2002; West-Eberhard 2003; and, in this volume, chapters 5 and 6). Developmental networks are commonly modular, and are usually stable during phenotypic evolution.

5. Generic and evolved mechanisms that generate phenotypic plasticity have played a major role in evolution, initiating morphological and behavioral transformations (Forgács and Newman 2005; Kirschner and Gerhard 2005; Newman and Müller 2006; and, in this volume, chapters 10, 11, 12, and 14).

6. Group selection, involving selection of interactions among cooperating group members, is common (Sober and Wilson 1998; and, in this volume, chapter 4). Since many organisms (including humans) contain symbionts and parasites that are transferred from one generation of the host to the next, it may be necessary to consider such *communities* as targets of selection (Zilber-Rosenberg and Rosenberg 2008). Many

patterns of macroevolutionary change are the outcome of selection at the species level and above (Jablonski 2005; chapter 13 in this volume).

7. The "Tree of Life" pattern of divergence, which was supposed to be universal, fails to explain all the sources of similarities and differences between taxa. Sharing whole genomes (through hybridization, symbiosis, and parasitism) and partial exchange of genomes (through various types of horizontal gene transfer) lead to weblike patterns of relations (Arnold 2006; Goldenfeld and Woese 2007). These weblike patterns are particularly evident in some taxa (e.g., plants, bacteria), and in special circumstances (e.g., during the initial stages that follow genome sharing or transfer). Co-evolution between viruses, and between viruses and their cellularized hosts, is an ongoing feature of evolution (Villarreal 2005).

In this chapter we are focusing mainly on the first two of these challenges and some aspects of the third, but epigenetic inheritance undoubtedly also has significant implications for all of the other challenges to the Modern Synthesis that we have listed.

The Epigenetic Turn

Epigenetic-oriented approaches to evolution all have the developing phenotype rather than the gene as their starting point, and focus on aspects of development that lead to flexibility and adjustment when the environment or the genome changes. Although their roots are old, these approaches became influential during the 1990s, and today are an important part of the alternative view of evolution that is taking shape. We call this revival, extension, and elaboration of epigenetic approaches to evolution the "epigenetic turn."

There are three main types of epigenetic research that are having an impact on evolutionary thinking. The first was pioneered more than 60 years ago by Waddington in Great Britain (e.g., Waddington 1957) and Schmalhausen in the Soviet Union (e.g., Schmalhausen 1949), both of whom took a view of evolution that was centered on the complementary aspects of developmental canalization and phenotypic plasticity. They studied the processes that *decouple* genetic and phenotypic variations, and suggested that the capacity to react should be the focus of evolutionary studies. They reasoned that through selection for the developmental capacity to respond to a new environmental stimulus in an adaptive way, a genetic constitution that facilitates adaptation can be built up. Waddington's experiments with *Drosophila* showed the effectiveness of such selection. After being somewhat neglected during the 1970s and 1980s, this type of epigenetic view of evolutionary change has recently been revived and expanded by Gilbert (2001), West-Eberhard (2003), Bateson (2006), Pigliucci (Pigliucci et al., 2006; chapter 14 in this volume), and others.

The second epigenetic approach to evolutionary change emphasizes the mechanisms that give rise to phenotypic plasticity. These include processes that arise from the fundamental physico-chemical properties of biological matter, which, when interacting with new environmental conditions, lead to new patterns of development that can be the basis of dramatic morphological innovations (Forgács and Newman 2005; chapters 11 and 12 in this volume), and to evolved plasticity mechanisms that are based on exploration and selective stabilization processes (Kirschner and Gerhart 2005; chapter 10 in this volume). Both generic and evolved open-ended plasticity help to provide explanations of evolutionary innovations, rapid evolutionary change, and convergent and parallel evolution.

The third type of epigenetic research, which is the main subject of this chapter, focuses on cell memory and cell heredity. Several strands of research have contributed to its recognition as an important factor in development and evolution. In the 1950s and 1960s, biologists such as Boris Ephrussi, David Nanney, Ruth Sager, and Tracy Sonneborn emphasized that something more than Mendelian genes is necessary to explain certain patterns of inheritance in microorganisms (e.g., see Ephrussi 1958; Nanney 1958; Sager and Ryan 1961; Sonneborn 1964). David Nanney (1958) described the many cellular mechanisms that produce persistent changes in cell characteristics as "epigenetic control systems." He recognized that differences between cells do not always depend on "the primary genetic material" (DNA), and described how inducible systems within and outside the nucleus can bring about heritable differences between cells. However, Nanney's ideas remained peripheral, and the framework he outlined did not become integrated with other studies of cell heredity and differentiation until the 1990s.

The same is true for the work of botanists, many of whom were aware of the implications of their research for understanding development and heredity (Matzke and Mittelsten Scheid 2007). The investigations by Brink and others of paramutation (reviewed in Brink 1973; Kermicle 1978), a type of induced heritable epigenetic variation, were not widely known and had little impact on the Mendelian view of heredity. Similarly,

McClintock's suggestion (reviewed in McClintock 1984) that the transposability of some genetic elements is developmentally regulated and may be heritable was not incorporated into the theoretical framework of geneticists and evolutionary biologists, most of whom were zoologists.

Zoologists were certainly interested in epigenetic phenomena, particularly cell and tissue determination and differentiation, but their approach was different. Ever since the early days of genetics, it had generally been assumed that the different cell types within an individual are genetically identical, so how their phenotypes are determined and maintained was an important problem with implications for medicine, particularly for cancer biology. Various techniques were used to study it. For example, Hadorn (1968) studied *Drosophila* imaginal disc cells by serially transferring them through the abdomens of adult females. He found that they retained their determined states for many generations, although occasionally they switched to a different cell type. The stability and potential reversibility of determined states in amphibians was studied by nuclear transplantation (reviewed in Gurdon 2006). It was found that when transplanted into enucleated eggs, the ability of the nuclei of embryonic and larval cells to support normal development became limited: as the cells of the donor embryos became differentiated, it was more and more difficult to reverse their developmental legacies and "reprogram" their nuclei. Another example of the stability of developmental decisions came from Mary Lyon's work on X-inactivation. She postulated, and it was subsequently confirmed, that early in the development of female mammals, one of the two X chromosomes in each cell is inactivated; which X is inactivated is a random process, but once the decision is made, the functional states of the X chromosomes are inherited by subsequent cell generations (Lyon 1961).

Studies such as those just outlined led to speculation about the nature of the systems that control differentiation and the inheritance of differentiated states (e.g., see Cook 1974), and in 1975 two theoretical papers were published that had an enormous impact on future research in the field. Holliday and Pugh (1975) and Riggs (1975) independently put forward the idea that DNA modification—adding small chemical groups to DNA bases, or removing them—underlies cellular differentiation and heredity. In particular, they suggested that methylation or demethylation of cytosines affects the activity of genes, and through the action of what is now known as maintenance methylase, patterns of methylation, and hence gene activity, can be maintained in cell lines. Methylation, they proposed, may function as a cellular inheritance system. The two papers explored various aspects of development such as mechanisms of determination, X-inactivation, cancer, and developmental clocks. A third paper, by Sager and Kitchin, published in the same year, suggested that DNA modifications mark chromosomes as targets for restriction by endonucleases that eliminate or inactivate parts of the genome. In none of the three papers was the word "epigenetics" used, but the processes described were what we now call epigenetic control mechanisms and epigenetic inheritance. The specific mechanism they suggested, DNA methylation, became central to studies of cellular epigenetic inheritance during the 1980s.

In the late 1980s, there was a revival of interest in the phenomenon known as genomic imprinting. Genomic imprinting had been described many years earlier by biologists working with insects (Crouse 1960), and it had also been recognized in mice and plants. The term describes situations in which the state of activity of a gene, chromosome, or whole set of chromosomes depends on the sex of the parent from which it came. For example, an allele that is expressed when inherited from the father is inactive when inherited from the mother. Following the development in the 1980s of molecular technologies that allowed foreign genes to be introduced into plants and animals, many cases were found in which the activity of the introduced transgene depended on the sex of the parent from which it was inherited. This difference in activity was often associated with methylation differences. Some studies of transgenes revealed phenomena that went beyond classical genomic imprinting: the transgenes became inactivated through methylation, and this inactive state was transmitted to descendants irrespective of parental sex (reviewed in Jablonka and Lamb 1995). In other words, sometimes the epigenetic change was stable. Work in plants, mainly on transgenes and transposable elements (Matzke and Matzke 1991; Fedoroff 1989; Jorgensen 1993), and on hormonal regulation (Meins 1989a, 1989b), pointed to the generality and wide scope of epigenetic control mechanisms. It also pointed to the likely existence of what Jablonka and Lamb (1989) called "the inheritance of acquired epigenetic variation"—to transgenerational, developmentally induced, epigenetic inheritance.

Since the early 1990s, the study of cell heredity and cell memory has flourished. Epigenetic inheritance—the inheritance of phenotypic variations in cells and organisms that do not depend on variations in DNA sequence (Holliday 1994, 2006; Jablonka et al. 1992; Jablonka and Lamb 1995, 2005)—has become a major aspect of developmental-molecular research. In fact, epigenetics is often identified as epigenetic inheritance. Wu and Morris (2001), Jablonka and Lamb (2002), Haig (2004), and others have all discussed the origins and changes in the concept of epigenetics, but since the usage of "epigenetics" and "epigenetic inheritance" is confusing, we need to define them as they are used in this chapter:

Epigenetics is concerned with the regulatory mechanisms (epigenetic control systems) that can lead to inducible, persistent, developmental changes. It includes the establishment of variant cellular states that are transmitted through cell division, and those that are dynamically maintained for a long time in nondividing cells (i.e., are responsible for cell memory). At higher levels of organization, epigenetic mechanisms generate the self-sustaining interactions between groups of cells that lead to physiological and morphological plasticity and persistence. Usually changes in DNA sequence are not involved, but in some cases, for example, in the mammalian immune system and in ciliate development, epigenetic control mechanisms generate regulated alterations in DNA.

Epigenetic inheritance is a component of epigenetics. It includes body-to-body (soma-to-soma) information transfer that can take place through developmental interactions between mother and offspring, through social learning, through symbolic communication, and through the interactions between the individual and its environment that are involved in niche construction (figure 7.1). It also includes *cellular epigenetic inheritance*, which is the transmission from mother cell to daughter cell of

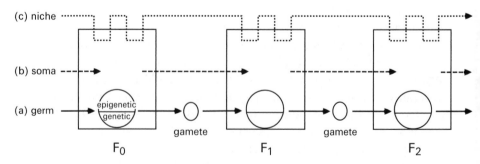

Figure 7.1
Epigenetic variations can be transmitted (a) through the germ line, (b) through the reconstruction of parental phenotypes during somatic development without the involvement of the germ line, and (c) through niche construction, in which the products of parental activities are used to reconstruct the same type of organism–environment relationship.

variations that are not the result of DNA differences or persistent inducing signals in the cells' environment. Cellular epigenetic inheritance can
occur within organisms, in mitotically dividing cell lineages in multicellular eukaryotes, and between organisms during cell division in prokaryotes and protists and during the meiotic divisions in the germ line
that give rise to sperm or eggs. Because epigenetic variations are often
developmentally induced, when epigenetic variations are transmitted
through the germ line, soft inheritance is possible. In this chapter, when
discussing the evolutionary aspects of cellular epigenetic inheritance,
we refer to between-organism, rather to within-organism, transmission
of epigenetic variations, although the mechanisms employed in the two
cases may be largely overlapping. It is also worth remembering that for
unicellular organisms, which during most of evolutionary history were
the only forms of life, epigenetic inheritance and the opportunities for
soft inheritance associated with it always occur between generations.

Cellular Epigenetic Inheritance: How Developmentally Induced, Cellular Epigenetic Variations Are Transmitted

Jablonka and Lamb (1989, 1995; Jablonka et al., 1992) suggested that
the mechanisms that lead to the transmission of cellular epigenetic variants should be understood and studied within a shared evolutionary
framework that incorporates their construction during development.
They described the processes that underlie cellular epigenetic inheritance as "epigenetic inheritance systems" (abbreviated to EISs by
Maynard Smith 1990), and called for the recognition of the Lamarckian
aspects of heredity and evolution that they bring about.

Four types of cellular EISs have now been described (Jablonka and
Lamb 2005): (1) systems based on self-sustaining regulatory loops; (2)
those that involve structural templating; (3) chromatin marking systems; (4) RNA-mediated inheritance. All can contribute to between-
generation epigenetic inheritance, and they interact. Their dual nature—
they are both developmental mechanisms and inheritance systems—
means that they have to be studied from both perspectives.

Self-Sustaining Metabolic Loops

The activity of genes and their products can be maintained by the organization of the metabolic circuits in which they participate. For example,
through positive feedback, the product of an inducible gene may act as
an activator of its own transcription. If the components of the circuit are

transmitted to daughter cells, the same patterns of gene activity may be reconstructed after cell division (figure 7.2). Such positive feedback can lead to two genetically identical cells, existing in the same environment, having alternative, heritable cell phenotypes.

The first feedback system of this type to be identified was the bi-stable lac operon of *Escherichia coli* (Novick and Weiner 1957). Subsequently, many other systems based on a similar cybernetic logic have been described in microorganisms. In the fungal pathogen *Candida albicans*, for example, an epigenetic switch underlies the transition between white and opaque, two cell states that are heritable for many generations and have different interactions with their human host. A certain level of the regulator protein Wor1 is necessary to establish the opaque state; once it is reached, Wor1 positively regulates its own transcription by binding to the regulatory region of its own DNA, thus activating its own synthesis. In this way, a stable self-sustaining feedback loop is formed, and the opaque state is maintained in the cell lineage (Zordan et al. 2006).

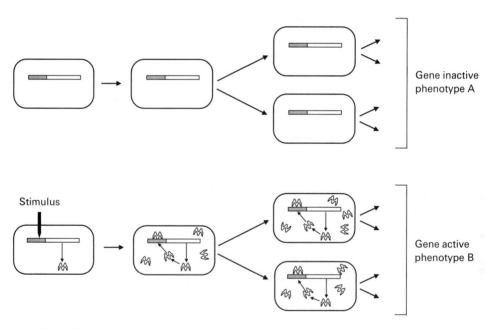

Figure 7.2
Epigenetic inheritance through self-sustaining loops. At the top, a gene having a control region (shaded) and a coding sequence (open) is inactive and transmits its inactive state. Below, the same gene is transiently activated by an external stimulus and produces a product which then associates with its control region and keeps it active; because the product is transmitted to daughter cells, the lineage retains the active state, even in the absence of the external stimulus.

Structural Templating

With this EIS, preexisting three-dimensional cellular structures act as templates for the production of similar structures, which then become components of daughter cells. This category covers a wide variety of mechanisms, including the inheritance of cortical structures in ciliates, prion-based inheritance, and the reconstruction of what Cavalier-Smith (2004) calls "genetic membranes": membranes whose assembly during reproduction is not spontaneous, but is dependent on the presence of preexisting membrane-templates.

Some of the first examples of this type of inheritance came from studies of ciliates such as *Paramecium*, where some variations in the structure of the cortex (e.g., an altered organization of the cilia), even when produced experimentally, can be inherited through many asexual and sexual generations (reviewed in Grimes and Aufderheide 1991). The molecular basis of this is not understood. Rather more is known about prions, proteins that are able to adopt alternative conformations that are self-templating (figure 7.3). Prions were originally associated with diseases affecting the mammalian nervous system, such as scrapie and kuru, but subsequent work with *Saccharomyces cerevisiae* and other fungi has shown that alternative protein conformations that are inherited can confer evolutionary advantages under different nutritional stress conditions (Shorter and Lindquist 2005).

Chromatin Marking

Chromatin marks result from small chemical groups (such as methyls) that are covalently bound to DNA, and modifiable, histone and non-histone proteins that are noncovalently bound. They are involved in the control of gene activity. Some chromatin marks segregate semi-conservatively and/or conservatively during DNA replication, nucleating the reconstruction of similar marks on the daughter molecules. The best-understood system is DNA methylation. In eukaryotes, the bases that are methylated are usually the cytosines (C) in CG doublets or CNG triplets (N can be any base); because of the symmetry of sites on the parental DNA duplex, patterns of cytosine methylation are replicated in a semi-conservative manner, hitchhiking on DNA replication (figure 7.4). Marks that involve modifications in the histones around which DNA is wrapped and the types of proteins that are attached to the nucleosomes are also inherited in cell lineages, although the mechanisms that bring about their reconstruction after DNA replication are far from clear. It is also unclear how marks, including methylation marks, are

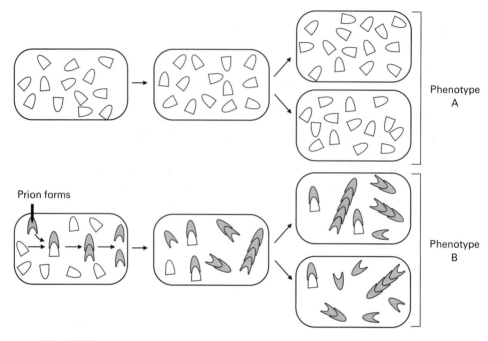

Figure 7.3
Epigenetic inheritance through structural templating. At the top, the cell produces a
normal protein; it is inherited by daughter cells, which continue to produce the protein;
the result is phenotype A. Below, a molecule of the protein adopts a prion conformation,
which interacts with normal protein molecules and converts them to its own structure.
When transmitted to daughter cells, the prion converts all newly formed proteins to its
own conformation, with the result that phenotype B is inherited.

reconstructed after the chromatin upheavals that go on during meiosis
and gamete formation. Nevertheless, there is good evidence that some
differences in marks are transmitted through many generations (see, for
example, Manning et al. 2006; Rangwala et al. 2006).

One of the most striking examples of this type of epigenetic inheritance was found in the toadflax, *Linaria vulgaris*. Over 250 years
ago, Carl Linnaeus described a morphological variant, Peloria, characterized by flowers that are radially rather than bilaterally symmetrical.
Later generations of botanists assumed that Peloria was a mutant
form. However, when Cubas and his colleagues (1999) studied *Lcyc*,
the *Linaria* version of the *cycloidea* gene that in related species is known
to control dorsoventral asymmetry, they found that the DNA sequences
of the normal and peloric forms are identical. What is different is not
the DNA, but the pattern of methylation: in the peloric variant, part

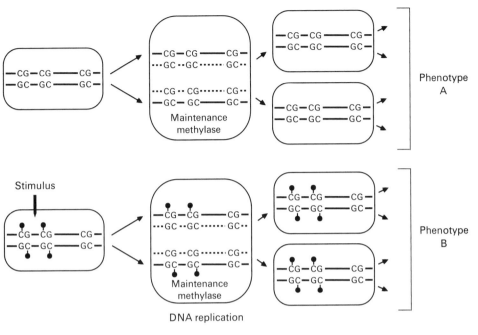

Figure 7.4
Epigenetic inheritance through methylation marks. The locus in the upper cell is unmethyl-ated, and this state is reproduced in daughter cells, giving phenotype A. In the cell below, a stimulus results in some CG sites in the same locus being methylated (•). Following DNA replication, the old strand remains methylated, but the new strand is initially unmethylated; a maintenance methylase recognizes CG sites that are asymmetrically methylated, and methylates the Cs of the new strand; the daughter cells thus inherit the pattern of methyla-tion and produce phenotype B.

of the gene is heavily methylated and transcriptionally silent. In other words, the Peloria phenotype is the result of an epimutation, not a mutation. Peloric strains are not totally stable, and occasionally branches with partially or even fully wild-type flowers develop on peloric plants, but it has been shown that the epigenetic marks on *Lcyc* are trans-mitted to progeny for at least two generations (Parker, personal communication).

Differences in DNA methylation and other chromatin components underlie many other cases of heritable epigenetic variations in plants, and similar epimutations have been found in animals (reviewed by Jablonka and Raz 2009). A hint of how induced heritable epigenetic changes can affect evolution by increasing selectable variation comes from work with an isogenic strain of *Drosophila melanogaster* carrying a mutant allele of the *Krüppel* gene, which affects eye morphology.

Ruden and his colleagues added geldanamycin, a drug that inhibits the activity of the heat shock protein Hsp90, to the food of larvae of this strain for a single generation. It enhanced the development of the abnormal eye phenotype in the adult. After selective breeding for the eye anomaly for six generations, the proportion of flies showing it had increased from just over 1% to more than 60% (Sollars et al. 2003). Since the strains used were isogenic, and therefore lacked genetic variability, the most reasonable interpretation of the results is that the variations that were selected were new, heritable, epiallelic differences induced by the drug treatment.

Experiments with mammals have also provided evidence for induced epigenetic variations that are inherited and affect fitness. One important study, which has worrying implications for medicine, is that by Anway and his colleagues, who injected pregnant female rats 8–15 days post coitus with vinclozolin, a fungicide that is also an androgen receptor antagonist (Anway et al. 2005, 2006a, 2006b). They found that the consequent abnormalities in the testis, immune system, and other tissues of male offspring were inherited for at least four generations. Fifteen different DNA sequences with altered methylation patterns in the F_1 males were transmitted from the F_1 to the F_3 generation.

DNA methylation has also been implicated in the patterns of inheritance seen with Fused, a dominant trait in the mouse. Carriers of the Fused gene (now known as $Axin^{Fu}$) have a very variable, kinked-tail phenotype. Many years ago, Belyaev and his group suggested that the rather strange patterns of inheritance found with Fused are manifestations of epigenetic, rather than purely genetic, phenomena (Belyaev et al. 1981a, 1983). Subsequently Rakyan and colleagues (2003) confirmed that the degree of expression of Fused is correlated with the extent to which a transposon-derived sequence in the Axin gene is methylated. Heavy methylation leads to the development of a normal tail, whereas a demethylated transposon element leads to abnormal RNA transcripts and a kinked tail.

RNA-Mediated Inheritance

With this EIS, silent transcriptional states are initiated and actively maintained through repressive interactions between small RNA molecules and the mRNAs or DNA to which they are complementary (Bernstein and Allis 2005). Transcriptional silence can be transmitted by cells and organisms through an RNA-replication system, and/or because the small RNAs interact with chromatin in ways that cause heritable modifications

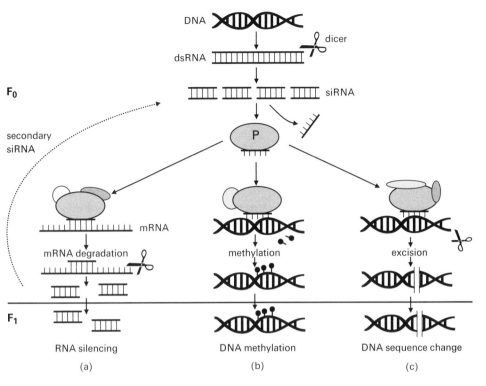

Figure 7.5
Epigenetic inheritance through gene silencing. By associating with various protein complexes (P), small RNAs cut from double-stranded RNA can silence genes (a) by causing degradation of the target mRNA with which they have sequence homology; this silencing can be inherited when the small RNAs are replicated by RNA polymerase and transmitted to daughter (and sometimes nondaughter) cells; (b) by pairing with and methylating homologous DNA sequences, so the chromatin marks are then inherited; (c) by pairing with DNA and causing sequences to be excised; the changed DNA sequences are then inherited.

of chromatin marks. RNA–DNA and RNA–RNA pairing interactions can also lead to targeted gene deletions and amplifications, which are then inherited (figure 7.5).

Transgenerational RNA-mediated inheritance has been studied in *Caenorhabditis elegans*, where injecting double-stranded RNA that targets specific *C. elegans* genes resulted in those genes becoming silent. The consequent induced morphological and physiological variations were transmitted for at least 10 generations, probably because the small RNAs interacted with chromatin and changed protein marks (Vastenhouw et al. 2006). An RNA-mediated EIS also seems to be responsible for the strange patterns of inheritance associated with the paramutable *Kit* alleles in mice, which result in white tail tips and paws (Rassoulzadegan et al. 2006). In this case there is evidence that suggests that small RNAs may be transmitted to the next generation in sperm.

Prevalence, Stability, and Induction of Cellular Epigenetic Variants

The existence of cellular epigenetic inheritance is beyond question, but how frequently epigenetic variants are transmitted between generations is unknown. In the absence of a molecular analysis, there is no way of distinguishing a stable epigenetic change from a genetic change, and very few molecular analyses of this type have been made so far. Nevertheless, in a recent survey Jablonka and Raz (2009) found over a hundred well-documented cases of epigenetic inheritance in 42 species. They included only four representative hybrids (out of the very many existing known cases), and excluded all cases of imprinting (where marks are dependent on the sex of the transmitting parent, and are therefore often transmitted for only one generation) and also many studies that were similar to those that had already been counted. The breakdown of their data shows:

• Twelve cases of epigenetic inheritance in bacteria. Most were of self-sustaining loops, but examples of chromatin marking and structural inheritance were also found.

• Nine cases in protists. Most were in ciliates, where structural inheritance (the transmission of cortical morphologies) is common, and all loci may be modified through the RNA-mediated EISs. The other two types of EISs were also found in protists.

• Nineteen cases in fungi, involving many phenotypes and loci. Examples of all four types of EISs were found.

• Thirty-eight cases in plants, involving many loci and many traits. Among the 38 cases, four were in plant hybrids, and in all of these, many loci were heritably modified. Genomic stresses such as hybridization and polyploidization, especially allopolyploidization, seem to induce genome-wide epigenetic changes, some of which are transmitted between generations through the chromatin-marking and the RNA-mediated EISs. No evidence was found for between-generation inheritance based on self-sustaining loops and structural templating.

• Twenty-seven cases in animals, some of which involved many loci. As with plants, stress seems to induce multiple epigenetic changes. Epigenetic variations were transmitted through the chromatin-marking and the RNA-mediated EISs; there was no evidence for between-organism cellular inheritance based on self-sustaining loops and structural templating.

Clearly, the evidence for transgenerational epigenetic inheritance is substantial, particularly if one bears in mind the nature of the data that are available. Most examples come from the model organisms *E. coli*, yeast, *Arabidopsis*, maize, rice, *Caenorhabditis*, *Drosophila*, and the mouse, so the data are obviously biased. Moreover, with all of the model animals, the segregation between germ line and soma occurs early in development, and epigenetic inheritance is expected to be more limited in these than in the nonrepresented animal taxa where segregation occurs late or not at all (Jablonka and Lamb 1995). For most phyla and the kingdom Archea, there are no studies of epigenetic inheritance at all. There are also few data that relate to epigenetic inheritance in viruses, and information on epigenetic inheritance in chloroplasts and mitochondria is scant.

In spite of the nature of the data and the many gaps, the evidence Jablonka and Raz (2009) have provided suggests that epigenetic inheritance is ubiquitous. This should not be surprising, because all organisms have chromosomes and chromatin, and theoretically heritable variations in chromatin marks could occur at any locus and be inherited. Similarly, the double-stranded nature of DNA and the possibility of transcription from both complementary strands mean that theoretically it is possible that, in organisms with RNA-mediated control systems, every DNA segment could form small double-stranded RNA molecules that lead to epigenetic silencing and its transmission. Furthermore, most proteins have the potential to form β sheets with spatial templating properties (Baxa et al. 2006), so the transfer of prion-like conformations in lineages of single-celled organisms may also occur quite frequently. The potential for transgenerational epigenetic inheritance is therefore present in all organisms. However, unlike DNA replication, which is largely insensitive to context, whether or not, and for how long, an epigenetic variation is inherited depends on other elements in the genome and on developmental conditions. This does not mean that the conditions that allow epigenetic inheritance are necessarily rare, however. We simply do not know how frequent they are. For example, many cases of epigenetic inheritance have already been found in fungi, but Benkemoun and Saupe (2006) suggest that there may be even more, because epigenetic inheritance may be behind the many "bizarre looking sectors or segregates that defy Mendelism" that people working with filamentous fungi often encounter but tend to discard.

Answering questions about the frequency with which epigenetic inheritance occurs is made more difficult by the wide range of stabilities displayed by epigenetic variations. Some seem to be relatively transient, lasting only two to four generations, but others are very stable, lasting tens of generations. Stability is very much dependent on the nature of the environmental conditions and the epigenetic control mechanisms which induce and maintain new epigenetic variants. New, heritable epigenetic variants can arise either as a result of developmental noise or as a result of changed conditions (Jablonka and Lamb 2008; Lamm and Jablonka 2008). Genomic and environmental stresses often seem to be involved. For example, the genomic stresses of hybridization and polyploidization are known to induce genetic and heritable epigenetic variation at many loci in plants (Pikaard 2003; Levy and Feldman 2004; Salmon et al. 2005), and a change from sexual to agametic reproduction leads to a heritable activation of some genes in sugar beet (Levites 2000; Levites and Maletskii 1999). In both plants and mammals, DNA damage through irradiation induces heritable epigenetic variation (Dubrova 2003; Molinier et al. 2006).

Evidence that environmental stresses can induce heritable epigenetic variations has been available for a long time but, perhaps because of its apparent "Lamarckian" implications, it did not receive much attention. In the 1950s it was found that nutritional stresses imposed during the development of flax can lead to heritable changes (reviewed in Cullis 2005). There were also strong hints that stress-induced hormonal changes in mammals result in heritable nongenetic changes. For example, in the 1970s and 1980s, Belyaev's group in the Soviet Union suggested that hormonal effects (in the serotonin system that controls aggression), brought about by selection for tameness, were involved in the heritable activation of the Star gene in silver foxes, which leads to white spotting (Belyaev et al. 1981b; Trut et al. 2004). They also found in the mouse that the penetrance of the Fused phenotype in the progeny of parents treated with hydrocortisone was heritably altered (Belyaev et al. 1983). At the present time, the best-understood case of hormonally mediated effects on transgenerationally transmitted epigenetic marks is the study mentioned earlier in which pregnant female rats were injected with the endocrine disruptor vinclozolin. Not only did the adult male offspring show a variety of heritable testis abnormalities and other diseases that were associated with altered levels of methylation, but mating preferences in the lineage were also affected. Females, whether from the ancestrally treated lineage or controls, preferred males whose ancestors had not been exposed to the chemical (Crews et al. 2007). Obviously, if environmental factors can induce germ-line-transmitted epigenetic effects on sexual selection, it has far-reaching evolutionary implications.

Environmental factors can induce heritable epigenetic variations either directly, in the germ line, or indirectly, through somatic mediation. The various possibilities are illustrated in figure 7.6. Jablonka and Raz's

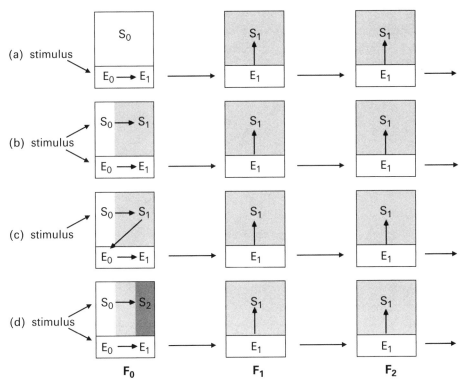

Figure 7.6
Direct and indirect induction of epigenetic variations. (a) Direct induction: the stimulus causes an epigenetic change from E_0 to E_1 in the germ line of the parental (F_0) generation; the phenotype of the parent is unaffected, but all subsequent generations have the S_1 phenotype. (b) Parallel induction: the stimulus causes the same epigenetic change, from E_0 to E_1, in both somatic and germ-line cells, and as a result both parent and offspring show the S_1 phenotype. (c) Somatic induction: the stimulus causes a change to S_1 in the somatic phenotype of the parent, which induces an epigenetic change from E_0 to E_1 in the germ line; because this is subsequently inherited, offspring have phenotype S_1. (d) Parallel induction with nonparallel effects: the stimulus causes a change in the somatic phenotype to S_2, and an epigenetic change to E_1 in the germ line, which is inherited and produces the S_1 phenotype in offspring.

(2009) survey shows that all possibilities—direct germ-line induction, somatically mediated induction with variable effects, and parallel induction of both germ line and soma—have been found.

Epigenetic Learning: Expanding the Scope of Studies of Cellular Epigenetics?

In studies of cellular epigenetic inheritance, it is usually assumed that if a mark—say a pattern of six methylated cytosine sites—is induced at a particular locus, it is reconstructed (with a certain error rate) in the descendants, where it has similar phenotypic effects (figure 7.7a). The dynamics of acquiring epigenetic marks (how many generations it takes for an inducer to produce a change that has a phenotypic effect), how quickly it is lost (how many generations are required for a mark to fade), and how the extent of marking relates to the phenotypic response are all issues that have not been addressed and studied systematically.

Ginsburg and Jablonka (2009) have suggested that the problem can best be approached by thinking about it in terms of *cellular epigenetic*

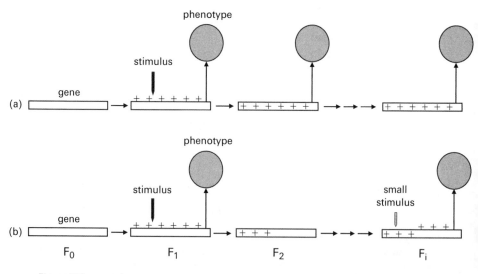

Figure 7.7
Cell learning through epigenetic inheritance. In (a), a stimulus alters a chromatin mark by adding six methyl groups (+), which leads to a phenotypic change; the mark persists (indicated by + symbols moving into the locus) and produces the same phenotype in subsequent generations. In (b), a stimulus alters a chromatin mark in the same way as previously, and it has a phenotypic effect, but the mark partially fades (only 3 + persist); in the next generation the induced phenotype is not produced, but because traces remain, a smaller stimulus elicits the response in a later generation (F_i).

learning. Epigenetic learning occurs when an inducing agent elicits a response that leaves a persistent epigenetic trace which later, upon subsequent induction, is the basis of a more effective response. So, for example, an inducing stimulus might cause a gene to become epigenetically marked in a way that affects the phenotype; in the absence of the inducer the mark decays, but when a second stimulus of the same type is applied, because a partial mark is already present, either a smaller stimulus is required to elicit the response or the response is faster (figure 7.7b). Variations on this theme, including the possibility of reactions involving more than a single gene and leading to a kind of "associative" epigenetic learning, have been described and discussed by Ginsburg and Jablonka (2009).

Epigenetic learning is not yet a part of the research program of epigenetics, but there are examples of simple learning in nonneural organisms such as the ciliates *Paramecium* and *Stentor* (Wood 1992; Armus et al. 2006) and the plant *Mimosa pudica* (Applewhite 1975), which can be interpreted in epigenetic terms. The molecular basis of this is at present unknown, but the ability to learn at this level may be adaptively important. Evolutionary models have shown that in fluctuating environments, epigenetic memory can be an advantage (Lachmann and Jablonka 1996; Balaban et al. 2004; Rando and Verstrepen 2007), but having epigenetic learning, rather than persistent memory, may often be selectively superior. With learning, the cost of a memorized response that is no longer adequate for present conditions (which is incurred when memory is perfect) is reduced, and the cost of development-from-scratch (which is incurred when "forgetting," or "resetting," is complete) is also reduced. If the molecular mechanisms that underlie epigenetic learning are the same as those that are the basis of cell memory and the EISs, it is not difficult to see how, through selection, small modulations in the conditions in which these mechanisms operate could lead to complex adaptive plastic responses. The mechanistic simplicity of such learning and the fitness benefits it is likely to confer suggest that a research project that investigates it could be fruitful.

Soma-to-Soma Transmission

Although epigenetic inheritance in cell lineages is currently receiving most attention, the soma-to-soma routes of transmission, which by-pass the germ line, are no less important for understanding the hereditary basis of evolution. Soma-to-soma transmission is an umbrella term for the many processes through which phenotypes are inherited because aspects of the niche in which development takes place are reconstructed in successive generations. It includes transmitting substances that affect development through feces ingestion, through the placenta and milk of mammals, and through the soma-dependent deposition of specific chemicals in the eggs of oviparous animals and plants (Avital and Jablonka 2000). In addition, maternal morphological features can constrain offspring development and lead to heritable and self-perpetuating developmental effects (Jablonka and Lamb 2007a, 2007c), as do socially learned behaviors that do not require the transfer of materials (Avital and Jablonka 2000). Ecological niche construction (Turner 2000; Odling-Smee et al. 2003), which includes developmental interactions among organisms that form coherent and persistent symbiotic communities (Zilber-Rosenberg and Rosenberg 2008), can also contribute to soma-to-soma inheritance.

New somatic phenotypes that are inherited by the following generations can be initiated in more than one way. They may be induced in the soma (figure 7.8a), but it is also possible for soma-to-soma transmission to be initiated by a germ-line mutation or epimutation that has somatic effects that are self-perpetuating after the mutation or epimutation has segregated away (figure 7.8b), or via niche construction (figure 7.8c). The frequency and diversity of soma-to-soma transmission can be seen from the following summary of some of the methods through which phenotypes are reconstructed in successive generations.

Maternal Morphological Constraints

Self-perpetuating phenotypes can arise through the effects that maternal morphology (for example, size) have on the development of offspring. This is especially true for viviparous animals, but affects other organisms too. In some insects and oviparous fish, for example, large mothers lay large eggs that develop into large females that will again lay large eggs, and this cycle continues for as long as the environment does not change too drastically (Mousseau and Fox 1998). For humans and rats, there is a positive correlation between environmentally influenced maternal size and offspring size: small mothers have small wombs, with reduced uterine perfusion, and this leads to small offspring. The small daughters of small females will tend to perpetuate the trend (Morton 2006; Gluckman et al. 2007). This means that in the same environment there could be two genetically identical lineages that differ in size because of transient environmental conditions that affected the nutritional state, and hence the size, of their maternal ancestors.

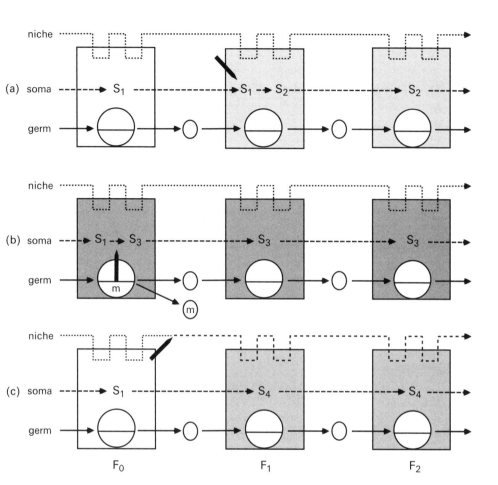

Figure 7.8
Different ways of inducing transmissible somatic changes: (a) an environmental stimulus (heavy arrow) induces a somatic change from S_1 to S_2 in the F_1 generation, and the changed phenotype is developmentally reconstructed in descendants; (b) a mutation (or epimutation) in generation F_0 has somatic effects changing S_1 to S_3, and this phenotype is developmentally reconstructed in the F_1 and later generations even though the causative mutation is not inherited; (c) because of some new activity, the organism changes its environment in a way that leads to a different organism–environment (niche) relationship and new phenotype, which are then reconstructed in subsequent generations.

Hormonal and Pheromonal Effects

In 1988 Campbell and Perkins reviewed the evidence for multigenerational effects of hormone treatments and exposure to drugs such as morphine, alcohol, and other chemicals. The work they described suggests soma-to-soma transmission, although germ-line-mediated transmission cannot be ruled out, and possibly both routes were involved. One clear case of soma-to-soma transfer of hormonal effects leading to similarity between parents and offspring is the transgenerational effect of the uterine concentration of testosterone on the sex ratio and behavior of Mongolian gerbils (Clark et al. 1993). In these small rodents, female embryos that develop in a uterus where most of their sibs are male are exposed to high testosterone levels. This has long-term effects: their development is delayed relative to that of females from a uterus with fewer males, they mate later, and their behavior is more territorial. Crucially, when they themselves become pregnant, they produce litters with more males than females, so their daughters are exposed to testosterone in utero, and the cycle is perpetuated.

Locust phase polyphenism, a dramatic example of epigenetic transmission between generations, may involve pheromonal switches. The solitary (nonswarming) and gregarious (swarming) phases differ markedly in morphological, physiological, and behavioral features. The switch from one phase to the other can occur during the lifetime of an individual, or cumulatively over several generations. Simpson and Miller (2007) have argued that gregarious mothers transmit their phenotypic characteristics to offspring by means of a pheromone secreted into the foam plug that protects the egg pods, although Tanaka and Maeno (2008) dispute this, claiming that phase-dependent differences are established in the ovary. In spite of this uncertainty about how transmission occurs, this is a clear case of soma-to-soma inheritance, and one with obvious adaptive significance.

Transmission through the Placenta, Milk, and Feces

In mammals, useful knowledge about what a mother has been eating can be passed to her offspring through substances present in her placenta, milk, and feces (Avital and Jablonka 2000). European rabbits provide a good example: the food preferences of the young are biased by the food information their mother transmits to them in the uterus, then through her milk, and finally through feces (Bilkó et al. 1994). Immunological information relevant to the current environment is also transmitted across the placenta and through milk. Evidence reviewed by Lemke and colleagues (2004) shows that transmitted maternal antibodies, which are the outcome of a mother's experiences of microbes and allergens, guide the maturation of the immune system of neonates in ways that enhance its effectiveness in later life; some of the effects can be carried over to grandchildren. Lemke and his colleagues describe this as "Lamarckian inheritance," because the acquired immunological phenotypes of the mother are transmitted to her offspring, and they point out that it has obvious selective advantages.

Information transmission through the placenta, milk, and feces is not always beneficial. Prion diseases and some other amyloidoses can be transmitted by these routes. Amyloidoses are diseases caused by conformational changes in proteins: proteins that are normally soluble form insoluble β sheets which induce further polymerization of β sheets from precursor proteins. Korenaga and his colleagues (2006) found that the milk of female mice with a particular type of amyloidosis transmitted factors that induced the same disease in their biological or fostered offspring. How common such prion-like behavior is with amyloidoses is unknown. Scrapie, a recognized prion disease of sheep, can be transmitted from mothers to offspring through milk, and there is a strong suggestion that it can also be transmitted laterally though saliva, feces, or urine (Konold et al. 2008).

Behaviorally Mediated Transmission

Soma-to-soma transmission in animals is often the result of their behavior: through social learning, ancestral patterns of feeding, mating, parenting, dispersal, predator avoidance, and other behaviors are actively reconstructed by descendants, leading to similarity between generations. Examples of this are numerous and well known: they range from transmission of local song dialects by songbirds and whales to the cultural differences found in traits such as nut-opening and ant-dipping in populations of chimpanzees (Avital and Jablonka 2000).

A fascinating insight into the kind of cellular epigenetic control mechanisms behind behaviorally mediated transmission is provided by Meaney and his colleagues' studies of maternal care in rats (Meaney 2001; Weaver et al. 2004). They found that some naturally occurring variations in a mother's style of caring not only influence the offsprings' responses to stress, they are also transmitted to them. The variations are in the amount of "licking and grooming" (LG) and "arch-back nursing" (ABN) that mothers give their pups during the first week after birth. Pups that receive a lot of licking and grooming are stress-resistant and nonneophobic, and the daughters that receive this type of care themselves become high LG and ABN mothers. Conversely, the pups of mothers who give their offspring less LG and ABN are more fearful and readily stressed, and, when adult, the females treat their offspring in the same way they were treated. Because the behavior is passed on, genetically identical rat lineages in the same environment can display different behaviors, depending on the history of their female ancestors.

Meaney and his team found that changes in gene expression in the hypothalamic-pituitary-adrenal system, which is known to underlie reactions to stress, accompany the behavioral variations. When compared with animals reared by less-caring mothers, offspring of high LG-ABN females have increased expression of the glucocorticoid receptor (GR) gene in the hippocampus. This is correlated with changes in DNA methylation and histone acetylation in the gene's promoter (Weaver et al. 2004). Once established, the state of the GR gene persists throughout life, and is reconstructed in the next generation through maternal behavior. The causal connection between the chromatin marks and the transmitted behavior was established by pharmacologically altering the epigenetic state of the gene in adults, using the methyl donor methionine and inhibitors of histone deacetylation. These treatments reversed the effects of previous maternal care (Weaver et al. 2005).

Maternal care studies in another mammal, the mouse, show how soma-to-soma transmission of a new variant can be initiated by a germline mutation (figure 7.8b). Curley and colleagues (2008) found that mouse mothers having a certain mutant gene gave a low quality of care to their offspring, causing them to be fearful and show decreased exploratory behavior. When these offspring reached adulthood, even though they did not carry the mutant gene, they, too, gave their offspring a low quality of care, with the result that the grandchildren of the original mutant females were also fearful and showed decreased exploratory behavior.

Transmission through Symbolic Systems

There is no doubt about the power of language and other symbolic systems to transmit information over many generations, and it is generally recognized that these systems have helped to shape human evolution (Richerson and Boyd 2005). They also provide the best examples of the type of soma-to-soma transmission shown in figure 7.8c. Although great apes seem to have the rudiments of a symbolic system, a capacity that is revealed when they are exposed to human language (Savage-Rumbaugh et al. 1998), we are not aware of any evidence showing that this is the basis of their various cultural traditions (Whiten et al. 2005). In contrast, human culture is dominated by symbolic systems, which lead to complex, cumulative, cultural evolution, and to far-reaching, rapid changes in the human niche. The invention of the airplane, for example, which depended on symbolic systems, enabled humans to fly, which has had far-reaching consequences for their social and individual lives.

Evolutionary Implications

The evidence presented in the previous sections shows that the transgenerational transmission of epigenetic variations through cellular inheritance and through routes that bypass the germ line is not a rarity. Therefore, if we are to understand heredity and evolution, we need to acknowledge these different types of information transfer between generations, and not focus exclusively on genetic transmission. Cellular epigenetic inheritance occurs in all organisms, although the relative importance of the various EISs differs among groups. Soma-to-soma transmission (in the sense that we are using it here) is specific to multicellular organisms, and some types of transmission are limited to certain taxa: obviously there is no neurally based behavioral inheritance in plants, and symbol-mediated inheritance is found almost exclusively in humans. Nevertheless, soft inheritance—the transmission of variations acquired during development—not only exists, it is found in every type of organism and seems to be common. It therefore has to be incorporated into evolutionary thinking.

We have discussed many of the evolutionary implications of epigenetic inheritance in previous publications (Jablonka and Lamb 2005, 2007a, 2007b, 2007c, 2008; Jablonka and Raz 2009), so here we will indicate only briefly how incorporating epigenetic inheritance affects various aspects of evolution.

Adaptation
Adaptation can occur through the selection of heritable epigenetic variations; genetic change is not necessary. This may be of particular importance when populations are small and have little genetic variability, such as during periods of intense inbreeding following population fragmentation. The discovery of extensive epigenetic variation in natural populations strengthens the view that it can play an important role in evolution (Bossdorf et al. 2008), and it is not difficult to imagine how some of the

epigenetic variations studied in the lab could be beneficial. For example, the inherited differences in the color and morphology of flowers that are known to be caused by different chromatin marks or RNA-mediated gene silencing might be beneficial if new pollinators or pests were introduced into their habitats. Adam and his colleagues (2008) have made a strong case, based on their experimental data, for epigenetic inheritance of stochastic variations in gene expression, rather than gene mutations, driving the evolution of antibiotic resistance in bacteria, although so far they have not pinpointed the mechanisms involved. Soma-to-soma transmission through social interactions and social learning is also undoubtedly behind some group-specific adaptive behaviors ("traditions") in animals (Avital and Jablonka 2000).

One of the reasons why recognizing epigenetic inheritance is so important for evolutionary thinking is that the dynamics of evolutionary change through inherited epigenetic variants are likely to be very different from those assumed in conventional population genetic models. While the population genetic models assume that mutations are rare events, and only one or very few individuals are likely to have a newly generated mutation, when epigenetic variations are induced, many individuals in a population may independently acquire a similar heritable phenotype at the same time. Moreover, epigenetic variations may be induced almost simultaneously at several different loci and coordinately affect several traits. How incorporating epigenetic inheritance can change an approach to evolutionary problems can be seen from Zuckerkandl and Cavalli's (2007) hypothesis for the origin of complex adaptations. They suggest that heritable epigenetic changes in "junk DNA," which spread across the genome, affecting the regulation of many genes, may be the answer to a problem that is difficult for Modern Synthesis theorists, namely, how in large animals, whose populations are relatively small, all the mutations necessary for a complex adaptation come together.

Another feature of epigenetic inheritance that makes it necessary to think again about the Modern Synthesis view of adaptation is that selection and mutation (epimutation) may not be independent. Epimutations may be induced by the selecting environment, and they sometimes revert if the environmental conditions change again.

Genetic Assimilation

Heritable non-DNA variations, even those that last for few generations, may enhance the effectiveness of genetic assimilation and accommodation processes, thereby accelerating adaptive evolution. One way in which this might happen is evident from True and Lindquist's (2000) study of pairs of yeast strains that differed only in whether or not they carried [PSI⁺], the prion form of a protein necessary for terminating mRNA translation. When present, the prion causes new and different proteins to be produced because translation goes beyond the normal end of the gene and stop-codons in the middle of nonfunctional genes are ignored. In some conditions, this was found to be beneficial: the prion strains grew faster. The presence of prions might therefore enable a lineage to adapt and to maintain the adaptation until such time as genetic changes took over. In this way, a heritable epigenetic variation—the prion conformation—produces phenotypic changes that pave the way for genetic changes. According to the theoretical model developed by Masel and Bergman (2003), the beneficial effects of such an epigenetically based system would lead to its selection even if the response is adaptive only once in a million years. It might be particularly important in asexual lineages, where the accumulation of mutations would be slow.

Soma-to-soma epigenetic inheritance can also facilitate genetic assimilation. Avital and Jablonka (2000) have argued that behaviorally transmitted information that is later partially or fully genetically assimilated is probably a major driver of animal evolution, and Dor and Jablonka (2000) have suggested that the evolution of the language faculty involved similar cultural and genetic processes that were mutually reinforcing. They argued that as language evolved culturally, and as it became an increasingly important element in the social lives of its speakers, the speakers came to be selected on the basis of their linguistic performance. The cultural invention and elaboration of language thus launched a process of genetic accommodation involving the selection of any genetic variants that contributed to linguistic ability, which in turn increased the possibilities of cultural linguistic evolution, and so on.

Reproductive Isolation

Heritable epigenetic variations may initiate post-zygotic reproductive isolation. Both the failure of some hybrid offspring to develop normally and hybrid sterility may be caused by incompatibilities in the chromatin marks on the two sets of parental chromosomes (Jablonka and Lamb 1995). These might arise by chance, or be the result of selection in periods when the parent populations were isolated.

Behavioral traditions, whether arising by chance or through selection, might initiate pre-zygotic isolation (Avital and Jablonka 2000). They

could reduce the likelihood of mating between members of two populations if they affected the preferred time or place at which courtship
occurs, or the song dialect used, for example. Transmissible differences
in food preferences or preferred habitats might also lead to partial pre-
zygotic reproductive isolation.

The Evolution of Development

Epigenetic inheritance has constrained the evolution of development.
There are several developmental phenomena—such as the difficulty of
reversing determined and differentiated cell states, the early segregation
and quiescent state of the germ line found in many animal groups, and
the massive changes in chromatin structure that occur during meiosis
and gamete production—that can be interpreted as indirect outcomes of
epigenetic inheritance. All could be the results of selection against transmitting chance epimutations and the parents' epigenetic "memories" to
the zygote, which needs to start its development from a totipotent epigenetic state. In some cases, as in the evolution of genomic imprinting,
selection may have favored the enhancement of germ-line-transmitted
epigenetic memories (Jablonka and Lamb 1995, 2005).

Macroevolutionary Change

Epigenetic control mechanisms may play a key role in many macroevolutionary changes, especially those that follow genetic exchanges
between species. Speciation through polyploidization and hybridization,
which are of central importance in plant evolution, probably depends on
them (Jorgensen 2004; Rapp and Wendel 2005; Arnold 2006). Following
auto- and allopolyploidization, there is a burst of selectable epigenomic
variation, which provides ample opportunities for adaptive change. It
seems that, just as McClintock (1984) argued, genomic stress reshapes
the genome.

    Because regulated genome rearrangements are found during the
development of so many different eukaryotes, it has been suggested that
the epigenetic control mechanisms that bring them about are very ancient
(Zufall et al. 2005). The role of chromatin marking and RNA-based
epigenetic mechanisms in silencing foreign viral genes and experimentally introduced genes in eukaryotes, which is now well established, has
also suggested that these are ancient mechanisms, and it has been argued
that their evolution was driven by their role in genome defense (Buchon
and Vaury 2006). Whether or not this is so, it increasingly looks as if the
epigenetic control mechanisms that silence genes have played a very

significant role in shaping evolutionary change throughout almost the whole of the history of life. Whenever there was complete or partial genome merging, not only through hybridization but also as a result of symbiogenesis or horizontal gene transfer, then by silencing some of the introduced genes and heritably altering patterns of gene expression, epigenetic control mechanisms allowed the new organism to survive. We therefore suggest that in response to genome disturbance through gene acquisition and damage caused by ecological stresses, the activities of epigenetic control mechanisms produce large-scale epigenetic variations that are inherited and lead to macroevolutionary changes. These epigenetic control mechanisms could underlie the systemic changes (genome repatterning) that Goldschmidt (1940) believed drives macroevolution (Lamm and Jablonka 2008). Certainly, what is already known suggests that a better understanding of macroevolution might come from looking at chromosomal changes and epigenetic systems rather than from studying differences in coding genes.

The Major Transitions

We have argued previously (Jablonka and Lamb 2006) that epigenetic inheritance and epigenetic control mechanisms have played a key role in all of the major evolutionary transitions identified by Maynard Smith and Szathmáry (1995). For example, as we indicated earlier in this chapter, structural templating mechanisms were probably important during the crucial periods of symbiogenesis when bacteria became integrated within the ancestors of the modern eukaryotic cell. The evolution of long eukaryotic chromosomes also necessitated the recruitment and evolutionary elaboration of the chromatin-marking epigenetic systems: as DNA sequences were linked together or were added by duplication, there was selection for organizing and packaging these long molecules in ways that protected them, allowed them to be replicated, and made then available for transcription following replication. The later transition to multicellularity is also impossible to understand without taking epigenetic inheritance into account, because for anything other than the simplest levels of organization, cell lineages have to remember their determined state. As we argued previously, the efficiency of cell memory, the stability of the differentiated state, and various features of development enumerated in the section "The Evolution of Development" were all shaped in part by the effects of epigenetic inheritance, and the epigenetic inheritance systems were, in turn, shaped by the evolution of development (Jablonka and Lamb 1995, 2005).

Soma-to-soma epigenetic inheritance in the form of social learning was probably involved in the establishment and evolution of animal social groups, another of Maynard Smith and Szathmáry's (1995) major transitions. Their final transition—to linguistic communities, the hallmark of human culture—involved the co-evolution of symbolic systems and hominid genes, with the former leading the latter. The last two transitions depended on a highly evolved nervous system, and we have suggested that the importance of neural activities in animal evolution means that the origin of neural communication, a new information-transmitting system, should be added to Maynard Smith and Szathmáry's list of major transitions (Jablonka and Lamb 2006).

An Extended Evolutionary Synthesis?

At the beginning of this chapter we pointed out that the Modern Synthesis denied the possibility of soft inheritance, and insisted that evolution is usually gradual. However, the mechanisms of epigenetic inheritance that we have discussed are simultaneously involved in the regulation of gene expression and production of phenotypes, as well as in the transmission of information between cells and organisms; they therefore enable soft inheritance. Moreover, probably because some cellular epigenetic control mechanisms have evolved in the context of defenses against genomic parasites, which they silence or eliminate (Bestor 1990; Cerutti and Casas-Mollano 2006), they are recruited and produce genome-wide epigenomic repatterning following the introduction of foreign genes by horizontal gene transfer, symbiogenesis, and hybridization. These are processes leading to reticulate evolution. The same or somewhat modified epigenetic mechanisms may also be recruited under conditions of continuous physiological stress, such as the nutritional stress in flax that causes changes in DNA methylation and in the number of ribosomal genes (Cullis 2005). Hence, both the mechanisms that allow soft inheritance during microevolution, and the epigenetic mechanisms that lead to macrovariations and instances of rapid evolutionary change, need to be incorporated in the emerging extended evolutionary synthesis.

Although the primacy we give to the developmental aspects of variation places our view firmly within the emerging Evo-Devo framework of evolution, our perspective differs from most others because it is focused on inheritance. Evo-Devo biologists often reject the gene-centered view that has dominated evolutionary theory since the 1940s, and argue convincingly that variations in genes should be regarded as inputs into developmental networks or units. However, the genome should be seen not just as a repository of genes that are inputs into development, but also as a developmental system with its own specific, inducible, variational mechanisms. A broader notion of heredity, based on the mechanisms of epigenetic inheritance at all levels of biological organization, could help to unite the different developmental approaches and transform our understanding of evolution.

References

Adam M, Murali B, Glenn NO, Potter SS (2008) Epigenetic inheritance based evolution of antibiotic resistance in bacteria. BMC Evolutionary Biology 8: 52.

Anway MD, Cupp AS, Uzumcu M, Skinner MK (2005) Epigenetic transgenerational actions of endocrine disruptors and male fertility. Science 308: 1466–1469.

Anway MD, Leathers C, Skinner MK (2006a) Endocrine disruptor vinclozolin induced epigenetic transgenerational adult-onset disease. Endocrinology 147: 5515–5523.

Anway MD, Memon MA, Uzumcu M, Skinner MK (2006b) Transgenerational effect of the endocrine disruptor vinclozolin on male spermatogenesis. Journal of Andrology 27: 868–879.

Applewhite PB (1975) Learning in bacteria, fungi and plants. In Invertebrate Learning, vol. 3: Cephalopods and Echinoderms. WC Corning, JA Dyal, AOD Willows, eds.: 179–186. New York: Plenum Press.

Armus HL, Montgomery AR, Jellison JL (2006) Discrimination learning in paramecia (*P. caudatum*). Psychological Record 56: 489–498.

Arnold ML (2006) Evolution through Genetic Exchange. New York: Oxford University Press.

Avital E, Jablonka E (2000) Animal Traditions: Behavioural Inheritance in Evolution. Cambridge: Cambridge University Press.

Balaban NQ, Merrin J, Chait R, Kowalik L, Leibler S (2004) Bacterial persistence as a phenotypic switch. Science 305: 1622–1625.

Bateson P (2006) The adaptability driver: Links between behavior and evolution. Biological Theory 1: 342–345.

Baxa U, Cassese T, Kajava AV, Steven AC (2006) Structure, function and amyloidogenesis of fungal prions: Filament polymorphism and prion variants. Advances in Protein Chemistry 73: 125–180.

Belyaev DK, Ruvinsky AO, Borodin PM (1981a) Inheritance of alternative states of the fused gene in mice. Journal of Heredity 72: 107–112.

Belyaev DK, Ruvinsky AO, Trut LN (1981b) Inherited activation-inactivation of the star gene in foxes: Its bearing on the problem of domestication. Journal of Heredity 72: 267–274.

Belyaev DK, Ruvinsky AO, Agulnik AI, Agulnik SI (1983) Effect of hydrocortisone on the phenotypic expression and inheritance of the *Fused* gene in mice. Theoretical and Applied Genetics 64: 275–281.

Benkemoun L, Saupe SJ (2006) Prion proteins as genetic material in fungi. Fungal Genetics and Biology 43: 789–803.

Bernstein E, Allis CD (2005) RNA meets chromatin. Genes and Development 19: 1635–1655.

Bestor TH (1990) DNA methylation: Evolution of a bacterial immune function into a regulator of gene expression and genome structure in higher eukaryotes. Philosophical Transactions of the Royal Society of London B326: 179–187.

Bilkó A, Altbäcker V, Hudson R (1994) Transmission of food preference in the rabbit: The means of information transfer. Physiology and Behaviour 56: 907–912.

Bossdorf O, Richards CL, Pigliucci M (2008) Epigenetics for ecologists. Ecology Letters 11: 106–115.

Brink RA (1973) Paramutation. Annual Review of Genetics 7: 129–152.

Buchon N, Vaury C (2006) RNAi: A defensive RNA-silencing against viruses and transposable elements. Heredity 96: 195–202.

Campbell JH, Perkins P (1988) Transgenerational effects of drug and hormonal treatments in mammals: A review of observations and ideas. Progress in Brain Research 73: 535–553.

Carroll SB (2005) Endless Forms Most Beautiful: The New Science of Evo Devo and the Making of the Animal Kingdom. New York: Norton.

Cavalier-Smith T (2004) The membranome and membrane heredity in development and evolution. In Organelles, Genomes and Eukaryote Phylogeny: An Evolutionary Synthesis in the Age of Genomics. RP Hirt, DS Horner, eds.: 335–351. Boca Raton, FL: CRC Press.

Cerutti H, Casas-Mollano JA (2006) On the origin and functions of RNA-mediated silencing: From protists to man. Current Genetics 50: 81–99.

Clark MM, Karpiuk P, Galef BG (1993) Hormonally mediated inheritance of acquired characteristics in Mongolian gerbils. Nature 364: 712.

Cook PR (1974) On the inheritance of differentiated traits. Biological Reviews 49: 51–84.

Crews D, Gore AC, Hsu TS, Dangleben NL, Spinetta M, Schallert T, Anway MD, Skinner MK (2007) Transgenerational epigenetic imprints on mate preference. Proceedings of the National Academy of Sciences of the USA 104: 5942–5946.

Crouse HV (1960) The controlling element in sex chromosome behavior in Sciara. Genetics 45: 1429–1443.

Cubas P, Vincent C, Coen E (1999) An epigenetic mutation responsible for natural variation in floral symmetry. Nature 401: 157–161.

Cullis CA (2005) Mechanisms and control of rapid genomic changes in flax. Annals of Botany 95: 201–206.

Curley JP, Champagne FA, Bateson P, Keverne EB (2008) Transgenerational effects of impaired maternal care on behaviour of offspring and grandoffspring. Animal Behaviour 75: 1551–1561.

Dor D, Jablonka E (2000) From cultural selection to genetic selection: A framework for the evolution of language. Selection 1: 33–55.

Dubrova YE (2003) Radiation-induced transgenerational instability. Oncogene 22: 7087–7093.

Ephrussi B (1958) The cytoplasm and somatic cell variation. Journal of Cellular and Comparative Physiology 52(suppl 1): 35–53.

Fedoroff NV (1989) About maize transposable elements and development. Cell 56: 181–191.

Forgács G, Newman SA (2005) Biological Physics of the Developing Embryo. Cambridge: Cambridge University Press.

Gilbert SF (2001) Ecological developmental biology: Developmental biology meets the real world. Developmental Biology 233: 1–12.

Ginsburg S, Jablonka E (2009) Epigenetic learning in non-neural organisms. Biosciences.

Gluckman PD, Hanson MA, Beedle AS (2007) Non-genomic transgenerational inheritance of disease risk. BioEssays 29: 145–154.

Goldenfeld N, Woese C (2007) Biology's next revolution. Nature 445: 369.

Goldschmidt RB (1940) The Material Basis of Evolution. New Haven, CT: Yale University Press.

Grimes GW, Aufderheide KJ (1991) Cellular aspects of pattern formation: The problem of assembly. Monographs in Developmental Biology 22: 1–94.

Gurdon JB (2006) From nuclear transfer to nuclear reprogramming: The reversal of cell differentiation. Annual Review of Cell and Developmental Biology 22: 1–22.

Hadorn E (1968) Transdetermination in cells. Scientific American 219 (Nov): 110–120.

Haig D (2004) The (dual) origin of epigenetics. Cold Spring Harbor Symposia on Quantitative Biology 69: 67–70.

Holliday R (1994) Epigenetics: An overview. Developmental Genetics 15: 453–457.

Holliday R (2006) Epigenetics: A historical overview. Epigenetics 1: 76–80.

Holliday R, Pugh JE (1975) DNA modification mechanisms and gene activity during development. Science 187: 226–232.

Jablonka E, Lachmann M, Lamb MJ (1992) Evidence, mechanisms and models for the inheritance of acquired characters. Journal of Theoretical Biology 158: 245–268.

Jablonka E, Lamb MJ (1989) The inheritance of acquired epigenetic variations. Journal of Theoretical Biology 139: 69–83.

Jablonka E, Lamb MJ (1995) Epigenetic Inheritance and Evolution: The Lamarckian Dimension. Oxford: Oxford University Press.

Jablonka E, Lamb MJ (2002) The changing concept of epigenetics. Annals of the New York Academy of Sciences 981: 82–96.

Jablonka E, Lamb MJ (2005) Evolution in Four Dimensions: Genetic, Epigenetic, Behavioral, and Symbolic Variation in the History of Life. Cambridge, MA: MIT Press.

Jablonka E, Lamb MJ (2006) The evolution of information in the major transitions. Journal of Theoretical Biology 239: 236–246.

Jablonka E, Lamb MJ (2007a) Bridging the gap: The developmental aspects of evolution. Behavioral and Brain Sciences 30: 378–392.

Jablonka E, Lamb MJ (2007b) The expanded evolutionary synthesis: A response to Godfrey-Smith, Haig, and West-Eberhard. Biology and Philosophy 22: 453–472.

Jablonka E, Lamb MJ (2007c) Précis of Evolution in Four Dimensions. Behavioral and Brain Sciences 30: 353–365.

Jablonka E, Lamb MJ (2008) The epigenome in evolution: Beyond the Modern Synthesis. VOGis Herald 12: 242–254.

Jablonka E, Raz G (2009) Transgenerational epigenetic inheritance: Prevalence, mechanisms, and implications for the study of heredity and evolution. Quarterly Review of Biology 84: 131–176.

Jablonski D (2005) Mass extinctions and macroevolution. Paleobiology 31: 192–210.

Jorgensen RA (1993) The germinal inheritance of epigenetic information in plants. Philosophical Transactions of the Royal Society of London B339: 173–181.

Jorgensen RA (2004) Restructuring the genome in response to adaptive challenge: McClintock's bold conjecture revisited. Cold Spring Harbor Symposia on Quantitative Biology 69: 349–354.

Kermicle JL (1978) Imprinting of gene action in maize endosperm. In Maize Breeding and Genetics. DB Walden, ed.: 357–371. New York: Wiley.

Kirschner MW, Gerhart JC (2005) The Plausibility of Life: Resolving Darwin's Dilemma. New Haven, CT: Yale University Press.

Konold T, Moore SJ, Bellworthy SJ, Simmons HA (2008) Evidence of scrapie transmission via milk. BMC Veterinary Research 4: 14.

Korenaga T, Yan J, Sawashita J, Matsushita T, Naiki H, Hosokawa M, Mori M, Higuchi K, Fu X (2006) Transmission of amyloidosis in offspring of mice with AApoAII amyloidosis. American Journal of Pathology 168: 898–906.

Lachmann M, Jablonka E (1996) The inheritance of phenotypes: An adaptation to fluctuating environments. Journal of Theoretical Biology 181: 1–9.

Lamm E, Jablonka E (2008) The nurture of nature: Hereditary plasticity in evolution. Philosophical Psychology 21: 305–319.

Lemke H, Coutinho A, Lange H (2004) Lamarckian inheritance by somatically acquired maternal IgG phenotypes. Trends in Immunology 25: 180–186.

Levites EV (2000) Epigenetic variability as a source of biodiversity and a factor of evolution. In Biodiversity and Dynamics of Ecosystems in North Eurasia, vol. 1, part 3: 73–75. Novosibirsk: Institute of Cytology and Genetics SB RAS.

Levites EV, Maletskii SI (1999) Auto- and episegregation for reproductive traits in agamospermous progenies of beet *Beta vulgaris* L. Russian Journal of Genetics 35: 802–810.

Levy AA, Feldman M (2004) Genetic and epigenetic reprogramming of the wheat genome upon allopolyploidization. Biological Journal of the Linnean Society of London 82: 607–613.

Lyon MF (1961) Gene action in the X-chromosome of the mouse (*Mus musculus* L.). Nature 190: 372–373.

Manning K, Tör M, Poole M, Hong Y, Thompson AJ, King GJ, Giovannoni JJ, Seymour GB (2006) A naturally occurring epigenetic mutation in a gene encoding an SBP-box transcription factor inhibits tomato fruit ripening. Nature Genetics 38: 948–952.

Masel J, Bergman A (2003) The evolution of the evolvability properties of the yeast prion [PSI+]. Evolution 57: 1498–1512.

Matzke MA, Matzke AJM (1991) Differential inactivation and methylation of a transgene in plants by two suppressor loci containing homologous sequences. Plant Molecular Biology 16: 821–830.

Matzke MA, Mittelsten Scheid O (2007) Epigenetic regulation in plants. In Epigenetics. CD Allis, T Jenuwein, D Reinberg, eds.: 167–189. Cold Spring Harbor, NY: Cold Spring Harbor Laboratory Press.

Maynard Smith J (1990) Models of a dual inheritance system. Journal of Theoretical Biology 143: 41–53.

Maynard Smith J, Szathmáry E (1995) The Major Transitions in Evolution. Oxford: Freeman.

Mayr E (1982) The Growth of Biological Thought. Cambridge, MA: Harvard. University Press.

McClintock B (1984) The significance of responses of the genome to challenge. Science 226: 792–801.

Meaney MJ (2001) Maternal care, gene expression, and the transmission of individual differences in stress reactivity across generations. Annual Review of Neuroscience 24: 1161–1192.

Meins F (1989a) A biochemical switch model for cell-heritable variation in cytokinin requirement. In The Molecular Basis of Plant Development. R Goldberg, ed.: 13–24. New York: Alan R Liss.

Meins F (1989b) Habituation: Heritable variation in the requirement of cultured plant cells for hormones. Annual Review of Genetics 23: 395–408.

Molinier J, Ries G, Zipfel C, Hohn B (2006) Transgeneration memory of stress in plants. Nature 442: 1046–1049.

Morton SMB (2006) Maternal nutrition and fetal growth and development. In Developmental Origins of Health and Disease. P Gluckman, M Hanson, eds.: 98–129. Cambridge: Cambridge University Press.

Mousseau TA, Fox CW (1998) Maternal Effects as Adaptations. New York: Oxford University Press.

Nanney D (1958) Epigenetic control systems. Proceedings of the National Academy of Sciences of the USA 44: 712–717.

Newman SA, Müller GB (2006) Genes and form: Inherency in the evolution of developmental mechanisms. In Genes in Development: Re-reading the Molecular Paradigm. E Neumann-Held, C Rehmann-Sutter, eds.: 38–73. Durham NC: Duke University Press.

Novick A, Weiner M (1957) Enzyme induction as an all-or-none phenomenon. Proceedings of the National Academy of Sciences of the USA 43: 553–566.

Odling-Smee FJ, Laland KN, Feldman MW (2003) Niche Construction: The Neglected Process in Evolution. Princeton, NJ: Princeton University Press.

Pigliucci M, Murren CJ, Schlichting CD (2006) Phenotypic plasticity and evolution by genetic assimilation. Journal of Experimental Biology 209: 2362–2367.

Pikaard CS (2003) Nucleolar dominance. In Nature Encyclopedia of the Human Genome, vol 4. DN Cooper, ed.: 399–402. London: Nature Publishing Group.

Rakyan VK, Chong S, Champ ME, Cuthbert PC, Morgan HD, Luu KVK, Whitelaw E (2003) Transgenerational inheritance of epigenetic states at the murine $Axin^{Fu}$ allele occurs after maternal and paternal transmission. Proceedings of the National Academy of Sciences of the USA 100: 2538–2543.

Rando OJ, Verstrepen KJ (2007) Timescales of genetic and epigenetic inheritance. Cell 128: 655–668.

Rangwala SH, Elumalai R, Vanier C, Ozkan H, Galbraith DW, Richards EJ (2006) Meiotically stable natural epialleles of $Sadhu$, a novel Arabidopsis retroposon. PLoS Genetics 2: 270–281.

Rapp RA, Wendel JF (2005) Epigenetics and plant evolution. New Phytologist 168: 81–91.

Rassoulzadegan M, Grandjean V, Gounon P, Vincent S, Gillot I, Cuzin F (2006) RNA-mediated non-mendelian inheritance of an epigenetic change in the mouse. Nature 441: 469–474.

Richerson PJ, Boyd R (2005) Not by Genes Alone: How Culture Transformed Human Evolution. Chicago: University of Chicago Press.

Riggs AD (1975) X inactivation, differentiation, and DNA methylation. Cytogenetics and Cell Genetics 14: 9–25.

Sager R, Kitchin R (1975) Selective silencing of eukaryotic DNA. Science 189: 426–433.

Sager R, Ryan FJ (1961) Cell Heredity. New York: Wiley.

Salmon A, Ainouche ML, Wendel JF (2005) Genetic and epigenetic consequences of recent hybridization and polyploidy in $Spartina$ (Poaceae). Molecular Ecology 14: 1163–1175.

Savage-Rumbaugh ES, Shanker SG, Taylor TJ (1998) Apes, Language, and the Human Mind. New York: Oxford University Press.

Schmalhausen II (1949) Factors of Evolution: The Theory of Stabilizing Selection. Philadelphia: Blakiston.

Shorter J, Lindquist S (2005) Prions as adaptive conduits of memory and inheritance. Nature Reviews Genetics 6: 435–450.

Simpson SJ, Miller GA (2007) Maternal effects on phase characteristics in the desert locust, $Schistocerca$ $gregaria$: A review of current understanding. Journal of Insect Physiology 53: 869–876.

Sober E, Wilson DS (1998) The Evolution and Psychology of Unselfish Behavior. Cambridge, MA: Harvard University Press.

Sollars V, Lu X, Xiao L, Wang X, Garfinkel MD, Ruden DM (2003) Evidence for an epigenetic mechanism by which Hsp90 acts as a capacitor for morphological evolution. Nature Genetics 33: 70–74.

Sonneborn T (1964) The differentiation of cells. Proceedings of the National Academy of Sciences of the USA 51: 915–929.

Tanaka S, Maeno K (2008) Maternal effects on progeny body size and color in the desert locust, *Schistocerca gregaria*: Examination of a current view. Journal of Insect Physiology 54: 612–618.

True HL, Lindquist SL (2000) A yeast prion provides a mechanism for genetic variation and phenotypic diversity. Nature 407: 477–483.

Trut LN, Plyusnina IZ, Oskina IN (2004) An experiment on fox domestication and debatable issues of evolution of the dog. Russian Journal of Genetics 40: 644–655.

Turner JS (2000) The Extended Organism: The Physiology of Animal-Built Structures. Cambridge, MA: Harvard University Press.

Vastenhouw NL, Brunschwig K, Okihara KL, Müller F, Tijsterman M, Plasterk RHA (2006) Long-term gene silencing by RNAi. Nature 442: 882.

Villarreal LP (2005) Viruses and the Evolution of Life. Washington DC: ASM Press.

Waddington CH (1957) The Strategy of the Genes. London: Allen and Unwin.

Weaver ICG, Cervoni N, Champagne FA, D'Alessio AC, Sharma S, Seckl JR, Dymov S, Szyf M, Meaney MJ (2004) Epigenetic programming by maternal behavior. Nature Neuroscience 7: 847–854.

Weaver ICG, Champagne FA, Brown SE, Dymov S, Sharma S, Meaney MJ, Szyf M (2005) Reversal of maternal programming of stress responses in adult offspring through methyl supplementation: Altering epigenetic marking later in life. Journal of Neuroscience 25: 11045–11054.

West-Eberhard MJ (2003) Developmental Plasticity and Evolution. New York: Oxford University Press.

Whiten A, Horner V, de Waal FBM (2005) Conformity to cultural norms of tool use in chimpanzees. Nature 437: 737–740.

Wilkins AS (2002) The Evolution of Developmental Pathways. Sunderland, MA: Sinauer.

Wood DC (1992) Learning and adaptive plasticity in unicellular organisms. In Encyclopedia of Learning and Memory. LR Squire, ed.: 623–624. New York: Macmillan.

Wu C-T, Morris JR (2001) Genes, genetics, and epigenetics: A correspondence. Science 293: 1103–1105.

Zilber-Rosenberg I, Rosenberg E (2008) Role of microorganisms in the evolution of animals and plants: The hologenome theory of evolution. FEMS Microbiology Reviews 32: 723–735.

Zordan RE, Galgoczy DJ, Johnson AD (2006) Epigenetic properties of white-opaque switching in *Candida albicans* are based on a self-sustaining transcriptional feedback loop. Proceedings of the National Academy of Sciences of the USA 103: 12807–12812.

Zuckerkandl E, Cavalli G (2007) Combinatorial epigenetics, "junk DNA," and the evolution of complex organisms. Gene 390: 232–242.

Zufall RA, Robinson T, Katz LA (2005) Evolution of developmentally regulated genome rearrangements in eukaryotes. Journal of Experimental Zoology, B: Molecular Development and Evolution 304: 448–455.

8 Niche Inheritance

John Odling-Smee

The Modern Synthesis has been a highly successful theory of evolution. However, due to some of its underlying assumptions, it is also the source of conceptual barriers that are currently making further progress in some areas stubbornly difficult.

One example is the long-standing inability of the Modern Synthesis to recognize *niche construction* as a co-causal process in evolution (see below). The Modern Synthesis's omission of niche construction not only restricts our understanding of the dynamics of the evolutionary process, it also makes it difficult to integrate evolutionary biology with several neighboring disciplines (Odling-Smee et al. 2003). For instance, it is still not possible to integrate ecosystem-level ecology with evolutionary biology (Jones and Lawton 1995). It is not possible to fully relate evolutionary biology to developmental biology ("evo-devo") correctly, nor to explain some recent data (West-Eberhard 2003; Laland et al. 2008). Nor, in our own species, is it possible to integrate our potent human cultural processes with human genetic evolution without introducing some unacceptable distortions (Odling-Smee et al. 2003).

This chapter seeks to demonstrate how the addition of niche construction to evolutionary theory alleviates or removes many of these problems. Since we have discussed the relationship between evolutionary theory, ecosystem ecology, and the human sciences before (Odling-Smee et al. 2003), I will concentrate primarily on the EvoDevo relationship.

Niche Construction Theory

Half a century ago a developmental geneticist, C. H. Waddington (1959), proposed the concept of an "exploitive system" in which animals choose and modify their environments, and by doing so, change some of the natural selection pressures they and their descendants confront. Later, a population geneticist, R. C. Lewontin (1983), proposed a similar idea: "Organisms do not adapt to their environments: they construct them out of the bits and pieces of the external world" (p. 280). Lewontin's proposal has since gathered momentum and is now called niche construction (Odling-Smee 1988; Odling-Smee et al. 2003; Laland and Sterelny 2006).

Figure 8.1 compares the modern synthetic theory of evolution (henceforth the MS) to niche construction theory (henceforth NCT). In the MS (figure 8.1a) natural selection pressures in autonomous environments, E, act on populations of diverse phenotypes to influence which individuals survive and reproduce and pass on their genes to the next generation through a single inheritance system, genetic inheritance. The adaptations of organisms are therefore consequences of autonomous selection pressures molding organisms to fit preestablished environmental templates. The templates are dynamic because processes that are independent of organisms change the environments to which organisms have to adapt, yet the changes that organisms bring about in their own environments are seldom thought to have evolutionary significance.

However, all organisms, through their metabolisms, movements, behavior, and choices, partly create and partly destroy their environments. In doing so, they transform some of the selection pressures in the environments that subsequently select them (Lewontin 1983; Odling-Smee et al. 2003). Therefore the adaptations of organisms cannot be exclusively consequences of organisms responding to autonomous selection pressures in environments. Sometimes they must involve organisms responding to selection pressures previously transformed by their own, or by their ancestors', niche-constructing activities.

When niche construction is added to the MS, it extends the "Synthesis." The evolution of organisms now depends on natural selection and niche construction (figure 8.1b). The transmission of genes by ancestral organisms to their descendants is influenced by natural selection, as in figure 8.1a. However, selected habitats, modified habitats, and modified sources of natural selection in those habitats are also transmitted by those same organisms to their descendants through a second general inheritance system, *ecological inheritance*. Ecological inheritance comprises the inheritance of selection pressures previously modified by niche-constructing organisms in an external environment (Odling-Smee et al. 2003). In NCT, the selective environments of organisms are therefore partly determined by independent sources of natural selection, for instance, by climate, or physical and chemical events, as usual. They are also partly determined by what organisms do, or previously did, to their own and each others' environments, by niche construction.

(a)

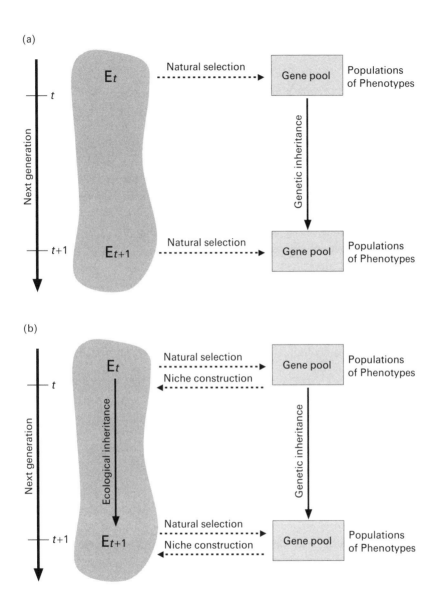

(b)

Figure 8.1
(a) The Modern Synthesis. (b) Niche construction theory.

There are innumerable examples of niche construction. Animals manufacture nests, burrows, webs, and pupal cases; plants modify fire regimes, levels of atmospheric gases, and nutrient cycles; fungi decompose organic matter; and bacteria fix nutrients (Turner 2000; Odling-Smee et al. 2003; Scwhilk 2003; Hansell 2005; Meysman et al. 2006). There are also examples of social niche construction in insects (Frederickson et al. 2005) and primates (Flack et al. 2006), and of cultural niche construction in humans (Feldman and Cavalli-Sforza 1989; Laland et al. 2000; Smith 2007). For decades ecologists have realized that organisms do alter their environments in ecologically significant ways, now called *ecosystem engineering* (Jones et al. 1994, 1997; Wright and Jones 2006; Erwin 2008).

Niche construction has also been modeled. Components of niche construction are implicit in many standard population genetic models of frequency- and density-dependent selection (Futuyma 1998); habitat selection (Hanski and Singer 2001); maternal inheritance (Kirkpatrick and Lande 1989); extended phenotypes (Dawkins 1982); indirect genetic effects and epistasis (Wolf 2000); and co-evolution (Thompson 1994, 2006). Other, less standard models have explicitly investigated how niche construction and ecological inheritance affect the dynamics of evolution (Laland et al. 1996, 1999, 2001; Odling-Smee et al. 2003; Ihara and Feldman 2004; Hui et al. 2004; Boni and Feldman 2005; Borenstein et al. 2006; Silver and Di Paolo 2006; Lehmann 2008). All the latter models find that niche construction is consequential because it changes what happens in evolution.

The Limitations of Standard Evolutionary Theory

In spite of these findings, the full significance of niche construction has, until recently, been neglected. Niche construction is an obvious process, so why had it been marginalized for so long? The answer probably lies with a seldom reconsidered assumption of the MS concerning the role of environments in evolution. I called it the "reference device" problem (Odling-Smee 1988). Lewontin (1983) described it with two pairs of equations.

Equations 8.1a and 8.1b summarize the MS:

$$\frac{dO}{dt} = f(O, E), \tag{8.1a}$$

$$\frac{dE}{dt} = g(E). \tag{8.1b}$$

The MS assumes that evolutionary change in organisms, dO/dt, depends on both organisms' states, O, and environmental states, E (1a). It also assumes that environmental change, dE/dt, depends exclusively on environmental states (1b). Therefore, organisms evolve in response to independent changes in their environments. Hence, environmental change becomes the implicit reference device relative to which the evolution of organisms is understood (Odling-Smee 1988).

Equations 8.2a and 8.2b summarize NCT:

$$\frac{dO}{dt} = f(O, E), \tag{8.2a}$$

$$\frac{dE}{dt} = g(O, E). \tag{8.2b}$$

Change in organisms, dO/dt, depends on organisms' states and environmental states, but environmental change, dE/dt, now depends on both environmental states and the environment-modifying activities of organisms (8.2b). In part, the environment is now co-evolving with its organisms. Hence, "the environment" cannot be used as an independent reference device for understanding the evolution of organisms. It is not independent.

The philosopher Peter Godfrey-Smith (1996) highlighted the same problem by describing the MS as an "externalist" theory. It is externalist because it seeks to explain the internal properties of organisms, their adaptations, exclusively in terms of properties of their external environments, natural selection pressures.

The MS obscures the point that to stay alive, organisms must be active as well as reactive. Organisms must gain resources from their external environments by genetically informed, or possibly brain-informed, fuel-consuming, nonrandom work. They must return detritus to their environments, and they must choose and perturb specific components of their environments when they do so (Odling-Smee et al. 2003). Organisms are therefore compelled to change some of the selection pressures in their environments. This point is captured by Lewontin's equation 8.2b. Equation 8.2b introduces a second "causal arrow" in evolution in addition to Darwin's first "causal arrow" of natural selection. Odling-Smee (1988) called this second causal arrow niche construction.

Now we can see why the MS neglects niche construction. One of the causal arrows in equations 8.2a and 8.2b, *natural selection*, is compatible with the externalist assumption of the MS because it is pointing in the "right" direction, from environments to organisms. It is conceptually straightforward to describe how external natural selection pressures in environments cause adaptations in organisms. But the second causal arrow, *niche construction*, is pointing in the "wrong" direction, from organisms to environments. Hence, niche construction is incompatible with the MS's externalist assumption, making it difficult or impossible for evolutionary biologists to describe changes in natural selection pressures caused by prior niche construction as evolutionarily causal. Instead, the MS is forced to explain away all observed instances of niche construction as nothing but phenotypic, or possibly extended phenotypic (Dawkins 1982, 2004), consequences of prior natural selection. The MS can recognize niche construction as a consequence of evolution, but it cannot recognize it as causal.

NCT overcomes this obstacle by describing the evolution of organisms relative to their niches instead of relative to their environments (Odling-Smee et al. 2003).

$$N(t) = h(OE) \tag{8.3}$$

In equation (8.3), $N(t)$ represents the niche of a population of organisms O at time t. The dynamics of $N(t)$ are driven by both population-modifying natural selection pressures in E, and by the environment-modifying niche-constructing activities of populations, O. Because niches always include two-way interactions between organisms and their environments (Chase and Leibold 2003), this step is sufficient to allow an "interactionist" (Godfrey-Smith 1996) theory of evolution, NCT (figure 8.1b), to be substituted for the MS (figure 8.1a). Niches are neutral. The *(OE)* niche relationship does not impose a bias in favor of natural selection and against niche construction, nor vice versa. Instead, it allows both the causal arrows in (8.2), natural selection and niche construction, to be modeled as *reciprocal* causal processes in evolution (Griffiths and Gray 2004; Laland and Sterelny 2006).

Ecological Inheritance

Multiple consequences flow from this revision. I will focus on those that affect the relationship between evolution and development.

When niche construction is added as a co-causal process in evolution, it not only contributes to the adaptations of organisms, it also generates ecological inheritances whenever the environmental consequences of the prior niche-constructing activities of organisms (e.g., the presence of burrows, mounds, and dams or, on a larger scale, changed atmospheric states, soil states, substrate states, or sea states—Dietrich et al. 2006; Meysman et al. 2006; Erwin 2008) persist or accumulate in environments as modified natural selection pressures, relative to successive generations of organisms.

Ecological inheritance is very different from genetic inheritance (Odling-Smee 1988). First, ecological inheritance is transmitted by organisms through the medium of an external environment. It is not transmitted by reproduction. Second, ecological inheritance seldom depends on the transmission of discrete replicators. Typically it depends on organisms bequeathing altered selective environments to their offspring by choosing, or physically perturbing, biological or non-biological components of their environments. Third, in sexual populations, genes are transmitted by two parents only, on a single occasion only, to each offspring. In contrast, an ecological inheritance is continuously transmitted by multiple organisms, to multiple other organisms, within and between generations, throughout the lifetimes of organisms. Fourth, ecological inheritance is not always transmitted by genetic relatives. It can be transmitted by other organisms in shared ecosystems that must be ecologically related, but need not be genetically related to the organisms receiving the inheritance.

Population genetic models show that the inclusion of ecological inheritance in NCT affects the dynamics of the evolutionary process. For example, niche construction can generate ecological inheritances to the point where modified natural selection overrides independent sources of selection, and drives populations down alternative evolutionary trajectories. Niche construction can initiate novel evolutionary episodes; it can influence the amount of genetic variation carried by populations; and it can generate unusual dynamics such as time lags and momentum effects (Laland et al. 1996, 1999, 2001; Schwilk and Ackerly 2001; Ihara and Feldman 2004; Borenstein et al. 2006; Hui et al. 2004; Silver and Di Paolo 2006; Lehmann 2008).

Niche Inheritance

NCT applies to development as well as to evolution by substituting *niche inheritance* for genetic inheritance. If in each generation each individual offspring inherits not only genes relative to its selective environment, but also an ecological inheritance, in the form of a modified local selective environment relative to its genes, then each offspring must actually inherit an initial organism-environment relationship, or "niche," from its ancestors. From equation (8.3), niche inheritance is given by $N(t_o) = h$ $[O,E]$, where $N(t_o)$ represents the state of an individual organism's inherited niche at the moment of origin, time t_o, of a new organism, the proviso being that "niche" now refers to the "personal" inherited developmental niche of an individual organism, and not to the evolutionary niche of its population (Odling-Smee 1988).

Minimally, each inherited niche for each offspring organism must include the inheritance of an initial environmental "address" in space and time as well as its inherited genes. Often that address will be influenced by parental choices, for example, simply by the timing of parental reproduction (Donahue 2005). In addition, in many species parents ensure that a resource package is also present at their offspring's initial address. For instance, phytophagous insects not only supply their offspring with eggs, they also choose specific host plants on which to lay their eggs, the chosen plants subsequently serving as energy and matter resources for their offspring. An organism's start-up niche may also include other environmental resources due to the niche-constructing activities of other, less closely related organisms. For example, when termites build a mound, they modify the temperature and humidity experienced by developing larvae by their collective niche construction (Hansell 2005).

NCT's niche inheritance is therefore richer than the MS's genetic inheritance, and its richness has two immediate implications. The first concerns how prior evolutionary processes (evo-) affect subsequent developmental (devo-) processes. Niche inheritance means evolution must contribute more to the development of individual organisms than just genes, because it also bequeaths modified selective environments. The second concerns how the prior development (devo-) of individual organisms may influence the subsequent evolution (evo-) of populations. Niche inheritance introduces some new ways in which prior devo- could affect subsequent evo-.

Classifying Niche Inheritance

Before exploring these implications, we need to expand the concept of niche inheritance to demonstrate how the various candidate inheritance systems currently being discussed by biologists (West-Eberhard 2003; Pigliucci and Preston 2004; Jablonka and Lamb 2005; Müller 2007; Wagner et al. 2007; Pigliucci 2008) correspond to different components of niche inheritance. For that, we need some descriptive dimensions.

Two dimensions can be derived from further unpacking the $[OE]$ niche relationship (Odling-Smee 2007). One concerns the relationship between the internal and external environments of organisms. It demarcates the two transmission channels through which the two principal components of niche inheritance, genetic and ecological, are inherited (table 8.1). A second dimension stems from the relationship between the two principal kinds of resources that organisms inherit, semantic information resources and energy and material resources (table 8.1). Both are essential to life. Both can serve as natural selection pressures. Both can be modified by niche construction.

The first transmission channel (channel 1) comprises the direct connection between the internal environments of parent organisms and the internal environments of their offspring through reproduction. Channel 1 is reducible to the mechanisms of cell division and cell fusion. It was discovered by Schwann and Virchow in the 19th century, and it gave rise to cell theory. Any kind of inheritance that travels between organisms, from cell to cell, or from O to O, directly during reproduction will be treated as a channel 1 type of inheritance.

The second transmission channel (channel 2) connects possibly multiple ancestral niche-constructing organisms to descendant organisms indirectly, through the modification of selection pressures in external environments. Any kind of inheritance that does not travel between organisms, from cell to cell directly, will be treated as a channel 2 type of inheritance. Channel 2 works in the way that ecological inheritance works, from O to E, and back to O again.

These definitions imply that it is easy to demarcate the internal and external environments of organisms, but that is not always true. For example, it would seem straightforward to define any individual organism, O, by whatever boundary exists between the organism and its environment, E. In a single-celled organism the boundary is its cell wall or membrane. O's external environment E is therefore everything outside that same boundary. Depending on the focus of interest, however, there can be complications. For example, in a multicelled organism the external environment of most cells includes all the other cells in the metazoan's body, as well as the metazoan organism's own external environment. Hence, demarcating between an internal and external environment depends on how an organism, or part of an organism, is defined. Assuming an organism, O, can be defined sufficiently clearly, then relative to that O, an external environment, E, can also be defined.

It may also be difficult to distinguish between the two kinds of heritable resources, semantic information versus energy and matter. The relationship between information and energy and matter is notoriously tricky (Bergstrom and Lachmann 2004), so several points need clarifying.

The distinction between information versus energy and matter in biology takes us back to the origin of life. Physically, organisms are highly improbable systems. To survive and reproduce, they must import energy and materials from, and must return detritus to, their environments. But organisms can do neither of these things unless they are adapted to their environments. Nor can they adapt without being sufficiently "informed" a priori by adaptive semantic information. Organisms need "meaningful" information to build and control adaptive phenotypes that can tap into energy and matter flows in their ecosystems, and dump detritus (Odling-Smee et al. 2003). Organisms cannot be sufficiently informed a priori, however, unless they, or their ancestors, also possess sufficient energy and material resources a priori to pay for the physical acquisition, storage, use, and transmission of adaptive semantic information. So what came first, physical resources or semantic information, metabolism or replication? This is the classic origin of life problem, and it is still not resolved (Fry 2000).

Equally, it is notoriously difficult to define semantic information. I will offer only a working definition: *Semantic information is anything that reduces uncertainty about selective environments, relative to the fitness interests of organisms.* This definition is strongly relativistic. Semantic information is "meaningful" only relative to particular selection pressures in the particular environments of particular organisms. Conversely, environmental resources are only resources, and can only serve as natural selection pressures, relative to the specific needs and traits of specific organisms. This is true regardless of whether selection pressures stem from positive resources such as food or water, or negative resources such as threatening ecological conditions. This definition also refrains from specifying any physical basis for the memory systems that actually carry semantic information within and between organisms. The physical carriers of semantic information could be DNA, RNA, or other molecules, or neurons, or even symbols entrenched in human artifacts (Aunger 2002). All that matters is that the semantic information carried by any physical carrier can potentially influence the fitness of organisms.

Now we can ask which transmission channel carries what kind of heritable resource. By strongly identifying genes with information, and by reserving channel 1 for genetic inheritance only, the MS encourages the idea that channel 1 transmits only genetically encoded semantic information. Similarly, our initial models of NCT restricted ecological inheritance to biotically modified energy and matter resources only (Laland et al. 1996, 1999). That inadvertently encouraged the idea that channel 2 transmits only modified physical resources. Both restrictions are misleading.

Nongenetic components of "start-up niches" can also be transmitted through channel 1, including physical resources. For instance, in insects and birds, the eggs supplied by mothers to their offspring carry some energy and matter resources in the form of cytoplasm and protein in egg yolks, as well as genetically encoded semantic information (Sapp 1987; Amundson 2005). Similarly, modified semantic information, as well as modified physical resources, can be transmitted through channel 2. Semantic information is transmitted when an ecological inheritance includes other organisms in an environment that are likewise the inheritors of, and carriers of, semantic information, and when the semantic information they carry has previously been modified by niche construction. This is a complex issue requiring further explanation.

One way to explain it is to return to the logic, but not the mathematicss, underpinning all our initial models of niche construction (Laland et al. 1996, 1999, 2001). These models were based on two-locus population genetic theory. They focused on two genetic loci, labeled E and A, in single populations, and assumed that (1) the population's capacity for niche construction is influenced by the frequency of alleles at the first, or E, locus; (2) the amount of some resource, R, in the population's environment depends wholly or in part on the niche-constructing activities of past and present generations of organisms; and (3) the amount of the resource, R, subsequently influences the pattern and strength of selection acting on alleles at the second, or A, locus, in the same population.

R can be any environmental condition or resource, provided it is possible for organisms to modify it by niche construction. R might therefore be an abiotic environmental component, for instance, a sediment or a water hole. Or R could be an artifact built by an animal, for example, a termite nest or a beaver dam. Or R could be a biotic component in the form of one or many other organisms in the environments of niche-constructing organisms. In the latter case these other organisms could belong to the same population as the niche-constructing organisms, or they could belong to different populations.

In spite of R's generality, we initially kept R as simple as possible, by assuming it referred exclusively to energy and matter resources. That assumption suffices for abiota, but not for biota, because, unlike abiota, other organisms contain two kinds of resources, physical resources (from hereon notated R_p) and semantic information (from hereon notated R_i). Also, both the R_p and R_i carried by other organisms can be altered, or exploited, or defended against by niche-constructing organisms. Hence, modified R_i states, as well as modified R_p states, can potentially become part of an ecological inheritance for descendant organisms.

If R is another organism, it might act only as a physical resource, R_p, say a food item, for a niche-constructing organism. Alternatively, a niche-constructing organism may "manipulate" another organism (e.g., the brood parasitism of cuckoos; Davies et al. 1998) or "copy" (e.g., social learning in animals; Fragaszy and Perry 2003) the semantic information, R_i, carried in either the genome or the brain of another organism by communicating with it. For example, a parasite, say a virus, may insert DNA into its host's DNA, and by doing so, manipulate the physiology or the behavior of its host (Combes 2001). When that happens, the parasite does not immediately gain any physical resource, R_p, by what, from hereon, I'll call *communicative niche construction*. What it gains is a degree of control over its host's phenotype by corrupting the semantic information in its host's genome or brain. Moreover, it has to pay a fitness cost for doing that, in the form of some expenditure of R_p. However, the parasite may subsequently use its capacity to control its host to gain R_p, and hence gain a subsequent fitness benefit, by causing its host to supply it with a physical resource. For example, galls produced by parasitized plants benefit the parasites at a cost to their hosts (Combes 2001; West- Eberhard 2003). In general, it should pay any organism to invest in communicative niche construction when the eventual benefit, measured in R_p, exceeds its initial cost, also measured in R_p (Bergstrom and Lachmann 2004; Odling-Smee 2007).

Benign forms of communication are also common, and obey the same rules. For example, it should "pay" a parent animal to transmit some of the semantic information in its brain to its offspring via social learning if that increases the parent's fitness. Or an offspring organism should pay the initial fitness cost of soliciting and copying some of the semantic information held by its parents, or peers, or possibly other organisms in its social group, if the extra information it gains subsequently improves its fitness.

Therefore, in spite of restricting ecological inheritance in our initial models to the transmission of modified R_p, that restriction is neither necessary nor desirable. Ecological inheritance can refer to both physical,

R_p, and semantic, R_i, resources, because both can be modified by niche construction. Hence, both channel 1 and channel 2 transmit both physical R_p and informational R_i resources between organisms, and between generations.

The Components of Niche Inheritance

We can now classify the principal subcomponents of niche inheritance. Several authors have proposed a variety of inheritance systems in evolution in addition to genetic inheritance (West-Eberhard 2003; Jablonka and Lamb 2005; Bird 2007). Jablonka and Lamb offer the most general scheme, so I will use theirs. It includes four different inheritances: (1) genetic, (2) epigenetic, (3) behavioral, and (4) symbolic.

Table 8.1 classifies all four of these inheritance systems by using the two transmission channels, channel 1 versus channel 2, and the two principal kinds of heritable resources, semantic information (R_i) versus energy and matter (R_p) to generate a 2×2 table. Each tabular Cell (differentiated from biological cells by uppercase Cs) is then assigned the types of inheritance proposed to belong to it. Illustrative examples are given in each Cell.

Many of the assignments are straightforward. For instance, naturally selected genes encode semantic information, or R_i (Maynard Smith 2000), and insofar as they are internally inherited, genetic inheritance clearly belongs to Cell 1a. Given that Jablonka and Lamb (2005: 147) define epigenetic inheritance as the "transfer of information from cell to cell," although it may be possible to define epigenetic inheritance more broadly, under their definition epigenetic inheritance also belongs to Cell 1a. Some examples discussed by Jablonka and Lamb under the heading of epigenetic inheritance are shown in Cell 1a. They include chromatin markings, methylation patterns, RNAi (interference), and some maternal effects, for example, the inheritance of maternal mRNA (Davidson 2006).

Cell 1b includes cytoplasmic inheritance and other heritable physical resources, or R_p, typically transmitted by mothers to their offspring, primarily through eggs. These resources include proteins and components of membranes (Davidson 2006).

Cell 2a refers to the conventional kinds of ecological inheritance we previously modeled, namely, the inheritance of *directly* modified selection pressures in the external environments of organisms, as a consequence of either the perturbation of energy and matter (R_p) resources by niche-

Table 8.1
Niche inheritance: The "evo-devo" and the "devo-evo" relationships

Transmission channel	What is transmitted	Type of inheritance system	Processes affected by the inheritance system
Channel 1 Internal environment	Semantic information R_i	1a Genetic inheritance, Epigenetic inheritance (e.g.) Chromatin marks, Methylation patterns, RNAi, Maternal effects	Development/**Weismann barrier**/Evolution Development → Evolution sometimes
	Energy/matter R_p	1b (e.g.) Cytoplasmic inheritance, Other maternal effects	Development → Evolution sometimes
Channel 2 External environment	Energy/matter R_p	2a Ecological inheritance of modified selective environments (e.g.) Perturbed physical environments	Development → Evolution
	Semantic information R_i	2b Ecological inheritance of modified selective environments (e.g.) Changed informational environments, Behavioral traditions, Symbolic	Development → Evolution

constructing organisms, or of the relocation of the niche-constructing organisms themselves in their external environments (Odling-Smee et al. 2003). In humans it includes one component of human cultural processes, the inheritance of material culture (cultural R_p).

Cell 2b refers to the second kind of ecological inheritance, the inheritance of modified selection pressures in the external environments of organisms as a consequence of prior communicative niche construction. Here, the modification of selection pressures is *indirect* because it involves two steps. The first step is, typically, the modification of semantic information, or R_i, by some kind of communicative niche construction. The second step is the subsequent modification of physical resources, or R_p, in an external environment, as a consequence of the first step. Jablonka and Lamb's third inheritance system, involving behavioral traditions and social learning in animals, belongs here. So does their fourth inheritance system, the inheritance of cultural knowledge or R_i, transmitted, via symbols, in humans (Cavalli-Sforza and Feldman 1981; Boyd and Richerson 1985; Laland et al. 2000).

Table 8.1 demonstrates that it is not always easy to allocate non-genetic inheritance systems to single Cells. For example, it is not easy to assign maternal effects to any single Cell in table 8.1. Mousseau (2006) points out that maternal effects are currently defined in diverse ways, and Mousseau himself defines them very broadly as ". . . all sources of offspring phenotypic variance due to mothers above and beyond the genes that she herself contributes" (p. 19). Thus, depending on how they are defined, it is possible to assign maternal effects, and probably paternal effects, too, to every Cell in table 8.1. If the scheme in table 8.1 is correct, maternal inheritance refers to more than one kind of inheritance, and probably needs clarifying.

Niche Regulation 1: How Evo- Affects Devo-

Table 8.1 allows us to reconsider the two questions at the heart of the evo-devo debate: (1) How does the prior evolution of populations affect the subsequent development of individual organisms? (2) How might the prior development of individual organisms affect the subsequent evolution of populations? Let's start with (1).

According to the MS, each new organism inherits a start-up set of genes from its ancestors, but little else. The new organism is then supposed to develop in the context of an independent external environment.

The ... ontogenetic process is seen as an *unfolding* of a form, already latent in the genes, requiring only an original triggering at fertilization and an environment adequate to allow "normal" development to continue. (Lewontin 1983: 276; italics in the original)

According to NCT, development is different. In NCT each offspring organism inherits a start-up niche from its parent(s) combining both a genetic inheritance and an ecological inheritance, and therefore several, or maybe all, of the subcomponents of niche inheritance in table 8.1. The organism then develops by responding to inputs from its local environment, and by emitting niche-constructing outputs to its environment that change some of the components in its own developmental environment. Development therefore ceases to be the unfolding of gene programs in the context of independent environments, and becomes a matter of active *niche regulation* by phenotypically plastic, niche-constructing organisms. Niche regulation starts at the moment of origin of a new organism, and continues for the rest of the organism's life (Oyama et al. 2001; West-Eberhard 2003; Pigliucci and Preston 2004; Sultan 2007). Hence:

... the organism [is] itself a *cause* of its own development. (Lewontin 1983: 279; italics in the original)

An arbitrary sequence of niche-regulating events in the life of a developing organism is illustrated in figure 8.2. The organism's phenotypic traits are symbolized by lowercase letters in O, while selection pressures, or stimuli in its environment, are symbolized by uppercase letters in E. The organism's [OE] niche relationship is assumed to be adaptive whenever there is a "match" between lowercase and uppercase letters, and maladaptive when there is a "mismatch."

At time t, O is tolerably well adapted to its environment E. Most of the letters match. However, there are two mismatches, between b and C at the top, and between j and Z at the bottom. At time $t + 1$, one of these mismatches disappears. Factor Z in the environment has now caused a "plastic" O to express the adaptive trait z instead of the maladaptive trait j. So O is now better adapted to its environment than it was before, thanks to its within-lifetime plastic capacity to respond to Z by changing its expression. For example, an animal might change its behavior in response to an environmental stimulus. Or a plant might change its morphology in response to light. Developmental plasticity stretches the MS by demanding a comprehensive reaction norm approach

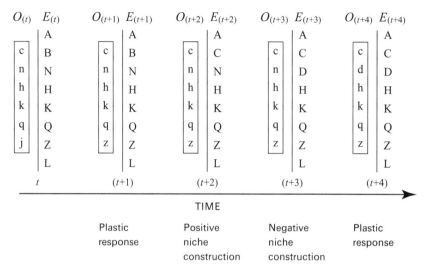

Figure 8.2
Niche regulation: An arbitrary sequence of niche-regulating events in the life of a developing, plastic, niche-constructing organism. The organism's phenotypic traits are symbolized
by lowercase letters in O. Selection pressures, or stimuli, in its environment are symbolized
by uppercase letters in E. (Based on Odling-Smee et al. 2003: 49, fig. 2.1).

(Schlichting and Pigliucci 1998), but does not go beyond it. The next step
does.

At time $t + 2$, O further improves its adaptation by positive niche
construction. O converts factor B in its environment to factor C by active
niche construction without changing itself. Trait c in O remains c. For
example, an animal might dig a burrow or move somewhere else, thus
evading predation, or a plant might kill a competitor by emitting an
allelopathic compound. The central point is that O achieves a new adaptive match not by changing itself in response to its environment, but by
changing its environment to suit itself. In effect, O causes its environment E to "adapt" to it. This action is described as positive niche construction if it enhances O's fitness (Odling-Smee et al. 2003).

At time $t + 3$, O damages its environment in some way by negative
niche construction, for example, by polluting it or overexploiting it.
Consequently, O impairs its own fitness by changing factor N in its environment to factor D, which generates a new mismatch between n and
D. Subsequently, at time $t + 4$, O responds to this mismatch with a conventional response to the environmental change it has itself caused, by
changing its own phenotypic expression again from n to d, thereby
restoring an adaptive match.

Ashby (1956) formulated some "rules" that apply to adaptive niche regulation in his *Introduction to Cybernetics*. Possibly because Ashby was primarily (but not exclusively) concerned with the control of human artifacts, for instance, autopilots, his work has largely been forgotten by biologists, but it is still relevant. Ashby's principal "rule" is the "law of requisite variety," which formally relates the minimal amount of variety any device, or organism, O, must deploy, relative to whatever variety it encounters in its heterogeneous environment, E, to "protect" an "essential variable." In organisms the key essential variable that O must protect is an adaptive [OE] niche relationship, represented in figure 8.2 by matching characters. Developing organisms must keep their niche relationships continuously adaptive by constantly matching environmental variance with adaptive variance of their own.

The MS proposes that developing organisms achieve adaptive niche relationships in one way only, by responding to events in their independent environments. NCT proposes that developing organisms achieve adaptive niche relationships in two ways. O can either express adaptive variant phenotypic traits in response to variant environmental states in E, or O can actively change some of the variant states in E by niche construction, thereby causing some variant E states to match its own variant states. Both kinds of niche regulation are consistent with Ashby's (1956) law of requisite variety. Modifying your own environment to suit yourself is just another way in which organisms can obey the law of requisite variety.

Niche Regulation 2: How Devo- May Affect Evo-

The final column on the right of table 8.1 raises the second question: Can the prior development of individual organisms affect the subsequent evolution of populations? We have just noted that NCT grants developing organisms their initial (OE) niche inheritances and an enhanced ability to regulate their niches adaptively. Do any of the inheritance systems in table 8.1 also have consequences for the "devo-evo" relationship?

Because the MS emphasizes genetic inheritance, and because it is not possible for organisms to inherit acquired characteristics directly from their ancestors by genetic inheritance ("hard" Lamarckism is not true), from the MS's point of view, within-lifetime developmental processes cannot have a direct causal influence on the subsequent evolution of populations. Development contributes to evolution only indirectly, through the differential transmission of genes by fit organisms between generations. Amundson (2005) showed how this historical disconnection between developmental and evolutionary biology is based on a series of putative barriers between devo- and evo-, the best-known being Weismann's (table 8.1). Weismann's barrier was originally derived from the early segregation of germ-line cells from soma in animals (but not plants), and it appeared to grant evolutionary biologists a license to ignore developmental biology (Maynard Smith 1982).

That license is currently being challenged by new molecular data (West-Eberhard 2003; Carroll 2005; Jablonka and Lamb 2005; Wagner et al. 2007; Pigliucci 2008). For instance, although controversial for decades, the idea that epigenetic variation can be transmitted across generations through epigenetic inheritance (Cell 1a, table 8.1) is now well supported (Bird 2002, 2007; Reik 2007). The peloric flower form of the toadflax *Linaria* is a widely known example. It is caused by an epigenetically inherited methylation pattern that has been stably inherited for over 200 years (Jablonka and Lamb 2005; Bird 2007). There is also good evidence that maternal inheritance (Cells 1a and 1b, table 8.1) can have evolutionary consequences. A mother's experience of her environment can lead to variations in her growth, condition, and physiological state that she can, in part, transmit to her offspring through cytoplasmic factors (Mousseau and Fox 1998; Mousseau 2006).

Others are better qualified than I to evaluate the devo-evo implications of the channel 1 inheritance systems in table 8.1, so I will not discuss them further. Instead, I will turn to the ecological inheritance systems in channel 2.

Cell 2a: Physical Ecological Inheritance

Niche construction originates from the niche-regulating activities of individual developing organisms. Niche construction is itself a developmental process, but it need not have consequences that go beyond the immediate development of an individual organism. For example, if the niche-regulating activities of an individual organism are merely idiosyncratic, or if the environmental consequences of its activities dissipate rapidly, then even though its niche-construction may affect its own fitness, in these circumstances it is unlikely to have wider ecological or evolutionary consequences.

That changes if the consequences of the developmental niche-constructing activities of multiple organisms, over many generations, combine and accumulate in the environments of populations. In Cell 2a, the relevant niche construction refers to the direct modification of energy and matter resources (R_p) in environments by niche-constructing organisms. If the consequences of this first type of niche construction accumulate in an environment, because they are collectively expressed by multiple organisms, they may have significant *ecosystem engineering* effects for populations (Jones et al. 1994, 1997; Wright and Jones 2006).

One often-cited example of ecosystem engineering is the building of dams in rivers by beavers. These niche-regulating activities of beavers modify beaver environments by changing and partly controlling a series of resource flows (used by other species as well as beavers), including nutrient flows, the buildup of sediments, and the availability of the river water itself (Naiman 1988). However, the ecosystem engineering consequences of beaver niche construction may still be exclusively ecological, unless they persist in the environments of populations for enough generations to allow any modified natural pressures they produce to cause the subsequent evolution of a population.

If that happens—if the ecosystem engineering consequences of niche construction do persist in the environments of populations for multiple generations—there can be evolutionary consequences. For instance, beaver dams create wetlands that can persist for centuries (Naiman 1988; Naiman and Rogers 1997; More 2006), long enough, relative to the short generational turnover times of many species in riparian ecosystems, for them to evolve in response to beaver-modified selection pressures. Another example is the niche-constructing activities of earthworms. Earthworms cause major changes in soils (Darwin 1881), and have apparently modified their environments by niche construction to suit themselves, instead of evolving new physiological adaptations.

... earthworms are essentially aquatic oligochaetes, poorly equipped physiologically for life on land. Yet there they are. (Turner 2000: 105)

So persistent ecosystem engineering effects caused by repeated niche construction, by many organisms, for many generations, can translate into evolutionarily significant ecological inheritances (Cells 2a and 2b, table 8.1) in the form of ancestrally modified selection pressures, relative to descendant organisms, in either a niche-constructing population or another population. When that happens, the developmental niche-constructing activities of individual ancestral organisms become collectively capable of causing evolutionary consequences in populations.

How likely is that chain of events? One cause that makes it likely comes from the other major component of niche inheritance, genetic inheritance (Cell 1a, table 8.1). If most organisms in a population inherit mostly the same genes, generation after generation, and if their inherited genes influence most organisms in the population to niche construct in similar ways, then genetic inheritance alone is likely to cause successive generations of organisms to act as "unidirectional biological pumps" that "pump" both biotic and abiotic ecological variables into new states by repetitive niche construction. For example, given time, niche-constructing organisms can drive large-scale abiotic ecosystem components such as an atmospheric gas, or soil chemistry, into thermodynamically out-of-equilibrium states that could never exist on a "dead" planet (Odling-Smee et al. 2003; Meysmann et al. 2006).

Other supplementary factors then determine how potent particular niche-constructing activities are. Most of these factors correspond to the set of ecological variables, listed by Jones et al. (1994, 1997), as responsible for determining the potency of ecosystem engineering. When applied to niche construction, Jones et al.'s list is as follows: (1) lifetime per capita activity of individual niche-constructing organisms; (2) density of the niche-constructing population; (3) length of time a population persists in the same place; (4) durability of the "construct" in the environment; (5) number and types of energy and matter flows that are modulated by niche construction; and (6) how many other species utilize, or are affected by, those flows.

The higher the value of each of these variables, the more ecologically potent the niche construction is likely to be. Hence, the more likely it is that the ecosystem engineering consequences of niche construction will translate into evolutionarily significant ecological inheritances for populations (Cells 2a and 2b, table 8.1). Also, it is not necessary for all these variables to be high simultaneously. Different species become potent niche constructors in different ways. Some organisms, such as beavers, are potent niche constructors because variables (1), (5), and (6) are high (More 2006). Others, for instance, photosynthesizing cyanobacteria, whose individual impacts on their environments (variable 1) are tiny, can also have a vast impact if they are present in sufficiently high densities (variable 2) over sufficiently large areas, for sufficient time (variable 3). They may also affect vast numbers of other species (variable 6). The Earth's aerobic atmosphere is a monument to the potency of bacterial niche construction.

When the consequences of niche construction do translate into ecological inheritances, there are various paths through which modified selection pressures can feed back to influence the evolution of populations. Via each feedback loop it is possible for the ecosystem engineering consequences of the niche construction originating from the niche-regulating activities of individual *developing* organisms to translate into ecological inheritances capable of influencing the *evolution* of populations. All these feedbacks are consistent with the logic of the two- locus population genetic models (above), in which changes in an environmental resource R, due to niche construction, influenced by the E locus, feed back in the form of a modified selection pressure(s) to the A locus. The following feedback paths are variations of this theme (Odling-Smee, et al. 2003).

The simplest (1) corresponds to Dawkins's (1982) "extended phenotype." The E locus expresses an extended phenotype, R, by "niche construction." R then generates a modified selection pressure that feeds back exclusively to the E locus itself, so only one genetic locus, E, is relevant (A = E). (2) corresponds to most of our original models. Both the E locus and the A locus are relevant. Suppose the E locus influences earthworm niche construction, which modifies the soil, R, which feeds back in the form of a modified selection pressure to the A locus in the same population of earthworms, to influence the subsequent evolution of earthworms.

The third case (3) differs in that the A locus is no longer in the same population as the E locus. Here, the niche-constructing activities of one population feed forward to affect natural selection in a second population. Organisms in the first niche-constructing population interact *directly* with organisms in a second population. The second population is therefore R, and the carrier of the A locus. Alternatively, two populations may interact *indirectly* through the modification by the first population of a separate ecosystem component, R, which feeds forward as a modified selection pressure for a second population. For example, a population of earthworms might change a biotic component of the soil and affect a population of plants. If the second population reciprocates in a similar manner, then (4) its reciprocal niche construction might cause the two populations to co-evolve as in standard models of co-evolution. A variant of (4) occurs if (5) the interactions of two co-evolving populations are mediated by an abiotic ecosystem component R. The modification of an abiotic R, say soil pH, by the niche-constructing activities of one population, say a population of litter-supplying plants, may make it possible for two populations, say earthworms and plants, to co-evolve indirectly through the mediation of a shared abiotic ecosystem component, R, in the soil.

Finally, (6) the co-evolutionary scenarios in (4) and (5) potentially extend to multiple populations that either co-evolved in the past or are still co-evolving today. Their collective co-evolution generates extra webs in ecosystems, in addition to energy and matter webs. Jones et al. (1997) called them "engineering" or "control" webs. Previously we suggested control webs could be modeled by "environmentally mediated genotypic associations," or EMGAs, in ecosystems, that can capture connections between distinct genotypes in networks of diverse populations, mediated by both biotic and abiotic ecosystem components (Odling-Smee et al. 2003).

Cell 2b: Informational Ecological Inheritance

The second kind of ecological inheritance generated by niche construction is in Cell 2b (table 8.1). It works in the same way, except that it also includes the modification of semantic information, R_i, by communicative niche construction, as well as the subsequent modification of energy and matter resources, R_p, by conventional niche construction. I'll focus on this extra step.

Why do organisms communicate? Organisms need R_i to grant them sufficient control over R_p to satisfy their energy and matter needs, and to protect themselves against threats (Odling-Smee et al. 2003), but there is a subtle difference between how organisms acquire R_i and R_p. Organisms can take raw R_p directly from their habitats, and dump raw detritus in their habitats by interacting with their environments. In contrast, organisms cannot take raw semantic information (R_i) from their habitats because environments do not contain "knowledge" except in the genomes or brains of other organisms. To acquire R_i, organisms must access one or more R_i-gaining processes, including those that depend on between-organism communication.

Common to all R_i-gaining processes is their ultimate dependence on the ability of organisms to register, or "remember," the outcomes of previous niche interactions in ways that subsequently permit organisms some chance of preparing for their futures adaptively, by using registered information in "fit" memories to constrain their phenotypes (see Szilard 1929 [1964]).

The primary R_i-gaining process is the evolutionary process itself. The genes inherited by individual organisms in each generation include some that are naturally selected as a consequence of the outcomes of the prior niche interactions of diverse ancestral phenotypes in populations. These selected genes necessarily reflect whatever selective biases were imposed on ancestral organisms by natural selection. The reexpression by contemporary organisms of whatever semantic information is registered in their inherited genes, amounts to an inductive gamble that both the selective environments and the fitness requirements of contemporary organisms are sufficiently similar to those of their ancestors to make whatever was adaptive before, adaptive again (Slobodkin and Rapoport 1974). The inheritance of genes by successive generations of organisms is the most fundamental kind of between-organism communication. All organisms depend on it (Waddington 1969; Campbell 1974; Odling-Smee et al. 2003).

Many organisms also have some capacity to "self-inform" through supplementary developmental R_i-gaining processes. Examples are the vertebrate immune system, and learning in animals. Instead of acquiring all their R_i from their ancestors a priori, via genetic inheritance, they "make" some R_i for themselves a posteriori by registering the outcomes of their own individual prior [OE] niche interactions (Odling-Smee 1983; Plotkin 1994; Dennet 1995).

The existence of both evo- and devo- R_i-gaining processes raises two questions. If developing organisms can self-inform, why do all organisms depend so heavily on ancestral R_i? Conversely, if evolution is the primary R_i-supplying process, how come it pays some organisms to invest in costly self-informing processes as well?

Two obstacles prevent organisms from gaining sufficient R_i through developmental R_i-gaining processes alone. First, to a large extent, organisms must be informed a priori to have any chance of surviving (Waddington 1969; Odling-Smee et al. 2003). No organism can inform itself a priori on the basis of "memories" of its own past before its past has occurred; that is why the origin of life problem is so difficult (Fry 2000). Only after life is present is it possible for organisms to be informed a priori by their ancestors. Parent(s) can then transmit an apparently free "gift" of "start-up" R_i to their offspring via genetic inheritance. The "gift" is deceptive because it includes a demand for the next generation to pass on R_i as a "gift" to their offspring in their turn. The "gift" is a "mortgage." The debt is repaid by reproduction in each generation.

Second, individual organisms cannot gain sufficient "data" during their lives to prepare themselves adaptively for their remaining futures simply by registering the outcomes of their own prior individual niche interactions. The maximum database provided by the "private" niche-sampling "experiences" of any individual organism is too meager to allow any organism to demarcate between "true" and "false" instances of causality, relative to the "causal texture" of its own niche, sufficiently reliably. However, unless the R_i carried by organisms does demarcate between true and false instances of causality, organisms cannot prepare for their own futures adaptively, except by chance (Odling-Smee et al. 2003). For this reason, all organisms have to depend primarily on the much richer "databases" collectively assembled by ancestors.

But the same argument also runs in reverse. Largely because the R_i supplied to individual organisms a priori by evolution has been collected by vast numbers of ancestors, relative to vast regions of past environmental space and time, it may not always be adaptive relative to the local niches of contemporary organisms. It may be out of date, or too "noisy," or misinformation relative to an individual's particular niche. Evolution resolves this dilemma between sufficiency and relevance in several ways. A fundamental way is through speciation. Speciation limits the R_i in ancestral databases, potentially making them more relevant to the particular niches of descendant organisms.

Other solutions depend on the plasticity of niche-regulating organisms. I'll discuss two of them more fully. First, plastic organisms can sometimes put some of the R_i they inherit to a novel use, and by doing so, change or enhance its meaning. For example, even though the semantic information carried by an inherited gene may have been selected for by a particular selection pressure in the environments of an organism's ancestors, there is no reason why a plastic, niche-constructing descendant organism should restrict its subsequent use to expressing an adaptation relative to that particular selection pressure only. An opportunistic organism may sometimes use old inherited R_i in a new way.

A difference between the MS and NCT is that the MS expects organisms to respond only to autonomous selection pressures in their environments. Organisms are not expected to initiate novel changes, except by chance. NCT, however, does expect organisms to initiate new changes nonrandomly by *inceptive niche construction* (Odling-Smee et al. 2003). If the consequences of inceptive niche construction subsequently scale up from individual developing organisms to populations, for similar reasons as in Cell 2a, then inceptive niche construction by developing organisms may become the source of evolutionary novelties.

That becomes more likely if the inceptive niche construction gets locked in, at a population level, by the subsequent modification of other natural selection pressures. Evolutionary trends, and the evolution of complex relationships such as host-parasite relationships (Combes 2001), or mutualisms (e.g., Martin et al. 2008), are hard to understand in terms of a sequence of supposedly disconnected natural selection pressures in autonomous environments. They could be easier to understand by realizing that successive selection pressures are rarely disconnected from each other, and can be connected by intermediate episodes of niche construction. Inceptive niche construction can initiate bouts of niche construction and natural selection, each modifying the other reciprocally, thereby causing a population to evolve in a novel direction.

A second solution is for plastic organisms to add to their inherited a priori "knowledge" by investing in supplementary R_i-gaining processes during their lives. I will discuss only animal learning. Presumably, learning by individual animals affects their differential survival and reproduction, but it can seldom have more far-reaching evolutionary consequences, because everything an individual learns is erased when it dies.

The ability of animals to learn, however, introduces the prospect of animals learning from each other by social learning. If individuals communicate some of what they learn to and from each other's memories within and between generations, and if, in the process, they generate an additional R_i ecological inheritance system (Cell 2b, table 8.1) by doing so, their collective memories may become evolutionarily significant in populations.

But why should animals communicate with each other's memories, given that communication incurs an initial fitness cost, not a benefit? In practice, the initial cost of exchanging R_i through social learning need not be high. Even though gaining "new" semantic information "de novo" is typically expensive (Odum 1988; Coolen et al. 2005), the transmission of old semantic information by making copies of it is often cheap, especially when the information is already in symbolic form. The subsequent benefit from gaining additional R_i through communication may then be considerable. Given that the fundamental function of R_i is to allow organisms a degree of control over physical resources (R_p), and given that only a little R_i can potentially control a lot of R_p, the extra control over R_p granted to animals that invest in extra R_i by communicating can far outweigh the initial cost of communicating.

That still fails to explain where the physical benefits, measured in R_p, ultimately come from. For that we also need *social niche construction* (Flack et al. 2005). Boehm and Flack (2009) describe the social niches of individual animals that live in groups by vectors of behavioral connections, primarily communication networks, between animals in overlapping social networks. An individual animal's behavioral connections may then provide it with social resources, food-sharing facilities, exchanges of goods, and access to social and ecological information. Cooperation among group members may also generate "public goods" that are potentially advantageous to all members of a group. In one example, Flack et al. (2006) demonstrated the advantages of social niche construction in pig-tailed macaques. The macaques benefited from stable social networks by building larger social networks, characterized by greater partner diversity and increased cooperation in their groups. These group benefits occurred, however, only in the presence of dominant animals that imposed social order by "policing" their groups. Without "policing," social disorder increased, the social networks splintered, and the benefits disappeared.

Costs to individuals also arise from social niche construction because cooperation in groups is often subverted by animals that "cheat" (Trivers 1971). Cheating ensures that "winners" and "losers" are typically found in every social group. Boehm and Flack (2009) discuss the emergence of power structures, built by social niche construction, in social groups, based on "perceived" and "legitimized" authority. Power structures enhance the benefits to some individuals, at a cost to others, in social groups.

Social learning can enter this scenario in several ways (Borenstein et al. 2008), and it can increase or depress benefits to different individuals in groups. However, that is not what primarily matters here. What matters is that social learning can transmit previously learned semantic information, R_i, from one generation to the next, through the ecological inheritances shown in Cell 2b (table 8.1). It therefore provides the mechanisms that underpin Jablonka and Lamb's (2005) third and fourth inheritance systems: the inheritance of social organization, communication networks, and behavioral traditions in animals, and the inheritance of cultural knowledge in humans.

The evolutionary potency of R_i ecological inheritance is particularly visible in humans (Smith 2007). A well-known example is the evolution of lactose tolerance among people who transmitted pastoralist agricultural traditions between generations in human social groups, and by doing so apparently did affect human genetic evolution (Holden and Mace 1997; Enattah et al. 2002; Burger et al. 2007).

Conclusion

Adding niche construction to evolutionary theory changes the predicted dynamics of the evolutionary process. It also facilitates the integration of evolutionary biology with other disciplines.

NCT connects evolution to ecosystem-level ecology in two main ways. First, because niche construction modifies abiotic as well as biotic environmental resources, it admits abiota into population genetic models in the guise of modified natural selection pressures. It thus overcomes one of the barriers separating evolutionary biology from ecosystem ecology: the difficulty of handling abiota in evolutionary models because abiota do not carry genes (O'Neill et al. 1986; Jones and Lawton 1995). Second, NCT proposes a new way of investigating the ecosystem engineers' claim that ecosystems incorporate "engineering" or "control" webs, in addition to energy and matter webs, by offering the EMGAs concept. NCT proposes that control webs in ecosystems should depend on the evolved adaptations of organisms, derived from selected genes and regulatory gene networks in developing organisms, and on weakly regulating EMGA-based gene networks among co-evolving populations in ecosystems (Odling-Smee et al. 2003).

In the human sciences the main innovation is the introduction by NCT of a human ecological inheritance system that incorporates both heritable material culture and heritable cultural knowledge (Odling-Smee 2007). The general significance of ecological inheritance is that it assigns to all phenotypes a second role in evolution. Phenotypes not only survive and reproduce differentially, they also modify their environments by niche construction. In the human case, that carries new philosophical as well as biological implications. Culturally, niche-constructing humans cannot be just "vehicles" for their genes (Dawkins 1989), nor the passive "playthings" of chance and necessity. Instead, we are bound to influence, but not control, our own future evolution, and the future evolution of other organisms in our shared ecosystems as well. That realization might eventually change not only how biologists, but also everyone else, perceives evolution and reevaluates their own existence.

Last, niche construction contributes to the "evo-devo" relationship by enhancing the contributions of prior evolution to the subsequent development of individual organisms, and by allowing plastic, niche-constructing developing organisms to affect the subsequent evolution of populations. For example, there are no barriers in channel 2 comparable to the Weismann barrier in channel 1 (table 8.1), to stop the consequences of prior niche construction by developing organisms from scaling up to affect the subsequent evolution of populations. That opens up a new "devo-evo" highway, which, apart from the pioneering work of Baldwin, Schmalhausen, and Waddington in the twentieth century, has seldom been explored. It probably cannot be explored further without extending the Synthesis.

Acknowledgments

I thank Kevin Laland and Lucy Odling-Smee for their valuable criticisms of an earlier draft of this chapter, and my editors, Gerd Müller and Massimo Pigliucci.

References

Amundson R (2005) The Changing Role of the Embryo in Evolutionary Thought. Cambridge: Cambridge University Press.

Ashby WR (1956) An Introduction to Cybernetics. New York: Chapman and Hall.

Aunger R (2002) The Electric Meme: A New Theory of How We Think. New York: Free Press (Simon and Schuster).

Balter M (2005) Are humans still evolving? Science. 309: 234–237.

Bergstrom CT, Lachmann M (2004) Shannon information and biological fitness. IEEE Information Theory Workshop: 50–54.

Bird A (2002) DNA methylation patterns and epigenetic memory. Genes and Development 16: 6–21.

Bird A (2007) Perceptions of epigenetics. Nature 447: 396–398.

Boehm C, Flack JC (2009) The emergence of simple and complex power structures through social niche construction. In The Social Psychology of Power. A Guinote, T Vescior, eds.: xx. New York: Guilford Press.

Boni MF, Feldman MW (2005) Evolution of antibiotic resistance by human and bacterial niche construction. Evolution 59: 477–491.

Borenstein E, Feldman MW, Aoki K (2008) Evolution of learning in fluctuating environments: When selection favors both social and exploratory individual learning. Evolution 62: 586–602.

Borenstein E, Kendal J, Feldman M (2006) Cultural niche construction in a metapopulation. Theoretical PopulationBiology 70: 92–104.

Boyd R, Richerson PJ (1985) Culture and the Evolutionary Process. Chicago: University of Chicago Press.

Burger J, Kirchner M, Bramanti B, Haak W, Thomas MG (2007). Absence of the lactase-persistence-associated allele in early Neolithic Europeans. Proceedings of the National Academy of Sciences of the USA 104: 3736–3741.

Campbell DT (1974) Evolutionary epistemology. In The Philosophy of Karl R Popper. PA Schilpp, ed.: 413–463. La Salle, IL: Open Court.

Carroll SB (2005) Endless Forms Most Beautiful: The New Science of Evo Devo and the Making of the Animal Kingdom. New York: Norton.

Cavalli-Sforza LL, Feldman MW (1981) Cultural Transmission and Evolution: A Quantitative Approach. Princeton, NJ: Princeton University Press.

Chase JM, Leibold MA (2003) Ecological Niches: Linking Classical and Contemporary Approaches. Chicago: University of Chicago Press.

Combes C (2001) Parasitism: The Ecology and Evolution of Intimate Interactions. Chicago: University of Chicago Press.

Coolen I, Ward AJW, Hart PJB, Laland KN (2005) Foraging nine-spined sticklebacks prefer to reply on public information over simpler social cues. Behavioral Ecology 16: 865–870.

Darwin C (1881) The Formation of Vegetable Mold Through the Action of Worms, with Observations on Their Habits. London: John Murray.

Davidson EH (2006) The Regulatory Genome: Gene Regulatory Networks in Development and Evolution. New York: Academic Press.

Davies NB, Kilner RM, Noble DG (1998) Nestling cuckoos, *Cuculus canorus*, exploit hosts with begging calls that mimic a brood. Proceedings of the Royal Society of London B265: 673–678.

Dawkins R (1982) The Extended Phenotype. Oxford: W. H. Freeman.

Dawkins R (1989) The Selfish Gene. 2nd ed. Oxford: Oxford University Press.

Dawkins R (2004) Extended phenotype—but not too extended. A reply to Laland, Turner and Jablonka. Biology and Philosophy 19: 377–396.

Dennet D (1995) Darwin's Dangerous Idea: Evolution and the Meanings of Life. London: Penguin.

Dietrich WE, Taylor Perron J (2006) The search for a topographic signature of life. Nature 439: 411–418.

Diggle S, Griffin AS, Campbell GS, West SA (2007) Cooperation and conflict in quorum-sensing bacterial populations. Nature 450: 411–414.

Donohue K (2005) Niche construction through phonological plasticity: Life history dynamics and ecological consequences. New Phytologist 166: 83–92.

Enattah NS, Sahi T, Savilhati E, Terwilliger JD, Peltonen L, Järvelä I (2002) Identification of a variant associated with adult-type hypolactasia. Nature Genetics 30: 233–237.

Erwin DH (2008) Macroevolution of ecosystem engineering, niche construction and diversity. Trends in Ecology and Evolution 23: 304–310.

Feldman MW, Cavalli-Sforza LL (1989) On the theory of evolution under genetic and cultural transmission, with application to the lactose absorption problem. In Mathematical Evolutionary Theory. MW Feldman, ed.: 145–173. Princeton, NJ: Princeton University Press.

Flack JC, de Waal FBM, Krakauer DC (2005) Social structure, robustness, and policing cost in a cognitively sophisticated species. American Naturalist 165: E126–E139.

Flack JC, Girvan M, de Waal FBM, Krakauer DC (2006) Policing stabilizes construction of social niches in primates. Nature 439: 426–429.

Fragaszy DM, Perry S, eds. (2003) The Biology of Traditions. Cambridge: Cambridge University Press.

Frederickson ME, Greene MJ, Gordon DM (2005) "Devil's gardens" bedevilled by ants. Nature 437: 495–496.

Fry I (2000) The Emergence of Life on Earth: A Historical and Scientific Overview. New Brunswick, NJ: Rutgers University Press.

Futuyma DJ (1998) Evolutionary Biology. 3rd ed. Sunderland, MA: Sinauer.

Godfrey-Smith P (1996) Complexity and the Function of Mind in Nature. New York: Cambridge University Press.

Griffiths PE, Gray RD (2004) The developmental systems perspective: Organism-environment systems as units of development and evolution. In Phenotypic Integration: Studying the Ecology and Evolution of Complex Phenotypes. M Pigliucci, K Preston, eds.: 409–431. New York: Oxford University Press.

Hansell MH (2005) Animal Architecture. Oxford Animal Biology Series. Oxford: Oxford University Press.

Hanski I, Singer MC (2001) Extinction-colonization and host-plant choice in butterfly metapopulations. American Naturalist 158: 341–353.

Holden C, Mace R (1997) Phylogenetic analysis of the evolution of lactose digestion in adults. Human Biology 69: 605–628.

Hui C, Li ZZ, Yue DX (2004) Metapopulation dynamics and distribution and environmental heterogeneity induced by niche construction. Ecological Modelling 177: 107–118.

Ihara Y, Feldman MW (2004) Cultural niche construction and the evolution of small family size. Theoretical Population Biology 65: 105–111.

Jablonka E, Lamb MJ (1995) Epigenetic Inheritance and Evolution: The Lamarckian Dimension. Oxford: Oxford University Press.

Jablonka E, Lamb MJ (2005) Evolution in Four Dimensions: Genetic, Epigenetic, Behavioral, and Symbolic Variation in the History of Life. Cambridge, MA: MIT Press.

Jones CG, Lawton JH, eds. (1995) Linking Species and Ecosystems. London: Chapman and Hall.

Jones CG, Lawton JH, Shachak M (1994) Organisms as ecosystem engineers. Oikos 69: 373–386.

Jones CG, Lawton JH, Shachak M (1997) Positive and negative effects of organisms as physical ecosystem engineers. Ecology 78: 1946–1957.

Kirkpatrick M, Lande R (1989) The evolution of maternal characters. Evolution 43: 485–503.

Laland KN, Odling-Smee FJ, Feldman MW (1996) The evolutionary consequences of niche construction. Journal of Evolutionary Biology 9: 293–316.

Laland KN, Odling-Smee FJ, Feldman MW (1999) Evolutionary consequences of niche construction and their implications for ecology. Proceedings of the National Academy of Sciences of the USA 96: 10242–10247.

Laland KN, Odling-Smee FJ, Feldman MW (2000) Niche construction, biological evolution, and cultural change. Behavioral and Brain Sciences 23: 131–175.

Laland KN, Odling-Smee FJ, Feldman MW (2001) Cultural niche construction and human evolution. Journal of Evolutionary Biology 14: 22–23.

Laland KN, Odling-Smee FJ, Gilbert SF (2008) Evo devo and niche construction: Building bridges. Journal of Experimental Zoology B310: 549–566.

Laland KN, Sterelny K (2006) Seven reasons (not) to neglect niche construction. Evolution. 60: 1751–1762.

Lehmann L (2008) The adaptive dynamics of niche constructing traits in spatially subdivided populations: Evolving posthumous extended phenotypes. Evolution 62: 549–566.

Lewontin RC (1982) Organism and environment. In Learning, Development and Culture. HC Plotkin, ed.: 151–170. New York: Wiley.

Lewontin RC (1983) Gene, organism, and environment. In Evolution from Molecules to Men. DS Bendall, ed.: 273–285. Cambridge: Cambridge University Press.

Martin F et al. (2008) The genome of *Laccaria bicolor* provides insights into mycorrhizal symbiosis. Nature 452: 88–92.

Maynard Smith J (1982) Evolution and the Theory of Games. Cambridge: Cambridge University Press.

Maynard Smith J (2000) The concept of information in biology. Philosophy of Science 67: 177–194.

Meysman FJR, Middleburg JJ, Heip CHR (2006) Bioturbation: A fresh look at Darwin's last idea. Trends in Ecology and Evolution 21: 688–695.

More JW (2006) Animal ecosystem engineers in streams. BioScience 56: 237–246.

Mousseau TA (2006) Box 2.1: Maternal Effects. In Evolutionary Genetics: Concepts and Case Studies. W Fox, JB Wolf, eds.: 19. New York: Oxford University Press.

Mousseau TA, Fox CW, eds. (1998) Maternal Effects as Adaptations. New York: Oxford University Press.

Müller GB (2007) Evo-devo: Extending the evolutionary synthesis. Nature Reviews Genetics 8: 943–949.

Naiman RJ (1988) Animal influences on ecosystem dynamics. BioScience 38: 750–752.

Naiman RJ, Rogers KH (1997) Large animals and system-level characteristics in river corridors. BioScience 47: 521–529.

Odling-Smee FJ (1988) Niche-constructing phenotypes. In The Role of Behavior in Evolution. HC Plotkin, ed.: 73–132. Cambridge, MA: MIT Press.

Odling-Smee FJ (2007) Niche inheritance: A possible basis for classifying multiple inheritance systems in evolution. Biological Theory 2: 276–289.

Odling-Smee FJ, Laland KN, Feldman MW (1996) Niche construction. American Naturalist 147: 641–648.

Odling-Smee FJ, Laland KN, Feldman MW (2003) Niche Construction: The Neglected Process in Evolution. Princeton, NJ: Princeton University Press.

Odum HT (1988) Self-organization, transformity, and information. Science 242: 1132–1139.

O'Neill RV, DeAngelis DL, Waide JB, Allen TFH (1986) A Hierarchical Concept of Ecosystems. Princeton, NJ: Princeton University Press.

Oyama S, Griffiths PE, Gray RD, eds. (2001) Cycles of Contingency: Developmental Systems and Evolution. Cambridge, MA: MIT Press.

Pigliucci M (2008) Is evolvability evolvable? Nature Reviews Genetics 9: 75–82.

Pigliucci M, Preston K, eds. (2004) Phenotypic Integration: Studying the Ecology and Evolution of Complex Phenotypes. New York: Oxford University Press.

Plotkin HC (1994) The Nature of Knowledge: Concerning adaptations, instinct, and the Evolution of Intelligence. London. Penguin.

Reader SM, Laland KN (2003) Animal Innovation. New York: Oxford University Press.

Reik W (2007) Stability and flexibility of epigenetic gene regulation in mammalian development. Nature 447: 425–432.

Sapp J (1987) Beyond the Gene: Cytoplasmic Inheritance and the Struggle for Authority in Genetics. New York: Oxford University Press.

Schlichting CD, Pigliucci M (1998) Phenotypic Evolution: A Reaction Norm Perspective. Sunderland, MA: Sinauer.

Schwilk DW (2003) Flammability is a niche construction trait: Canopy architecture affects fire intensity. American Naturalist 162: 725–733.

Schwilk DW, Ackerly DD (2001) Flammability and serotiny as strategies: Correlated evolution in pines. Oikos 94: 326–336.

Silver M, Di Paolo E (2006) Spatial effects favour the evolution of niche construction. Theoretical Population Biology 70: 387–400.

Slobodkin LB , Rapoport A (1974) An optimal strategy of evolution. Quarterly Review of Biology. 49: 181–200.

Smith BD (2007) Niche construction and the behavioral context of plant and animal domestication. Evolutionary Anthropology 16: 188–199.

Sultan SE (2007) Development in context: The timely emergence of eco-devo. Trends in Ecology and Evolution 22: 575–582.

Szilard L (1929) Über die Entropieverminderung in einem thermodynamischen System bei Eingriffen intelligenter Wesen. Zeitschrift für Physik 53: 840–856. Translation (1964)

On the decrease of entropy in a thermodynamic system by the intervention of intelligent beings. Behavioural Science 9: 301–310.

Tomasello M, Carpenter M, Call J, Behne T, Moll H (2005) Understanding and sharing intentions: The origins of cultural cognition. Behavioral and Brain Sciences 28: 675–735.

Thompson JN (1994) The Coevolutionary Process. 2nd illus. ed. Chicago: University of Chicago Press.

Thompson JN (2006) Mutualistic webs of species. Science 312: 372–373.

Trivers RL (1971) The evolution of reciprocal altruism. Quarterly Review of Biology 46: 35–57.

Turner JS (2000) The Extended Organism: The Physiology of Animal-built Structures. Cambridge. MA: Harvard University Press.

Vandermeer J (2008) The niche construction paradigm in ecological time. Ecological Modelling 214: 385–390.

Waddington CH (1959) Evolutionary systems: Animal and human. Nature 183: 1634–1638.

Waddington CH (1969) Paradigm for an evolutionary process. In Towards a Theoretical Biology, vol. 2. CH Waddington, ed.: 106–128. Edinburgh: Edinburgh University Press.

Wagner GP, Pavlicev M, Cheverud JM (2007) The road to modularity. Nature Reviews Genetics 8: 921–931.

West-Eberhard MJ (2003) Developmental Plasticity and Evolution. New York: Oxford University Press.

Wolf JB (2000) Indirect genetic effects and gene interactions. In JB Wolf, ED Brodie III, MJ Wade, Epistasis and the Evolutionary Process. New York: Oxford University Press.

Wright JP, Jones CG (2006) The concept of organisms as ecosystem engineers ten years on: Progress, limitations, and challenges. BioScience 56: 203–209.

9 Chemical, Neuronal, and Linguistic Replicators

Chrisantha Fernando and Eörs Szathmáry

The Modern Synthesis can be extended laterally and vertically. Lateral extensions transfer the thought patterns and the methodology of evolutionary theory to different, previously nonevolutionary disciplines. Examples include replicator and systems chemistry, linguistics, cultural selection (Boyd and Richerson 2005), memetic replicators (Aunger 2002; Dawkins 1976), somatic selection (Edelman 1994), and microeconomics. Vertical extensions deepen our knowledge in traditional areas of evolution research such as EvoDevo, niche construction, epigenetic inheritance (Jablonka and Lamb 2005), and multiple levels of selection (Okasha 2006). In this chapter we first review the relevance of evolutionary thinking in certain areas of chemistry. Then we present, as the major novel contribution, a lateral extension to neuroscience, in that we outline a truly evolutionary approach to brain function in the higher vertebrates. Finally, some relevant aspects of language will be considered.

Evolution by natural selection is perhaps the most important process acting in populations of living systems. This is one of the reasons why it is so tempting to equate units of evolution (i.e., an abstract generalization that makes no reference whatsoever to any particular level of biological organization) to units of life. Another reason is that units of evolution can be much more readily defined. There are a few known alternative formulations of the concept of units of evolution; here we stick to the version outlined by Maynard Smith (1986): such units must multiply, show heredity across generations (like begets like), and heredity should not be exact. If some of the hereditary traits affect the chance of reproduction and/or survival of the units, evolution by natural selection can take place in a population of such units. The combination of survival and reproduction (translating into the expected number of descendants) is called fitness. The above characterization of Darwinian dynamics is deliberately general: note that it is not restricted to cover living systems only. (As a matter of fact, some living systems do not—sometimes cannot—multiply: mules and neurons normally do not reproduce.) Hence it is potentially applicable to molecules and cultural traits as far as the criteria really apply.

A general point about definitions is that they cannot be falsified. They have to be internally consistent, of course, but there can be an arbitrary number of such definitions for life, for example. It is the use of the alternative definitions that makes the difference: some definitions are found helpful because they categorize natural phenomena in a way that is conducive to further insights. There is always an ingredient of arbitrariness in definitions: we have to live with this fact.

Chemical Origin of Evolvability and Systems Chemistry *in Statu Nascendi*

Biology has its roots in chemistry (Von Kiedrowski 2001). Gánti (e.g., 2003) emphasized that contemporary living systems always have (1) some metabolic subsystem, (2) some systems for heritable control, and (3) some boundary system to keep the component together. We consider it unlikely that a chemical system satisfying all the constraints from this abstraction could have appeared just out of chemical chaos. This observation led to the formulation of the concept of infrabiological systems (Szathmáry 2005; Fernando et al. 2005). Infrabiological systems always lack one of the key components just listed. For example, in the original formulation of Gánti (1971), a model of minimal life did not include a boundary system. The combination of a metabolic cycle and a membrane was also conceived by Gánti (1978), and called a self-reproducing microsphere. In contrast, Szostak et al. (2001) conceived a protocell-like entity with a boundary and template replication but no metabolic subsystem. Such systems show a crucial subset of interesting biological phenomena. The three subsystems can be combined to yield three different doublet systems (figure 9.1).

The emerging field of systems chemistry deals with the analysis and synthesis of coupled autocatalytic systems (e.g., Kindermann et al. 2005; Ludlow and Otto 2008). Chromosomes made of DNA come in different lengths. They can harbor a small or a large number of genes. During replication of the bacterial chromosome, it makes perfect sense to say that replication is half complete when one half is already present in two copies. This sharply contrasts with the following example. Imagine a molecule A, which reacts with a number of compounds to yield two molecules of A after one turn of the cycle. Molecular systems of this

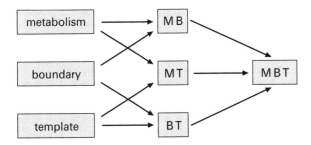

Figure 9.1
Elementary combinatorics of infrabiological systems (Fernando et al. 2005). The chemoton is a minimal biological system comprising three qualitatively different subsystems (metabolism, membrane, and template). The analyis and synthesis of such systems is the aim of the emerging field of systems chemistry (Kindermann et al. 2005; Ludlow and Otto 2008). (From Fernando et al. 2005)

kind do exist; examples include the formose reaction (figure 9.2), the reductive citric acid cycle (which is almost the exact reverse of the citric acid cycle and is used for carbon fixation by some bacteria), and the Calvin cycle (fixing carbon dioxide in plants). One, or a few, autocatalysts are sufficient to seed the system, and parts (the chemical moieties) of the autocatalytic molecules are held together by covalent bonds (and are thus sterically constrained). They are also stoichiometric in the sense that the elementary steps are simple chemical reactions (transformations). Two questions must be asked about such systems: (1) Are they feasible as autonomous replicators (self-replicators)? (2) Is there hereditary information stored in them? We discuss these questions in turn.

It is important to emphasize that all sufficiently well described metabolic networks contain at least one (sometimes several) autocatalytic metabolic seeds without which the cell cannot start running, despite the presence of all genes and enzymes (Kun et al. 2008). However, the reductive citric acid cycle and the Calvin cycle (which is in fact a complex network) are not autonomous, in the sense that they require the operation of enzymes that are not produced by them. This is in contrast to the formose reaction, which does not require enzymes.

Heredity requires alternative types of cycles. Currently, there are only hypothetical suggestions, put forward by Wächtershäuser (1988, 1992): they are various extensions of the (equally hypothetical) "archaic" reductive citric acid cycle. Even if alternative forms of such systems can exist, most changes will be mere fluctuations and will not lead to hereditary alterations ("mutations" in the general sense). It is expected that the system will flip from the basin of one attractor into that of another attrac-

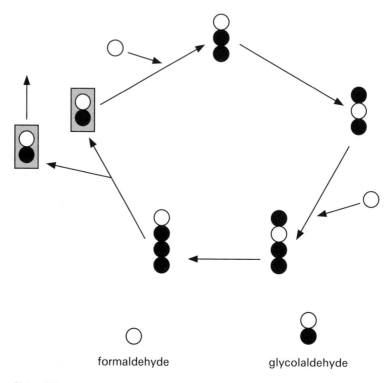

formaldehyde   glycolaldehyde

Figure 9.2
The autocatalytic core or seed of the formose reaction (Fernando et al. 2005). Each circle represents a chemical group including one carbon atom. Black and white circles denote different internal chemical structures.

tor very rarely; hence there will be infrequent "macromutations" only (Wächtershäuser 1988). Nevertheless, such macromutations may have been of paramount importance in chemical evolution. Models suggest that chemical evolution can occur by natural selection if a generative chemical system capable of producing satellite autocatalytic cycles is enclosed in a compartment. Variation is by "chemical avalanches" (macromutations), as suggested by Wächtershäuser, and selection is at the compartment level (Fernando and Rowe 2007, 2008). This idea is open to experimental test. Such systems are holistic replicators (Maynard Smith and Szathmáry 1999). If one looks at the core of the formose reaction (figure. 9.2), one sees that there is no real sense in which one could say that replication is "halfway through," in sharp contrast to a piece of RNA or DNA. This is because replication here is not template replication (copying) that rests on a modular polymerization of monomers.

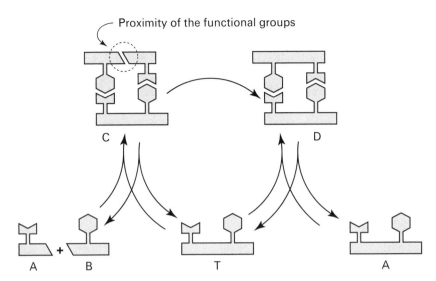

Figure 9.3
Scheme of simple modular self-replication (from Bag and Von Kiedrowski 1996). A and B, building blocks; T, template; C, catalytic complex; D, duplex. Note the reversible and irreversible reactions.

The first modular type of self-replicator (figure 9.3) was synthesized by Von Kiedrowski (1986). The palindromic arrangement of the template ensures that the copy will be identical to the template, despite complementary base pairing. There is now a large number of such experimentally produced replicators (for a review, see Von Kiedrowski 1999). A common criterion for the replication process is that the two strands (template and copy) must spontaneously separate. Since they are held together by hydrogen bonds (also necessary for replication), the strands cannot be too long; otherwise they will stick together for too long a time. Long pieces of nucleic acids can be replicated in the cell because enzymes of the replicase complex also ensure the unwinding of the strand—this cannot be assumed in nonenzymatic systems. These artificial replicators hence must generally be rather short. Although replication is modular, heredity is still limited because of size limitation. These replicators have only a didactic relevance to evolution since they are not feasible in prebiotic environments. Chemical evolutionists nevertheless do believe in prebiotically feasible counterparts.

The difficulty of long-strand replication has been analyzed in depth by Fernando et al. (2007). A novel stochastic model of nucleic acid chemistry was developed to allow rapid prototyping of chemical experiments designed to discover sufficient conditions for template replication. Experiments using the model brought to attention a robust property of nucleic acid template populations, the tendency for elongation to outcompete replication. Externally imposed denaturation-renaturation cycles did not reverse this tendency. For example, it has been proposed that fast tidal cycling could establish a TCR (tidal chain reaction) analogous to a PCR (polymerase chain reaction) acting on nucleic acid polymers, allowing their self-replication (Lathe 2003). However, elongating side reactions that would have been prevented by the polymerase in the PCR still occurred in the simulation of the TCR. The same finding was found with temperature and monomer cycles. It is only the short nucleic acids (like the ones similar to the structure in figure 9.3) that replicate readily, until they are mopped up by ever-elongating long polymers that show no sign of replication under the investigated conditions. Thus the chemical origin of evolvability is still obscure. A possible way out could be the self-construction of active replicase enzymes from smaller parts (Hayden et al. 2008). Alternatively, an early replicase ribozyme may have been a restriction ribozyme capable of cutting itself out of elongating strands.

Unlimited heredity has arisen from limited heredity on multiple occasions: in the origin of life, in the adaptive immune system, and in language. The next section proposes that unlimited heredity also arose in the form of symbolic neuronal replicators inhabiting the brains of humans, and that this was a prerequisite for language.

Neuronal Replicators in Cognition

We propose that the scope of natural selection should be further widened to cognitive phenomena, in order to address the following open questions. How can the remarkable creativity of human thought and problem solving be explained, namely, how does cognitive search work? How does the brain solve the delayed reinforcement problem, that is, on what basis do we decide to behave, given that sometimes external rewards are far removed in time or even absent, and environments are non-Markovian? How does long-term memory formation and retrieval interact with working memory? How do we form causal models of the environment? How can perceptual and conceptual operations be applied to sensory data in a position-independent manner? How is language transmitted?

We hypothesize that natural selection occurs in the brain at rapid timescales (e.g., overnight), and that this contributes to the unconscious generation of ideas, a process we equate with a kind of structured heuristic search. Furthermore, we propose that the human brain contains mechanisms for generating fitness functions based not only on the presence of extrinsic reward objects but also on internal value functions, such as the first derivative of predictability, causal coherence, and information-based criteria.

Just as fitness landscape terminology can be used to describe an evolutionary search, so it can also be used to describe a cognitive search through a problem space. We show how the *efficacy* of natural selection is increased by neuronal mechanisms that are capable of transforming a random search into a structured search, thus highlighting that there exists a continuum between blind variation and thought, the underlying process that structures thought being the evolution of evolvability in neuronal replicators (Toussaint 2003).

Many aspects of cognition appear to involve structured heuristic search rather than random search or pure hill-climbing; for example, insight and transformation problems (Chronicle et al. 2004) are in many respects more complex versions of problems such as the Stroop Task (Dehaene et al. 1998), the Wisconsin Card Sorting Task (Dehaene and Changeux 1991), and the Tower of London Task (Dehaene and Changeux 1997). These latter tasks involve choosing from a set of behaviors on the basis of positive and negative reward feedback. The search space of these tasks is typically small. More complex cognitive problems, in contrast (e.g., puzzles such as the nine-dot problem), have a larger search space that limits the effectiveness of exhaustive and random search. Also, they do not provide explicit feedback, making it necessary for the subject to depend on internally generated success criteria. We propose that the process of solving such problems involves natural selection of neuronal replicators in the brain. We suggest plausible mechanisms for neuronal replicators, neurophysiological evidence for these mechanisms, and which candidate behavioral tasks may be explained by natural selection. The characterization given here is a "law of qualitative structure" in that it describes a framework in which more detailed knowledge may potentially be obtained (Newell and Simon 1976).

The neuronal replicator hypothesis states that several types of replicator exist in the human brain: synaptic replicators, topological neuronal replicators (i.e. groups of synapses) (Fernando et al., 2008), and dynamical neuronal replicators. The hypothesis permits multiple realizability (Marr 1982), that is, it does not commit itself to a particular implementation of a neuronal replicator, although several are suggested here.

A crucial aspect of the extended evolutionary synthesis is a focus on the evolution of evolvability (G. B. Müller 2007; Pigliucci 2008). In neuronal replicators, simple yet powerful mechanisms exist to structure the exploration distribution of a neuronal replicator, that is, the phenotypic distribution of variants produced upon replication (Toussaint 2003). This permits a neuronal replicator system to modify a search based on the outcomes of previous searches, in a way that is not easily possible at the genetic replicator level. In addition, Lamarckian evolution takes place in neuronal replicators because neuronal changes due, for example, to reinforcement learning can be copied directly. Finally, neuronal topology copying has other functions that may be important in causal inference algorithms (Gopnik and Schulz 2004) for forming internal models (Craik 1943) and emulators (Grush 2004).

Precursors of the Neuronal Replicator Hypothesis

There is a long history of application of selectionist and Darwinian dynamics to intrabrain processes, from early thoughts on the evolution of ideas (James 1890 [1950]), to memetics (Dawkins 1976, 1982), neural selectionism (Changeux et al. 1973; Edelman 1987; Marr 1969; Young 1979), synaptic replicators (Adams 1998), synfire chains (Abeles 1991), and hexagonal replicators (Calvin 1996). These theories attempt to explain neural information processing, conscious thought, and human problem solving. All have been influential for the neuronal replicator hypothesis (see Fernando et al., 2008 for a full review). To this we may add the philosophical tradition of evolutionary epistemology, with some of its roots in Mach (1897, 1910), and the full exposition by Campbell (1974), Perkins (1995), Popper (1972), and Simonton (1995). An outline of these influences is given below, except for the philosophical part, which will be dealt with elsewhere (cf. Dennett 1981).

Psychologists such as William James were already writing about the selection of ideas (James 1890 [1950]). A century later, Monod discussed mutation and recombination as applied to ideas (Monod 1971). Dawkins revitalized the concept by introducing the "meme" (Dawkins 1976, 1982). The memetic paradigm has arguably remained in a poised state, at least for two reasons: first, it could not demonstrate a phenomenon that conclusively required memes; and second, it could not demonstrate a physical basis for memes (Aunger 2002). Typically, memes were not considered to replicate *within* a single brain; however, Aunger (2002) is a notable exception because he believes that historically intra-brain neuromeme replication would have preceded interbrain neuromeme transmission, a view to which we adhere.

Influenced by Donald Hebb's notion of neuronal assemblies (Hebb 1949), intrabrain selectionist mechanisms were proposed by Edelman and Changeux for neuronal groups (Changeux 1985; Changeux et al. 1973; Edelman 1987).[1] These mechanisms were intended to explain, among other things, executive function (Dehaene and Changeux 1997) and perceptual categorization (Edelman 1990). In Changeux's mechanism, the inherent dynamics of neural networks produces transient pre-representations, some of which get stabilized by resonance with perceptual inputs. Resonance is thought to arise in loops between cortex and thalamus. Dehaene and Changeux (1997) write that "in the absence of specific inputs, prefrontal clusters activate with a fringe of variability, implementing a 'generator of diversity.'" However, this idea appears to be poorly extendable due to one version of the curse of dimensionality (Belman 1957): If there is a large space to search, how can adaptive pre-representations be produced sufficiently rapidly? Changeux addresses this by allowing heuristics to act on the search through pre-representations; notably, he allows recombination between neuronal assemblies, writing that "this recombining activity would represent a 'generator of hypotheses,' a mechanism of diversification essential for the geneses of pre-representations." However, it is not clear how these heuristics are optimized. Changeux never mentions multiplication of pre-representations, that is, he does not consider pre-representations to be units of evolution. Recent models of neuronal networks exhibit structures and dynamics that correspond to Changeux's pre-representations, that is, polychronous groups (Izhikevich 2007); these are groups of neurons defined by the fact that they fire in temporal correspondence with each other. It is not clear how polychronous groups can be used to produce adaptive behavior.

Edelman (1987) proposed a theory similar to Changuex's in order to explain how an organism "decides" how to behave, given a set of sensory inputs and reward feedbacks as it interacts in a situated and embodied manner with an environment. To test his theory, Edelman has implemented what he calls a Darwinian system within the computer controller of a robot. In Edelman's theory, elements of a primary repertoire of neuronal groups within the brain are thought to compete with each other for stimulus and reward resources. This results in selection of a secondary repertoire of behaviorally proficient groups (Izhikevich et al. 2004). The theory has been the subject of a large number of models and variants that attempt to explain a wide range of behavioral and cognitive phenomena at various levels of abstraction (Gisiger et al. 2000), such as category formation (Edelman 1987), reinforcement learning using spike-time-dependent plasticity modulated by dopamine reward (Izhikevich 2007), and visual-motor control in a robotic brain-based device (Krichmer and Edelman 2005).

Again, Edelman's neuronal groups are not units of evolution, for there is no multiplication (Crick 1989, 1990). The most modern version of Edelman's neuronal group selection is presented by Izhikevich et al. (2004). In Izhikevich's model, neurons are arranged in a recurrent network with axonal condition delays and weights being modified by spike-time-dependent plasticity (STDP). Polychronous groups (PCGs) with stereotypical temporal firing patterns self-organize when a particular firing set of neurons is activated in a spatiotemporal pattern, resulting in the convergence of spikes in downstream neurons. STDP subsequently reinforces these causal relationships. Because the same neuron can occur in many groups, and because delays produce an effectively infinite-sized dynamical system (subject to temporal resolution constraints), the number of PCGs far exceeds the number of neurons in the network, allowing a very high memory capacity for stored PCGs. In a group of 1,000 neurons, approximately 5,000 PCGs are found, with a distribution of sizes. If inputs are random, the PCGs are formed and lost transiently, over the course of minutes and hours. Structured input can stabilize certain PCGs. Izhikevich attempts to understand the functioning of polychronous groups within Edelman's framework of neuronal group selection (Izhikevich 2006). Michod explains that in neuronal group selection, synaptic change rules replace replication as a mechanism of variability of the "unit of selection": there is correlation between the parental and offspring states of the same neuronal group even without multiplication (Michod 1988). Here "parental" is used to mean the group at time t, and "offspring" refers to the same group at time $t + x$. However, no mechanism is described showing how a beneficial trait of one PCG could be transmitted to another PCG. We claim that this absence of multiplication is a fundamental limitation of Edelman's Neural Darwinism.

To summarize, because neither Edelman's nor Changeux's model includes multiplication, we argue that they lack the algorithmic capabilities of natural selection. The problem of transmission of a favorable trait from one group to nongroup material (or another group) is a ubiquitous bind in selectionist theories. Replication is the most natural way to envisage this transfer of function operation.

What advantage has Darwinism over selectionism? First, adaptations can be transmitted between structures. This allows search to be undertaken by modifying one's offspring and not oneself, an important factor if most variants are harmful. Second, replication allows good solutions to obtain more search resources. Third, replication allows selection to act on variability properties, that is, the evolution of evolvability (Pigliucci 2008), by hitchhiking of neutral variability-structuring neuronal changes (mutations) that are selected because they shape the exploration distribution of variants (Kirschner and Gerhart 1998; Toussaint 2003). This is because, if there is variation in variability, then selection can act on that variation (Wagner and Altenberg 1996).

Consider how the evolution of evolvability is possible. A minimal natural selection algorithm lacking the capacity for the evolution of evolvability (except by neutral drift) is a $1 + 1$ evolutionary strategy (ES) in which a parent makes one mutated offspring (Beyer 2001). A $1+1$ ES is so called because it involves a "population" of only one parent producing only one offspring. If the offspring is immediately superior to the parent, the offspring replaces the parent. If it is not, the offspring is destroyed and the parent makes another offspring. One iteration is a generation. Imagine two offspring are produced that have equal fitness, yet differ in the following feature only: the first offspring is capable of producing fit offspring itself, whereas the second offspring can produce only low-fitness offspring. The $1+1$ ES cannot "see" the grandchildren's average fitness, and so both offspring will have equal likelihood of being selected. The minimal selective scenario required for selection for the more evolvable first offspring would have been a $1+1+1$ ES in which the grandparent is potentially replaced after assessing the grandchildren, rather than replacing the parent after assessing the children. If a population of genotypes has nontrivial neutrality, that is, if there exist two offspring with the same fitness but different exploration distributions, selection on variability properties is possible.

James Baldwin already in 1898 was applying concepts of structured variability to how "valuable thought-variations" could be generated by a process of trial and error, as a basis of intelligence (Baldwin 1898, 1909). The Baldwin Effect is an example of how developmental exploration can structure intergenerational genetic variability, but clearly Baldwin was concerned with structuring variability in other domains. Nontrivial neutrality may be a necessary condition for open-ended evolution (Bedau 1998). This leads us back to the meme. No one disputes that social communication involves duplication of information, but

Robert Aunger asks, "are people the exclusive agents behind this process," or "is an information-bearing replicator" acting as an agent?" (Aunger 2002). The concept of an agent is notoriously difficult to formalize (Barandiaran and Ruiz-Mirazo 2008), and we do not define it here, but one empirical finding is clear: entities that are known to possess agenthood are produced by the subset of units of evolution capable of nontrivial neutrality and whatever other features are necessary for these units to be capable of open-ended evolution. Self-referential units of evolution are distinct from simple units of evolution in being "reproducers" (Szathmáry and Maynard Smith 1997), that is, in specifying the means of their own production. Self-referential replication is one mechanism that produces the phenomenon of nontrivial neutrality. The memetic question can be reformulated: *Are there self-referential units of evolution that copy themselves between brains?* The best candidates may be linguistic replicators (Maynard Smith and Szathmáry 1999).

We suggest that self-referential replicators allow structured heuristic search, which is crucial in many aspects of cognition, and that the brain contains many such agents that compete and cooperate for extrinsic and intrinsic reward resources.

The idea of neuronal replicators has already arisen twice before (Adams 1998; Calvin 1996). The most promising suggestion is by Paul Adams, who describes synapses as replicating. This would make them the minimal neuronal units of evolution. Synapses replicate by increasing the amount of quantal release from the presynaptic to the postsynaptic neuron, according to a Hebbian rule. Mutations are noisy quantal Hebbian learning events where a synapse is made to contact a nearby postsynaptic neuron rather than to enhance the connection to the current postsynaptic neuron (Adams 1998). Synapses compete with each other for correlation resources and other reward resources if, for example, dopamine acts to modulate Hebbian learning. Adams demonstrates how vectors of synaptic weights can be selected, and how error-correction mechanisms in cortical layer VI can adjust the synaptic mutation rates. As described in Fernando et al. (2008), there is a mathematical isomorphism between Hebbian learning and Eigen's replicator equations, a standard model of natural selection dynamics in chemical and ecological systems. Hebbian learning can be said to *select* between weights on the basis of correlations in activity. Synaptic weights can be said to multiply in proportion to the product of presynaptic and postsynaptic activity. The synaptic equivalent of mutation is a shifting of synaptic weight resources between synapses afferent on the same postsynaptic neuron. Specifically, the Oja version of Hebbian learning is (Oja 1982):

$$\tau_w \frac{d\mathbf{w}}{dt} = v\mathbf{u} - \alpha v^2 \mathbf{w}, \tag{9.1}$$

where $\mathbf{w}$ is the synaptic weight vector, v is the output rate, $\mathbf{u}$ is the input rate vector, and the rest are constants. This is isomorphic to Eigen's replicator equation (Eigen 1971):

$$\frac{dx_i}{dt} = A_i Q_i x_i + \sum_{j \neq i}^{N} m_{ij} x_j - \frac{x_i}{c} \sum_{j=1}^{N} \sum_{k=1}^{N} m_{ij} x_j, \tag{9.2}$$

where x_i is the concentration of sequence i, m_{ij} is the mutation rate from sequence j to i, A_i is the gross replication rate of sequence i and Q_i is its copying fidelity, N is the total number of different sequences, and formally $m_{ii} = A_i Q_i$ (Eigen 1971).

The second proposal for neuronal replicators appeared in *The Cerebral Code* (Calvin 1996). Calvin outlined an algorithm for the copying of neural dynamics (spatiotemporal patterns) between hexagonally shaped regions of the cortex. Pyramidal cells in the superficial layers of the cortex have a circular field of excitatory efferents projecting to a standard length of 0.5mm from the central cell. Due to geometric constraints on a 2-D surface, this means that if six such cells are arranged in a hexagon and have the same receptive field, they form self-reexciting loops. If the cells behave as nonlinear relaxation oscillators (like fireflies), they can entrain each other and fire spikes in synchrony. The system may start with two cells firing in synchrony, which recruits a third cell to form an equilateral triangle, and so on. Due to long-term potentiation (LTP), the underlying synaptic weights would change to reinforce this synchrony in the future. A fundamental limitation of Calvin's proposal is that he does not explain how the circuitry between cells is to be replicated between hexagons.

Copying of neural receptive fields is already known to occur in the brain. The formation of topographic maps (from an initial random connection-weight matrix) by using spike-time-dependent plasticity (STDP) and lateral inhibition has been demonstrated (Song and Abbott 2001). This is effectively copying of the receptive fields in one layer onto a second, parallel layer. J. M. Young et al. (2007) have extended this work to explain the formation of new receptive fields in de-afferented regions of the cat's visual cortex. The extent of one-to-one topographic copying of receptive fields can be tuned by altering a neuronal gain parameter in the lesioned area. At high gain, instead of obtaining a one-to-one copying of adjacent receptive fields (copying with no selective amplification and low information loss), the neurons in the entire de-afferented area receive the receptive field of just one of the adjacent neurons (copying with selective amplification and high information loss). Nonselective copying can be obtained not just with STDP but with Hebbian learning as well, with gain being adjusted according to activity-dependent scaling, for example (Van Rossum et al. 2000). Calvin's proposal does only half the job of copying, just as hydrogen bond formation in DNA replication does only half the job of semi-conservative replication; phosphodiester bond formation is required to re-create the original topology of the parent strand; similarly, in neuronal topology copying, an extra mechanism would be required, one that Calvin has not described.

Finally, concepts from synfire chains can be modified to include replication of neuronal dynamics. A synfire chain is a feed-forward network of neurons with several layers (or pools). Each neuron in one layer feeds many excitatory connections to neurons in the next pool, and each neuron in the receiving layer is excited by many neurons in the previous one. When activity in such a cascade of layers is arranged like a packet of spikes propagating synchronously from layer to layer, it is called a synfire chain (Abeles 1982, 1991). There have been reports in the literature about observations of precisely repeating firing patterns (Abeles and Gat 2001). An excellent summary is provided by Abeles et al.[2] One can see that almost trivially, replication of the spike packet is possible simply if the synfire chain branches into two or more. If we have branches like this, then one can use them for the spread of spike packets gated by reward. If we imagine a lattice where every arrow between the neuronal groups can work both ways, but in a reward-gated fashion, then fitter and fitter packets can fill up the lattice (Fernando and Szathmáry, 2009). The snag, for the time being, is the limited heredity potential due to the limited information a spike packet can propagate. Recombination is easy to imagine when two roughly equally fit packets are transmitted to the same neuron group. The topology of the synfire network influences the outcome of selection. Neuronal evolutionary dynamics could turn out to be the best application field of evolutionary graph theory (Lieberman et al. 2005). It has been shown that some topologies speed up, whereas others retard, adaptive evolution. The brain could well influence the replacement topologies by gating, thereby realizing the most rewarding selection topologies (Fernando et al., 2008).

Mechanisms of Neuronal Replication

Neuronal Topology Replicators

A mechanism for copying of neuronal topology from one neuronal topographic layer to another is outlined. The mechanism utilizes topographic map formation (Song and Abbott 2001; Willshaw and Von der Malsburg 1976), coupled with spike-time-dependent plasticity (STDP; Markram et al. 1997), neuronal resetting (Crick and Mitchison 1995), topological error correction, and activity reverberation limitation. Neuronal topology copying works by using a neuronal implementation of a causal inference algorithm, in the sense that the offspring layer "observes" the activities of the parental layer in order to infer the connectivity of the parental layer.

Figure 9.4 shows a minimal example of neuronal topology copying (Fernando et al., 2008). It works by first establishing a topographic map between the parental and the offspring layers (Song and Abbott 2001; Willshaw and Von der Malsburg 1976). Spike-time-dependent plasticity (Markram et al. 1997) in the offspring layer is then used to infer the underlying topology of the parental layer, on the basis of activity received from the parental layer neurons as they are randomly sparsely activated.

To improve copying fidelity of neuronal topology, error-correction neurons are hypothesized that measure the difference in activity between corresponding neurons in parent and offspring layers (Adams and Cox 2002). On the basis of this discrepancy of activity, they modify the afferents to the offspring neuron accordingly. Two types of error-correction neurons are hypothesized, false-positive and false-negative error correctors. Activity reverberation limitation is required to allow the correct inference of Markov-equivalent causal graphs, and thus the copying of larger networks. This is implemented using inhibitory neurons that allow spikes to pass only if its associated excitatory neuron was principally depolarized from outside its layer.

The simplest way to demonstrate natural selection using the above copy operation is by using the 1+1 Evolutionary Strategy (1+1 ES; Beyer 2001) shown in figure 9.5. A 1+1 ES is a simple evolutionary algorithm that works as follows. If the offspring does not have fitness higher than the parent, then the offspring is erased and another attempt at copying the parent can be made (not shown). If the offspring has fitness higher than the parent, then the parent is erased and the offspring becomes the new parent and makes a new offspring in what was previously the parental layer (see parts 5, 6, 7, 8).

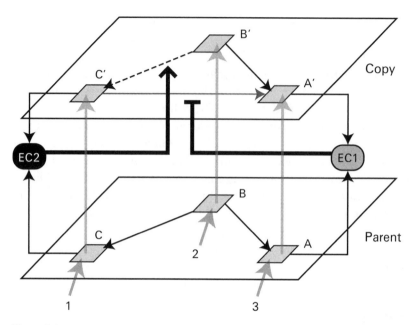

Figure 9.4
An outline of the neuronal topology replication mechanism with error correction. The
parental layer is on the bottom. The offspring layer is on the top. In this example, each
layer has three neurons. Topographic collaterals (vertical arrows) connect parent layer to
offspring layer. Copying is the reproduction of the intralayer topology of the parent to the
offspring. Error correction mechanisms are shown. STDP operates in the offspring layer.
There are two error correction mechanisms; EC1 (right) is a false-positive error correction
mechanism implemented using an "observer" neuron (EC1) that negatively neuromodu-
lates neuron A′ in the copy layer on the basis of differences in firing between the parental
(A) and copy (A′) layer neuron. We assume C is undergoing stimulation (1) when EC1
acts. EC2 (left) is a false-negative error correction mechanism implemented using an
"observer" neuron that positively neuromodulate inputs that pass to a poorly firing neuron
(C′) in the copy layer from the neuron that is undergoing interventional stimulation (in
this case we assume B is undergoing stimulation (2)) when EC2 acts. EC1- and EC2-type
neurons are required for each neuron pair—A, A′, B, B′ and C, C′—and their neuromodulatory outputs must pass *widely* to all synapses in the child layer.

Note that neuromodulation is critical in the function of the 1+1 ES
circuit in three ways. First, the direction of copy making depends on
modulation to open and close vertical up and down gates at different
times. Second, neuromodulation is necessary to switch on and off STDP-
based plasticity in L0 and L1 alternately. Third, a mechanism is necessary to *reset* the layer (i.e., reduce weights in the layer) that is to be
overwritten.

Figure 9.6 shows an example of an evolutionary run in which a 10-node
neural network is evolved using the above algorithm. The replicators in the
1+1 ES are units of evolution as defined previously. This corresponds to a

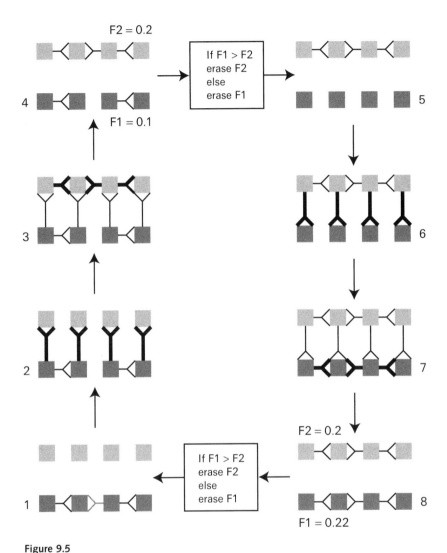

Figure 9.5
A neuronally implemented 1+1 ES using the STDP-based copying mechanism. (1) The circuit to be copied exists in the lower layer L0. The black connections in L0 show the original circuit. (2) Horizontal UP connections are activated (e.g., by opening neuromodulatory gating). These are the equivalent of the h-bonds in DNA copying. (3) A copy of the topology of L0 is made in L1, using STDP and error correction. (4) The layers are functionally separated by closing neuromodulatory gating of the UP connections. The fitness of each layer is assessed independently. (5) The layer with the lowest fitness is erased or reset (i.e., strong synaptic connections are reduced). In the above diagram we see that L1 fitness > L0 fitness, so L0 experiences weight unlearning. (6) DOWN vertical connection gates are opened. (7) STDP in layer 0 copies the connections in L1. (8) After DOWN connections are closed, fitness is assessed and the cycle continues.

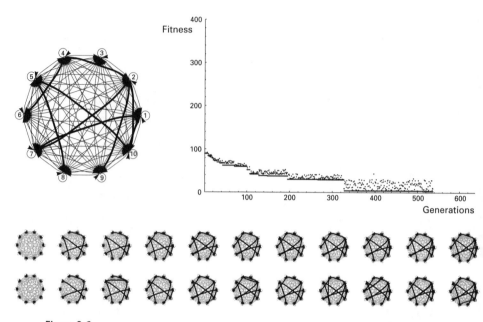

Figure 9.6
An example evolutionary run in which a particular desired topology is selected for. The neuronal copying algorithm is capable of sustaining evolution by natural selection to optimize the topology of a 10-node motif. Desired topology is on the top left (randomly initialized with 10% connectivity). Fitness (Euclidean distance between desired and actual topology) of parent (line) and offspring (dots) over 600 generations is shown on the top right. Bottom graphs show 11 parent–offspring pairs taken at intervals of 50 generations.

"genome size" of 90 synapses (self-synapses are not permitted). The network is selected for a particular topology of strong weights, the fitness being the Euclidean distance between the actual and the desired topology.

The network was initialized with all weak weights (fully connected; see bottom left). Explicit mutation operators were used after each copying event. A mutation involved the modification of a weight to be either very strong or very weak. In this implementation, the copying operation was often so effective that without an explicit mutation operator, there was insufficient variability for rapid evolution of the desired topology. The above model is capable of copying a sparsely connected neuronal topology of any size from one layer to another, if the two layers are connected topographically, and STDP and anisotropic reverberation limitation operates in the offspring layer. However, the rate of topology copying would be limited by the rate of synapse formation and resetting.[3] In the next section, dynamical neuronal replicators are described that are capable of orders of magnitude faster evolution.

Dynamic Neuronal Replicators

The minimal unit of dynamical neuronal replication consists of a bidirectional coupled and gated pair of bistable neurons (Izhikevich 2007a), as in figure 9.7. We propose that dynamical replicators operate in working memory (Baddeley and Hitch 1974), for which bistability (Wang 1999) and recurrence (Zipser et al. 1993) have previously been proposed as mechanisms.

Grouping these pairs together, one can make two layers coupled initially by a topographic map. The parental layer has neurons initialized

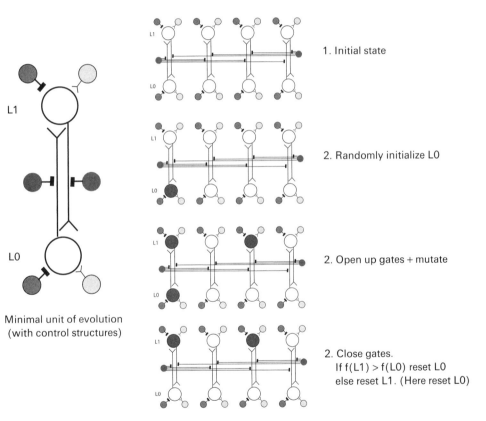

1. Initial state

2. Randomly initialize L0

2. Open up gates + mutate

2. Close gates.
If f(L1) > f(L0) reset L0
else reset L1. (Here reset L0)

L1

L0

Minimal unit of evolution
(with control structures)

Figure 9.7
An outline of how dynamical neuronal replicators can implement a 1+1 evolutionary stratergy. The two bistable neurons are shown as large circles (light = not firing, dark = firing). They are coupled bidirectionally by gated axons. Each neuron can be reset and activated by external control. A vector of such pairs is shown on the right. Initially the vector is all off (not spiking). Layer 0 (parental layer) is randomly initialized, here the neuron on the left becomes active. The up gates are opened allowing that neuron to activate its corresponding neuron in layer 1 (offspring layer). The gates are closed and the fitness of each layer is assessed. The least fit layer is reset (becoming the new offspring layer), and is overwritten by the fitter layer (the new parental layer) in the next copying operation.

randomly as follows. If a bistable neuron is given some depolarizing current, it begins to fire repeatedly, whereas if a bistable neuron is given some hyperpolarizing current (in the correct phase), it stops firing. The state of neurons in the offspring layer is reset (i.e., all neurons are hyperpolarized to switch off spiking). Activity gates are opened for a brief period from the parental layer to the offspring layer, allowing the spiking neurons in the parental layer to switch on the corresponding neurons in the offspring layer. Activity gates between layers are then closed. Thus the vector of activities in the parental layer is copied to the offspring layer. As in figure 9.5, a 1+1 ES can be implemented if the two layers have their fitness assessed, the higher fitness layer is defined as the parent, and the offspring layer is reset. Figure 9.8a shows the result of selecting for a particular desired vector of activity using the above protocol.

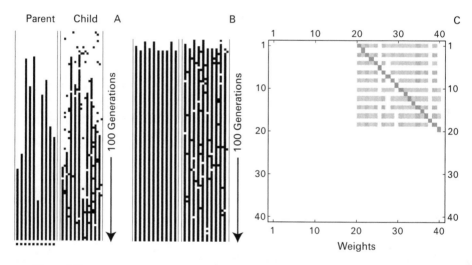

Figure 9.8
An example of dynamical replicators evolved to produce a desired activity vector with and without Hebbian learning. (A) shows that within 100 generations (each generation lasting 7 seconds), the desired activity vector can be selected for (bottom left sequence). The parental layer activities are shown on the left (i.e., the layer from which a copy is made) and the offspring layer activities are shown on the right. If the offspring layer is less fit than the parent, then it is overwritten by another copy made from the parent until it is fitter than the parent, at which point it takes on the parental role. (B) shows the much faster discovery of the desired activity vector when the copying operation was biased by a previously found solution using Hebbian learning. Hebbian learning was undertaken on the copying connections between the parent and the offspring layer neurons, after the previous run had found the correct solution. This was able to bias future exploration during the copy operation towards the previously found solution. (C) shows the weight vector connecting parental layer to offspring layer that arises due to Hebbian learning. Layer 0 contains neurons indexed 0 to 19, and layer 1 contains neurons indexed 20–39. Note that the diagonal weights are the initially strong topographic weights. These are not adjusted by Hebbian learning.

How might dynamical and topological neuronal replicators interact? The consolidation of memories into long-term stores is known to involve the changing of synaptic weights in the hippocampus and cortex (Abraham and Robins 2005; Nadel et al. 2007; Nadel and Moscovitch 1997; Nadel et al. 2000). The multiple-trace theory of memory consolidation suggests the conversion of solutions evolved in working memory, using dynamical replicators, into more permanent topological replicators that compete for synaptic resources.

Structuring Search Using Hebbian Learning

A remarkable capacity for structuring exploration distributions emerges if instead of limiting between-layer connections to a topographic map, one starts with a strong one-to-one topographic map and allows all-to-all Hebbian connections to develop once a local (or global) optimum has been reached. Imagine that evolution has been run to the end point of figure 9.8a, so that the optimum activity vector has been obtained and is present in both the parent layer and the child layer. Hebbian learning is then permitted between these two vectors (for all synapses except the original one-to-one topographic connections). If the activity vectors are then reset and a new evolutionary run is started, then copying will be biased by the Hebbian learning that took place in previous evolutionary runs. An active neuron in the parental layer will tend to activate not only the corresponding one-to-one topographic neuron, but also other neurons in the offspring layer that were previously active when the optimal solution had been found. Oja's rule is used to control the Hebbian between-layer synapses. Figure 9.8b shows that if Hebbian learning is permitted, then later evolutionary searches can converge faster, because they have learned from previous evolutionary searches. The Hebbian weight vector that evolved is shown in figure 9.8c.

Richard Watson and colleagues in Southampton have described a set of search problems that are particularly well suited to neuronal copying biased by Hebbian learning (Watson 2006), and have proposed that "symbiotic evolution" can effectively solve these problems (Watson et al. 2009). Waston's principle of using Hebbian learning to learn local-optima using a multiple restart hill climber is general; however, the neuronal replicator may be the most plausible implementation of it (Watson et al. 2009). In these problems, there is interdependency between problem variables because the fitness contribution of one variable is contingent upon the state of other variables. An archetypical

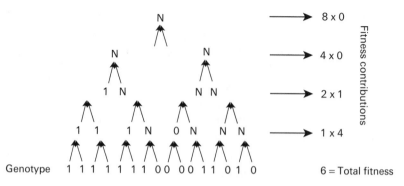

Genotype   1 1 1 1 1 1 0 0 0 0 1 1 0 1 0   6 = Total fitness

Figure 9.9
An example of the HIFF problem. The genotype is divided into pairs. A transfer function is applied recursively to pairs as follows: {0,0} à 0, {1,1} à 1, all other pairs produce a NULL (N). Fitness is given for every 0 or 1 produced. Fitness at each level is scaled exponentially, and summed to give total fitness of a genotype. To see that the fitness landscape has many local optima, imagine adding 1s from left to right to a genotype of all 0s.

example is the Hierarchical IF-and-only-IF problem (HIFF; Watson et al. 1998), illustrated in figure 9.9.

The lowest level of fitness contributions comes from looking at adjacent pairs in the vector and applying the transfer function and the fitness function. The transfer function is {0,0} → 0, {1,1} → 1, and all other pair types produce a NULL (N). The fitness function for each level just sums the 0 and 1 entries. The second level is produced by applying the same transfer function to the output of the first transfer function. The fitness contribution of this next layer is again the number of 0s and 1s in this layer, multiplied by 2. This goes on until there is only one highest-level fitness contribution. The fitness landscape arising from the HIFF problem is pathological for a hill climber because there is a fractal landscape of local optima, which means that the problem requires exponential time to solve. Figure 9.10 (left) shows the dynamical neuronal replicator with Hebbian learning applied to the HIFF problem. The Hebbian learning of local optima allows the problem to be solved by biasing variability, such that partial solutions are not forgotten. In the next section we argue that a wide range of problems in cognition involving generative creativity are of this type, and that rapid natural selection of dynamic neuronal replicators biased by Hebbian learning offers a possible mechanism for their solution. Figure 9.10 (right) shows a control for the above experiment in which there is no Hebbian learning. The HIFF problem is not solved, and there is no improvement after each reinitialization of the 1+1 ES.

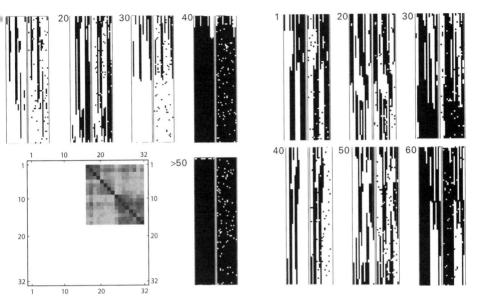

Figure 9.10
(Left) A 16-bit HIFF problem. Parental activity vectors (spanning one epoch consisting of 100 generation/copying events along the y-axis, and 16 bits along the x-axis) are shown on the left of each column (black = on, white = off), and offspring activity vectors are shown on the right of each column of results. The number next to each column shows how many Hebbian learning and resetting loops (i.e., epochs) have elapsed. Hebbian learning is only permitted in the last 10 generations of each epoch (i.e., when it is most likely that a local-optimum will have been found). It takes 30 epochs of Hebbian learning to find the all 0s solution rapidly. After 50 generations, all initial conditions lead to one of the global optima, i.e., the all 1s optimum. The final weight matrix is shown on the bottom left. (Right) Control case, as before, except with no Hebbian learning. There is no improvement each time the 16-bit activity vector is reinitialized at each epoch. The best solution that is found is the half 0's/half 1's optimum.

Structured Search in Evolution and Cognition

Phenotypic variability can be highly structured (Toussaint 2003). Unstructured search may be very inefficient in complex problem domains (Richards 1977). The same applies to unstructured hill climbing in any optimization problem with multiple interdependencies (Watson 2006). Ross Ashby's homeostat is an example. It becomes very slow in reaching a stable state, given a perturbation to a large and highly connected network (Ashby 1960). Many tasks for testing prefrontal cortex function do not require a structured search. In fact, often they can be solved by random search. Consider Rougier et al.'s (2005) model of a classification task. The system receives reward if it classifies a set of feature vectors according to a particular element of that vector. The network is given a cue instructing which element it should base the classification upon. A slightly more complex version involves feature matching, that is, doing XOR (the non-linear exclusive OR logic function which returns 1 if given inputs [0,1] or [1,0] but returns 0 if given inputs [0,0] or [1,1]) on the relevant element pair in two feature vectors. Reward has the effect of gating a stochastic search, stabilizing the activities in prefrontal cortex if an unexpected reward is obtained, and destabilizing them if the system does not get an expected reward. Similarly, in temporal difference learning (Sutton and Barto 1998) the network undertakes a random search, biased by instantaneous differences in predicted reward. A cue defines distinct environments in which a different behavior is rewarded. Rougier et al.'s task is fundamentally the same as the Stroop Task, in which the subject must either name the word "red" or name the color in which the word "red" is painted, or the Winsonsin Card Sorting Task (WCST), in which the subject must sort the cards on the basis of a particular feature dimension. Although switching is rapid, it can be achieved by random (unstructured) search in the space of classification dimensions. The authors admit that the system does not deal with generativity: "Generativity also highlights the centrality of search processes to find and activate the appropriate combination of representations for a given task." They describe the mechanism as "random sampling with delayed replacement."

Similar approaches have been taken by Changeux's group (Dehaene et al. 1998) for the Stroop Task, but the language used to describe the process is different. They talk of stabilization (selection) of internal pre-representations by external stimuli. Configurations in a global workspace are stabilized by internal reward and attention signals (Dehaene et al. 1998). Chialvo and Bak's paper "Learning from Mistakes" (Chialvo and Bak 1990) and variants (Bosman et al. 2004) also describe a similar stochastic hill-climbing type of algorithm. Seung has explicitly described the process of reward-biased search in spiking neurons as stochastic hill climbing (Seung 2003).

Selective attention in complex tasks (Desimone and Duncan 1995) and solving insight problems (Chronicle et al. 2004) may require a mechanism for structuring search. This is because for many problems it would take too long to exhaustively search all possible solutions, and hill climbing may be inefficient due to local optima. Rougier et al. write that "an important feature of future research will be to identify neural mechanisms that implement more sophisticated forms of search" (Rougier et al. 2005).

We do not yet know what neural mechanisms underlie human creativity in problems such as the Nine-Dot Problem: "Draw four continuous straight lines, connecting all the dots without lifting your pencil from the paper" (MacGregor et al. 2001).

• • •
• • •
• • •

The difficulty of finding the right solution depends upon the representation of solutions, the variability operators, and the selection algorithm. It is likely that for most representations of the Nine-Dot Problem the fitness landscape is rugged (Perkins 1995). That is, move operators will result in getting stuck on local minima. Some problem representations may be so poor as not to contain the correct solution. Several authors describe that insight problems require "restructuring of the initial problem representation" (Chronicle et al. 2004) or sculpting the response space (Frith 2000) to encompass the goal state (Öllinger et al. 2006) because of strong constraints that "prevent one from considering and evaluating the correct solutions" (Reverberi et al. 2005; Knoblich et al. 2005). For a problem to be an insight problem, performance cannot be explained by innate, random, or exhaustive search; for example, it has not been conclusively demonstrated that New Caledonian crows use insight when bending wires to lift peanuts in buckets out of bottles. Simpler search methods may be sufficient for this task (Weir et al. 2002).

Most people's phenomenal concomitants to solving this problem are something like undertaking trial-and-error search with constraint relaxation, that is, increasing the space of possible solutions, as well as hill climbing to improve some intermediate goal function. However, one must be careful not to assume that the brain works in the same way as the mind thinks. "The ghost has been chased further back into the machine, but it has not been exorcised" (Fodor 1983: 127). The moment of insight is instantaneous, and the correct solution does not seem to be arrived at consciously (Sternberg and Davidson 1995; Metcalfe and Wiebe 1987). What neuronal process may underlie performance in these tasks (Kaplan and Simon 1990; Simon and Reed 1976)?

It is proposed that a role for neuronal natural selection is to structure exploration distributions in order to bias cognitive search. An exploration distribution is defined by both the representation of solution space *and* the move set. The example of the HIFF problem shows how the exploration distribution of a dynamical neuronal replicator can be structured by Hebbian learning. Hebbian learning alters only the move set, not the underlying representation of solution space, not the selection method, and not the fitness function. The neuronal replicator hypothesis proposes that the brain configures all these processes to implement an effective natural selection algorithm for a given search problem. The frame problem (Pylyshyn 1987) then becomes the problem of how to choose an initial population of solutions. An example is provided below of an idealized method by which a neuronal natural selection algorithm could solve an "insight" problem.

Consider how to apply the neuronal replicator approach to the 10-coin problem; 10 coins are arranged in a triangle, and the aim is to make the triangle face the opposite direction by moving only three coins.

Figure 9.11 shows a possible representation of this problem. Assume that when the problem is defined to the subject verbally, the subject's brain is capable of representing the positions of the coins as activations of a vector of neurons. The figure shows a simple representation consisting of a square neuronal matrix with coin positions shown. This representation is capable of emulating various properties of the real coins, such as their relative spatial location, and constraints about how they can be arranged. Assume that there is a population of at least two solutions. A solution is a probabilistic neuronal description of three coin moves, implemented, for example, as a matrix of neurons with different firing rates. When a solution is applied to the coin emulation, the emulation is transformed into the final state. Another neuronal system is then capable of assigning a subjective utility to this final state. The assignment of subjective utility is not trivial, and depends on previous experience. Two subjective utility functions are used here. In the first, a simplifying assumption is made that subjective utility corresponds to the Hamming distance from the correct solution. In the second, subjective utility is the maximum number of coins corresponding to an inverted triangle convolved over the final state of coins after three moves. Performance using both functions was very similar. A potential solution consists of six probabilistic transformation matrices representing which coin is to be moved and where it is to be moved. Other potential solution representations may arise that are more or less likely to be fit; however, we do not consider the mechanism by which an initial population of neuronal replicators is first formed. A mechanism applies a solution to the emulator's representation of the coin positions by choosing a valid coin to move, and making sure it moves to a valid place. Because the solutions are initialized randomly according to a uniform distribution, many different coin moves

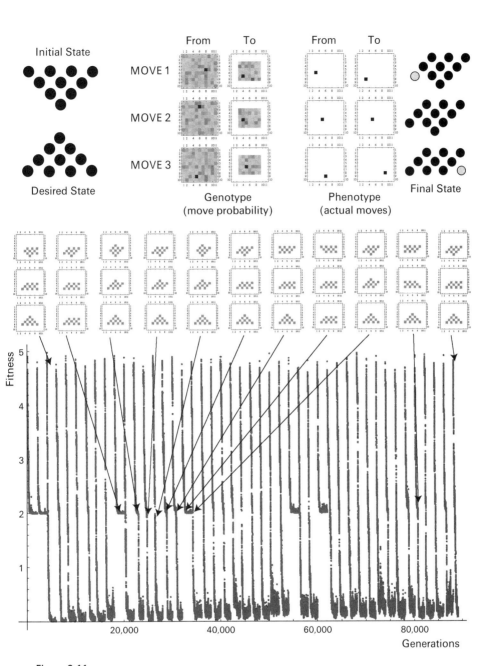

Figure 9.11
(Top left) The initial and the desired states of the 10 coins. (Top right) An example of the genotype, the chosen move, and the emulated coin positions for 3 moves that result in a partial solution. (Middle and bottom) An evolutionary run consisting of reinitialization of the 2 genotypes every 3000 generations. Hebbian learning is not used between generations. Most runs result in the global optimum being reached. Some runs get stuck with an error of 2 (i.e., one coin in the wrong place). The intermediate and final coin positions are shown in the boxes above the fitness plot.

can be produced by the same solution. Therefore, to accurately determine the fitness of a solution, the subjective utility of many final states arising from many (in this case 1,000) applications of the rule must be summed (Cooper 2001). Over time, the solution representation becomes more deterministic and fewer runs through the emulator are required to obtain a fitness assessment. Figure 9.11 shows the above algorithm implemented with a population size of 2. At each round, the fitness of both solutions is calculated, and the fitter solution replaces the less fit solution, with mutation. Mutation is defined as choosing each site in the transformation matrices with a probability of 1/10 and either doubling it or halving it, and then renormalizing that matrix. Sometimes the system can find the correct solution by a process of hill climbing, but sometimes it gets stuck in local optima. The local optima occur due to null neutral moves (i.e., where the initial and final positions of a coin are the same in one of the moves) and due to circular neutral moves (i.e., where a coin contributing to fitness in the correct inverted triangle is moved to another place where it provides equal fitness to the inverted triangle).

The ability to solve the problem by hill climbing alone calls into question whether this problem (as formulated above) is really an insight problem. The model presented says nothing about how such goal functions are chosen and modified. A more realistic model of human performance would include an algorithm for generating predicted goal states. The Hebbian learning mechanism is not required to find the solution to the 10-coins problem here; however, it can *save* the solution and allow this saved solution to bias mutation on another round of natural selection from random initial conditions.[4] This process of Hebbian learning of local optima with subsequent biasing of exploration distributions may be one of the means by which associations can be learned and applied (Simonton 1995).

An objection to the neuronal plausibility of the above stochastic hill-climbing type of algorithms may be that many runs through an emulator were required in order to obtain an accurate fitness assessment of a rule (e.g., 1,000 emulations per rule). Clearly, such a large number of applications of a probabilistic rule are not made consciously. In fact, consciously we are only aware of testing deterministic rules a relatively small number of times (e.g., we move a coin physically, then see what we can do next). Failure to find the correct solution may arise due to various limitations imposed upon the above algorithms; for example, emulation of a rule may be possible only consciously or by use of real coin moves in the external environment, in which case far fewer assessments of a probabilistic rule can be carried out. In this circumstance, fitness assessment of a probabilistic rule may be so noisy that no increase in population fitness is possible from a random initial probabilistic rule initialization. Where fewer emulations are allowed, a rule must either be defined deterministically, or the fitness function must be redesigned to decrease the variance in the outcomes of emulations.

This method of solving the coin problem differs from those discussed in Chronicle et al. (2004). There it is assumed that the only configurations tested are those that are physically moved in the real world. Here it is assumed that internal emulation of the effect of coin movements takes place unconsciously in the brain when one is told the problem. Evidence for unconscious mechanisms in insight tasks comes from an experiment showing that sleep doubles the rate at which explicit knowledge is gained about a hidden correlation in a stimulus-response task (Wagner et al. 2004). In reality, a complex interaction between what are described as conscious and unconscious processes is likely. In addition, people may vary greatly in their capacity for emulation.

The heuristic search hypothesis of Newell and Simon (1976) assumes that "solutions to problems are represented as symbol structures. A physical symbol system exercises its intelligence in problem solving by search—that is, by generating and progressively modifying symbol structures until it produces a solution structure." The physical symbol system hypothesis has been heavily criticized (Brooks 1990). But we note that there appears to be a parallel in the requirement for "symbols" both in the brain and in organismal natural selection, as became evident in the problem of how to maintain information and diversity by blending inheritance (Gould 2002: 622). The only solution was to allow symbolic (i.e., particulate Mendelian) inheritance (Fisher 1930). Symbols seem to be a crucial requirement for natural selection with unlimited heredity, irrespective of its implementation.

How Is Fitness Defined in the Brain?

How does the brain determine the criteria by which to select neuronal replicators? Midbrain dopamine systems signal the error in predicted reward (Izhikevich 2007b; Oudeyer et al. 2007); however, dopamine also signals more complex value functions, such as prediction error (Horvitz 2000), and some have claimed that the brain tries to minimize prediction error (Friston and Stephan 2007). But then why seek novelty, why play? Oudeyer et al. (2007) propose that the brain attempts to maximize learning progress (i.e., to produce continual increases in predictivity). Other formulations of intrinsic fitness functions exist (Lungarella and Sprons 2005), such as maximization of information flow in the sensorimotor loop (Klyubin et al. 2007), and maximization of mutual information between the future and the past (Bialek et al. 2001), and these have been tested in robotic control (Der et al. 2008). Alternatively, co-evolutonary approachs suggest that separate populations of neuronal replicators in the brain may have different functions, some units operating as fitness functions for other units that act as solutions or as predictors or as perceptual agents (De Jong and Pollack 2003). In an almost evolutionary system, the function being maximized in Copycat (a program for solving analogy-based insight problems) is the activation of a conceptual network (Hofstadter and Mitchell 1995), which is effectively a neuronal network with neuromodulation (Sporns and Alexander 2002) and gating (Steriade and Pare 2007).

We return now to Aunger's claim that neuronal replicators (neuromemes) escaped the confines of the single brain and became capable of being copied between brains (Aunger 2002).

Language and Neuronal Replicators

Prions are an example of molecular *phenotypic* replicators. Prions can have alternative conformations; molecules with bad conformation (phenotypes) transform peptides with the right conformation into ones with bad conformation (Mestel 1996). There is a direct phenotype-to-phenotype transmission, without modular copying of constituents, which is in sharp contrast to the case of RNA, for which the phenotypes are correlated because the parental sequence is replicated. Interestingly, further evolution of this initially purely selfish system has been co-opted by yeast where it transmits a certain phenotypic trait to the read-through of all three nonsense codons (Patino et al. 1996). The sequences of prions are coded for by genes. It is intuitively clear that such molecular replicators can exist in a few alternative states only; hence they belong to the class of limited hereditary replicators.

Can there be phenotypic replicators with unlimited heredity? Memes are proposed within this category (Dawkins 1976), although the fact that they are typically phenotypic replicators was recognized only recently (Maynard Smith and Szathmáry 1999). Consider, for example, the neuronal correlates of an understanding of Newton's Second Law. When teachers teach it to their students, there is no copying involved whatsoever. Copying would require the transmission of the synaptic configuration of the neural network storing the piece of information in question. There are reasons to believe that such a copying would produce no

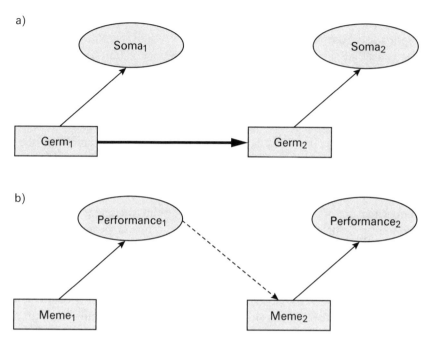

Figure 9.12
Memes and Lamarckian inheritance. (a) The Weissmanist segregation of soma and germ line. (b) Transfer of memes passes through the performance level, which is mostly absent in the molecular world (Szathmáry 2002).

meaningful result. Instead, the emerging hypotheses in the students are tested according to performance (phenotype), until performance in students and teacher is sufficiently similar (figure 9.12).

Why are we able to sustain an indefinitely large number of memes? We think the answer is human language. Language is a cultural inheritance system with indefinitely large semantic coverage (Maynard Smith and Szathmáry 1995). It is also digital, since an indefinitely large number of sentences can be generated with a limited alphabet. Although it is true that alphabets were superimposed on languages long after their invention, the number of basic phonemes we use in any language is a finite set. We construct words using this set. Even the number of words does not exceed 10^5.

It is relevant to ask whether the neuronal structures underlying language are true replicators or not. Given what we wrote above about neuronal replicators, we believe the answer is affirmative. Memes are genotypic replicators inside brains and phenotypic replicators between brains.

Accepting that memes within brains are also digital replicators, whose replication from brain to brain is phenotypic, could one think of a didactic molecular analogy? Here is one (Szathmáry 2000). There are two individuals, A and B. Take protein X from individual A. Suppose you want to enable individual B to develop a molecule with the same phenotypic effect (enzymatic function, for example). If gene transfer from A to B is not allowed, one then must have (1) some generative mechanism for proteins in B, and (2) some method for the assessment of phenotype. This comes very close to an immune system in B. The crucial difference is that the task now is to produce "antibody" Y, in individual B, that shares crucial phenotypic properties with "antigen" X, from individual A. Although both molecules would have sequences, it is most unlikely that they would be close to one another in protein space (cf. Maynard Smith 1970). In all probability the pleiotropic effects of the two proteins would differ. This is why cultural heredity is bound to be inexact and why cultural evolution is faster than biological evolution. By analogy, the transmission of language may involve the phenotypic transmission of linguistic constructions, inferred by a process of natural selection within the brain of the receiver.

Indeed, there is evidence that compositionality in language may be an adaptation for phenotypic transmissibility. Henry Brighton in his Ph.D. thesis (2003) describes an evolutionary model that was intended to demonstrate that languages were constrained by the fact that they had to be copyable. The basic task involves a conceptual space of a "speaker" modeled as a 2-D surface, onto which 100 randomly positioned points are assigned. Each point is classified into one of five classes. This constitutes the initial "concept" space. Then, in series, a further 100 unclassified points are presented to the speaker. The speaker classifies each point based on the class of the already existing point that is closest to the new point. The "listener" has access *only* to the 100 newly classified points, which are painted onto the listener's 2-D space. The listener then becomes the speaker, and the process iterates. The question is, how well can the pattern of classes over the 2-D surface be transmitted?

Because the sampling of the class space is stochastic, because the class distribution of learners is biased, and because new classes cannot be reintroduced once lost, eventually, after many generations, only one class of points exists, having taken over the whole surface. Heredity of more than one class type is not possible with this algorithm. The system has reached one of five global asymptotic stable points. How can the system be modified to allow many classes to be stably transmitted?

An extension of the above experiment was devised to test how a *mapping* from one 2-D space (M) to another 2-D space (S) could be inherited. The first speaker starts by being given a random one-to-one map from 100 points in M to 100 points in S. As before, 100 new points in M are presented to the speaker. The speaker decides what to say by finding the three closest points in M to the new point, which leads him to the corresponding three points in S. The speaker then uses some interpolation of the positions of S_1, S_2, and S_3 to produce a new point, s. It is these 100 new s points that are heard by the listener, who uses them to construct her set of 100 points in M, and the process iterates.

Brighton defines compositionality as the degree of correlation between distance between points in M and the distances between the corresponding points in S. A compositional map is a topographic map with $c = 1$, and a random M $\rightarrow$ S map has $c = 0$. For cases with the system started with a random map and with a fully compositional map, the above algorithm could sustain only weak compositionality, $c = 0.3$, with agents often having different M $\rightarrow$ S maps between generations. Thus a mapping could not be stably inherited using the above algorithm. This was shown to be because there was no error-correcting mechanism acting to counter small rounding errors in the interpolation calculation.

To solve this problem, Brighton introduced the *obverter procedure*, originally also devised to study the evolution of communication. The speaker uses "introspection," that is, when trying to produce a signal to represent a point in M, the speaker works out which signal, if received by itself, would maximize the probability of its inferring the correct point in M. The speaker is thus using himself as a model of the listener. An alternative method could be to use other techniques to infer the M $\rightarrow$ S mapping of the listener. By doing this (and by using new points to improve the classification of subsequent new points, "production memory"), a highly compositional ($c = 0.92$) *and* highly stable mapping was achieved.

What is the relevance of these dynamics to phenotypic copying? A simple "autistic" phenotypic transmission algorithm may be insufficient; a predictive mechanism may be necessary to ensure that stable and unlimited M $\rightarrow$ S maps are copied accurately between brains.

Conclusion

In this chapter we have ventured into the domain of bona fide neuronal replication, in the hope of extending the synthesis even further in breadth.

We are happy to point out that while doing so, we are standing on the shoulders of Maynard Smith, Changeux, and Edelman. Without the influence of these scholars we would have been unlikely to arrive even at an initial research program on neuronal evolution proper. We also hope, at least in the long run, to be able to contribute to the depth of the theory with our research program. From a general point of view it is likely that this program will shed new light on evolvability, exploration distributions in evolution, and the interplay of Lamarckian and Darwinian mechanisms, but strictly within the constraints of genetically based evolution of the component brain mechanisms. We do not know what all this will deliver. Maybe the particular neuroevolutionary mechanisms presented in this chapter will not survive, but the general research program as such will, and in the future some neurobiologists will also be evolutionary biologists at the same time.

The replication mechanisms are proposed to exist at the neuronal level and not at the level of thought. The existence of noise is a major constraint to neuronal copying of all kinds. Just as in the study of the origin of life, it is a nontrivial question to understand how unlimited heredity could have arisen in the brain when noise is taken into account (Szathmáry and Maynard Smith 1997). However, before the question of origins is addressed, it remains to be seen whether empirical evidence can be found for topological or dynamical neuronal replicators.

Acknowledgments

Thanks to a Career Development Fellowship at the MRC National Institute for Medical Research, Mill Hill, London, and to a Marie Curie Inter-European Grant to work at Collegium Budapest, Hungary. Partial support of this work has generously been provided by the Hungarian National Office for Research and Technology (NAP 2005/KCKHA005) and by the Regional Knowledge Centre of Eötvös University (National Office for Research and Technology grant no. RET2.4/2005). The work has also been supported by the COST Action CM0703 on systems chemistry. Thanks to Luc Steels, Richard Watson, Richard Goldstein, James McDonald, Anil Seth, and Zoltán Szatmáry for useful discussions.

Notes

1. Dawkins (1971) also proposed neuronal elimination as a basis of memory.
2. (http://www.scholarpedia.org/article/Synfire_chain).

3. An interesting parallel development is von Kiedrowski's (Von Kiedrowski et al. 2003) "connectivity copying" mechanism whereby selectable nano-robots can be constructed in the chemical domain.

4. Introducing Hebbian learning to the copy operator in a manner analogous to the previous approach with the HIFF problem allows evolution to structure the exploration distribution of variants based on the structure of solutions found in previously explored optima. A 660×660 Hebbian copying matrix is updated to represent the within genotype correlations after an optimum has been reached (a genotype is 660 units in length) and is used to bias the copy operation in subsequent evolution. However, since the system can find a global optimum without Hebbian learning, this system is largely redundant (see figure 9.12).

References

Abeles M (1982) Local Cortical Circuits: An Electrophysiological Study. Berlin: Springer.

Abeles M (1991) Corticonics: Neural Circuits of the Cerebral Cortex. New York: Cambridge University Press.

Abeles M, Gat I (2001) Detecting precise firing sequences in experimental data. Journal of Neuroscience Methods 107: 141–154.

Abraham WC, Robins A (2005) Memory retention—the synaptic stability versus plasticity dilemma. Trends in Neurosciences 28: 73–78.

Adams P (1998) Hebb and Darwin. Journal of Theoretical Biology 195: 419–438.

Adams PR, Cox KJA (2002) A new interpretation of thalamocortical circuitry. Philosophical Transactions of the Royal Society of London B357: 1767–1779.

Ashby WR (1960) Design for a Brain: The Origins of Adaptive Behavior. New York: Wiley.

Aunger R (2002) The Electric Meme: A New Theory of How We Think. New York: The Free Press.

Baddeley AD, Hitch GJ (1974) Working memory. In The Psychology of Learning and Motivation: Advances in Research and Theory. GH Bower, ed.: 8: 47–89. New York: Academic Press.

Bag BG, Von Kiedrowski G (1996) Templates, autocatalysis and molecular replication. Pure and Applied Chemistry 68: 2145–2152.

Baldwin JM (1898) On selective thinking. The Psychological Review 5: 4.

Baldwin JM (1909) The influence of Darwin on theory of knowledge and philosophy. Psychological Review 16: 207–218.

Barandiaran X, Ruiz-Mirazo K (2008) Introduction to the special issue: Modelling autonomy. Biosystems 91: 295–304.

Bedau MA (1998) Four puzzles about life. Artificial Life 4: 125–140.

Belman RE (1957) Dynamic Programming. Princeton, NJ: Princeton University Press.

Beyer H-G (2001) The Theory of Evolution Strategies. Berlin: Springer.

Bialek W, Nemenman I, Tishby N (2001) Predictability, complexity and learning. Neural Computation 13: 2409–2463.

Bosman RJC, Van Leeuwen WA, Wemmerhove B (2004) Combining Hebbian and reinforcement learning in a minibrain model. Neural Networks 17: 29–36.

Boyd R, Richerson PJ (2005) The Origin and Evolution of Cultures. Oxford: Oxford University Press.

Brighton H (2003) Simplicity as a driving force in linguistic evolution. Ph.D. thesis, University of Edinburgh.

Brooks RA (1990) Elephants don't play chess. Robotics and Autonomous Systems 6: 3–15.

Calvin WH (1996) The Cerebral Code: Thinking a Thought in the Mosaics of the Mind. Cambridge, MA: MIT Press.

Campbell DT (1974) Evolutionary epistemology. In The Philosophy of Karl Popper. PA Schillpp, ed.: 412–463. LaSalle, IL: Open Court.

Changeux J-P (1985) Neuronal Man: The Biology of Mind. Princeton, NJ: Princeton University Press.

Changeux J-P, Courrège P, Danchan A (1973) A theory of the epigenesis of neuronal networks by selective stabilization of synapses. Proceedings of the National Academy of Sciences of the USA 70: 2974–2978.

Chialvo DR, Bak P (1990) Learning from mistakes. Neuroscience 90: 1137–1148.

Chronicle EP, MacGregor JN, Ormerod TC (2004) What makes an insight problem? The roles of heuristics, goal conception, and solution recording in knowledge-lean problems. Journal of Experimental Psychology: Learning, Memory, and Cognition 30: 14–27.

Cooper WS (2001) The Evolution of Reason: Logic as a Branch of Biology. Cambridge: Cambridge University Press.

Craik K (1943) The Nature of Explanation. Cambridge: Cambridge University Press.

Crick F, Mitchison G (1995) REM sleep and neural nets. Behavioral and Brain Research 69: 147–155.

Crick FHC (1989) Neural Edelmanism. Trends in Neurosciences 12: 240–248.

Crick FHC (1990) Reply. Trends in Neurosciences 13: 13–14.

Dawkins R (1971) Selective neurone death as a possible memory mechanism. Nature 229: 118–119.

Dawkins R (1976) The Selfish Gene. Oxford: Oxford University Press.

Dawkins R (1982) The Extended Phenotype: The Gene as the Unit of Selection. Oxford: W.H. Freeman.

De Jong ED, Pollack JB (2003) Learning the ideal evaluation function. Genetic and Evolutionary Competition 2003: 203. No. 2723 of LNCS.

Dehaene S, Changeux J-P (1991) The Wisconsin Card Sorting Test: Theoretical analysis and modeling in a neuronal network. Cerebral Cortex 1: 62–79.

Dehaene S, Changeux J-P (1997) A hierarchical neuronal network for planning behavior. Proceedings of the National Academy of Sciences of the USA 94: 13293–13298.

Dehaene S, Kerszberg M, Changeux J-P (1998) A neuronal model of a global workspace in effortful cognitive tasks. Proceedings of the National Academy of Sciences of the USA 95: 14529–14534.

Dennett DC (1981) Brainstorms. Cambridge, MA: MIT Press.

Dennett DC (1995) Darwin's Dangerous Idea. New York: Simon & Schuster.

Der R, Güttler F, Ay N (2008) Predictive information and emergent cooperativity in a chain of mobile robots. In Proceedings of Alife XI: 166–172. Cambridge, MA: MIT Press.

Desimone R, Duncan J (1995) Neural mechanisms of selective visual attention. Annual Review of Neurosciences 18: 193–222.

Edelman GM (1987) Neural Darwinism: The Theory of Neuronal Group Selection. New York: Basic Books.

Edelman GM (1990) The Remembered Present: A Biological Theory of Consciousness. New York: Basic Books.

Edelman GM (1994) The evolution of somatic selection: The antibody tale. Genetics 138: 975–981.

Eigen M (1971) Selforganization of matter and the evolution of biological macromolecules. Naturwissenschaften 58: 465–523.

Fernando C, Rowe J (2007) Natural selection in chemical evolution. Journal of Theoretical Biology 247: 152–167.

Fernando C, Rowe J (2008) The origin of autonomous agents by natural selection. Biosystems 91: 355–373.

Fernando C, Santos M, Szathmáry E (2005). Evolutionary potential and requirements for minimal protocells. Topics in Current Chemistry 259: 167–211.

Fernando C, Karishma, KK, Szathmáry E (2008) Copying of neuronal topology by spike-time dependent plasticity and error-correction. PLoSONE 3: e3775.

Fernando C, Szathmáry E. (2009) Natural selection in the brain. In Towards a Theory of Thinking: Building Blocks for a Conceptual Framework. B Glatzeder, V Goel, A von Müller (Eds). New York: Springer.

Fernando C, Von Kiedrowski G, Szathmáry E (2007) A stochastic model of nonenzymatic nucleic acid replication: "Elongators" sequester replicators. Journal of Molecular Evolution 64: 572–585.

Fisher RA (1930) The Genetical Theory of Natural Selection. Oxford: Clarendon Press.

Fodor JA (1983) The Modularity of Mind. Cambridge, MA: MIT Press.

Friston KJ, Stephan KE (2007) Free-energy and the brain. Synthese 159: 417–458.

Frith CD (2000) The role of dorsolateral prefrontal cortex in the selection of action, as revealed by functional imaging. In Control of Cognitive Processes: Attention and Performance. S Monsell, J Driver, eds. Cambridge, MA: MIT Press.

Ganti T (1971) The Principles of Life (in Hungarian). Budapest: Gondolat. (English translation 1986. Budapest. OMIKK)

Ganti T (2003) The Principles of Life. Oxford: Oxford University Press.

Gisiger T, Dehaene S, Changeux J-P (2000) Computational models of association cortex. Current Opinion in Neurobiology 10: 250–259.

Gopnik A, Schulz L (2004). Mechanisms of theory formation in young children. Trends in Cognitive Sciences 8: 371–377.

Gould SJ (2002) The Structure of Evolutionary Theory. Cambridge, MA: Belknap Press of Harvard University Press.

Grush R (2004) The emulation theory of representation: Motor control, imagery, and percption. Behavioral and Brain Sciences 27: 377–442.

Hayden EJ, Von Kiedrowski G, Lehman N (2008) Systems chemistry on ribozyme self-construction: Evidence for anabolic autocatalysis in a recombination network. Angewandte Chemie (international ed.) 47: 8424–8428.

Hebb DO (1949) The Organization of Behavior: A Neuropsychological Theory. New York: John Wiley.

Hofstadter DR, Mitchell M (1995) The Copycat Project: A model of mental fluidity and analogy-making. In Fluid Concepts and Creative Analogies: Computer Models of the Fundamental Mechanisms of Thought. DR Hofstadter and the Fluid Analogies Research Group, eds.: 205–267. New York: Basic Books.

Horvitz J-C (2000) Mesolimbocortical and nigrostriatal dopamine responses to salient non-reward events. Neuroscience 96: 651–656.

Izhikevich EM (2006) Polychronization: Computation with spikes. Neural Computation 18: 245–282.

Izhikevich EM (2007a) Dynamical Systems in Neuroscience: The Geometry of Excitability and Bursting. Cambridge, MA: MIT Press.

Izhikevich EM (2007b) Solving the Distal Reward Problem through linkage of STDP and dopamine signaling. Cerebral Cortex 17: 2443–2452.

Izhikevich EM, Gally JA, Edelman GM (2004) Spike-timing dynamics of neuronal groups. Cerebral Cortex 14: 933–944.

Jablonka E, Lamb MJ (2005) Evolution in Four Dimensions: Genetic, Epigenetic, Behavioral, and Symbolic Variation in the History of Life. Cambridge, MA: MIT Press.

James W (1890 [1950]). The Principles of Psychology. New York: Dover.

Kaplan C, Simon HA (1990). In search of insight. Cognitive Psychology 22: 374–419.

Kindermann M, Stahl I, Reimold M, Pankau WM, Von Kiedrowski G (2005) Systems chemistry: Kinetic and computational analysis of a nearly exponential organic replicator. Angewandte Chemie (international ed.) 44: 6750–6755.

Kirschner M, Gerhart J(1998) Evolvability. Proceedings of the National Academy of Sciences of the USA 95: 8420–8427.

Knoblich G, Öllinger M, Spivey MJ (2005) Tracking the eyes to obtain insight into insight problem solving. In Cognitive Processes in Eye Guidance. G Underwood, ed. Oxford: Oxford University Press.

Klyubin AS, Polani D, Nehaniv CL (2007) Representations of space and time in the maximization of information flow in the perception-action loop. Neural Computation 19: 2387–2432.

Könnyű B, Czárán T, Szathmáry E (2008) Prebiotic replicase evolution in a surface-bound metabolic system: Parasites as a source of adaptive evolution. BMC Evolutionary Biology 8: 267.

Krichmer JL, Edelman GM (2005) Brainbased devices for the study of nervous systems and the development of intelligent machines. Artificial Life 11: 63–77.

Kun Á, Papp B, Szathmáry E (2008) Computational identification of obligatorily autocatalytic replicators embedded in metabolic networks. Genome Biology 9: R51.

Lathe R (2003) Fast tidal cycling and the origin of life. Icarus 168: 18–22.

Lieberman E, Hauert C, Nowak W (2005) Evolutionary dynamics on graphs. Nature 433: 312–316.

Ludlow RF, Otto S (2008) Systems chemistry. Chemical Society Reviews 37: 101–108.

Lungarella M, Sprons O (2005) Information self-structuring: Key principle for learning and development. In Proceedings of the 14th IEEE International Conference on Development and Learning: 25–30.

MacGregor JM, Ormerod TC, Chronicle EP (2001) Information-processing and insight: A process model of performance on the nine-dot and related problems. Journal of Experimental Psychology: Learning, Memory, and Cognition 27: 176–201.

Mach E (1897) Contributions to the Analysis of Sensations. Chicago: Open Court.

Mach E (1910) Popular Scientific Lectures. 4th edition. Chicago: Open Court.

Markram H, Lubke J, Frotscher M, Sakmann B (1997) Regulation of synaptic efficacy by coincidence of postsynaptic APs and EPSPs. Science 275: 213–215.

Marr D (1969) A theory of cerebellar cortex. Journal of Physiology 202: 437–470.

Marr D (1982) Vision: A Computational Investigation into the Human Representation and Processing of Visual Information. San Francisco: W.H. Freeman.

Maynard Smith J (1970) Natural selection and the concept of a protein space. Nature 225: 563–564.

Maynard Smith J (1986) The Problems of Biology. Oxford: Oxford University Press.

Maynard Smith J, Szathmáry E (1995) The Major Transitions in Evolution. Oxford: Oxford University Press.

Maynard Smith J, Szathmáry E (1999) The Origins of Life. Oxford: Oxford University Press.

Mestel R (1996) Putting prions to the test. Science 273: 184–189.

Metcalfe J, Wiebe D (1987) Intuition and insight and noninsight problem solving. Memory and Cognition 15: 238–246.

Michod RE (1988) Darwinian selection in the brain. Evolution 43: 694–696.

Monod J (1971) Chance and Necessity: An Essay on the Natural Philosophy of Modern Biology. New York: Knopf.

Müller GB (2007) Evo-devo: Extending the evolutionary synthesis. Nature Reviews Genetics 8: 943–949.

Muller HJ (1966) The gene material as the initiator and the organizing basis of life. American Naturalist 100: 493–517.

Nadel L, Campbell J, Ryan L (2007) Autobiographical memory retrieval and hippocampal activation as a function of repetition and the passage of time. Neural Plasticity: 90472.

Nadel L, Moscovitch M (1997) Memory consolidation, retrograde amnesia and the hippocampal complex. Current Opinion in Neurobiology 7: 217–227.

Nadel L, Samsonovich A, Ryan L, Moscovitch M (2000) Multiple trace theory of human memory: Computational, neuroimaging, and neuropsychological results. Hippocampus 10: 352–368.

Newell A, Simon HA (1976) Computer science as empirical inquiry: Symbols and search. Communications of the Association for Computing Machinery 19: 113–126.

Oja E (1982) Simplified neuron model as a principal component analyzer. Journal of Mathematical Biology 15: 267–273.

Okasha S (2006) Evolution and the Levels of Selection. Oxford: Oxford University Press.

Öllinger M, Jones G, Knoblich G (2006) Heuristics and representational change in two-move matchstick arithmatic tasks. Advances in Cognitive Psychology 2: 239–253.

Oudeyer P-Y, Kaplan F, Hafner VV (2007) Intrinsic motivation systems for autonomous mental development. IEEE Transactions on Evolutionary Computation 11: 265–286.

Patino MM, Liu J-J, Glover JR, Lindquist S (1996) Support for the prion hypothesis for the inheritance of a phenotypic trait in yeast. Science 273: 622–626.

Perkins DN (1995) Insight in minds and genes. In The Nature of Insight. RJ Sternberg, JE Davidson, eds.: 495–533. Cambridge, MA: MIT Press.

Pigliucci M (2008) Is evolvability evolvable? Nature Reviews Genetics 9: 75–82.

Popper KR (1972) Objective Knowledge: An Evolutionary Approach. Oxford: Oxford University Press.

Pylyshyn ZW, ed. (1987) The Robot's Dilemma: The Frame Problem in Artificial Intelligence. Norwood, NJ: Ablex.

Reverberi C, Toraldo A, D'Agostino S, Skrap M (2005) Better without (lateral) frontal cortex? Insight problems solved by frontal patients. Brain 128: 2882–2890.

Richards RJ (1977) The natural selection model of conceptual evolution. Philosophy of Science 44: 494–501.

Rougier NP, Noelle DC, Braver TS, Cohen JD, O'Reilly RC (2005) Prefrontal cortex and flexible cognitive control: Rules without symbols. Proceedings of the National Academy of Sciences of the USA 102: 7338–7343.

Seung SH (2003) Learning in spiking neural networks by reinforcement of stochastic synaptic transmission. Neuron 40: 1063–1073.

Sievers D, Von Kiedrowski G (1994) Self-replication of complementary nucleotide-based oligomers. Nature 369: 221–224.

Simon HA, Reed SK (1976) Modeling strategy shifts in a problem-solving task. Cognitive Psychology 8: 86–97.

Simonton DK (1995) Foresight in insight? A Darwinian answer. In The Nature of Insight. RJ Sternberg, JE Davidson, eds. Cambridge, MA: MIT Press.

Song S, Abbott L (2001) Cortical development and remapping through spike timing-dependent plasticity. Neuron 32: 339–350.

Sporns O, Alexander, WH (2002) Neuromodulation and plasticity in an autonomous robot. Neural Networks 15: 761–774.

Steriade M, Pare D (2007) Gating in Neuronal Networks. Cambridge: Cambridge University Press.

Sternberg RJ, Davidson JE, eds. (1995) The Nature of Insight. Cambridge, MA: MIT Press.

Sutton SR, Barto AG (1998) Reinforcement Learning: An Introduction. Cambridge, MA: MIT Press.

Szathmáry, E. (2000) The evolution of replicators. Philosophical Transactions of the Royal Society of London B355: 1669–1676.

Szathmáry E (2002) Units of evolution and units of life. In Fundamentals of Life. G Pályi, L Zucchi, L Caglioti, eds.: 181–195. Paris: Elsevier.

Szathmáry E (2005) Life: in search of the simplest cell. Nature 433: 469–470.

Szathmáry E (2006) The origin of replicators and reproducers. Philosophical Transactions of the Royal Society of London B361: 1761–1776.

Szathmáry E (2007) Coevolution of metabolic networks and membranes: The scenario of progressive sequestration. Philosophical Transactions of the Royal Society of London B362: 1781–1787.

Szathmary E, Maynard Smith J (1997) From replicators to reproducers: The first major transitions leading to life. Journal of Theoretical Biology 187: 555–571.

Szostak JW, Bartel DP, Luisi PL (2001) Synthesizing life. Nature 409: 387–390.

Toussaint M (2003) The evolution of genetic representations and modular adaptation. Ph.D. thesis, Institut für Neuroinformatik, Ruhr-Universität, Bochum, Germany.

Van Rossum MCW, Bi GQ, Turrigiano GG (2000) Stable Hebbian learning from spike-timing-dependent plasticity. Journal of Neuroscience 20: 8812–8821.

Von Kiedrowski G (1986) A self-replicating hexadeoxy nucleotide. Angewandte Chemie (international. ed.) 25: 932–935.

Von Kiedrowski G (1999) Molekulare Prinzipien der artifiziellen Selbstreplikation. In Gene, Neurone, Qubits & Co.: Unsere Welten der Information. D Ganten, ed.: Stuttgart: S. HirzelVerlag.

Von Kiedrowski G (2001) Chemistry: A way to the roots of biology. ChemBioChem. 2: 597–598.

Von Kiedrowski G, Eckhardt L, Naumann K, Pankau WM, Reinhold M, Rein M (2003) Toward replicatable, multifunctional, nanoscaffolded machines: A chemical manifesto. Pure and Applied Chemistry 75: 609–619.

Wächtershäuser G (1988) Before enzymes and templates: Theory of surface metabolism. Microbiological Reviews 52: 452–484.

Wächtershäuser G (1992) Groundworks for an evolutionary biochemistry: The iron^sulfur world. Progress in Biophysics and Molecular Biology 58: 85–201.

Wagner GP, Altenberg L (1996) Complex adaptations and evolution of evolvability. Evolution 50: 967–976

Wagner U, Gais S, Haider H, Verleger R, Born J (2004) Sleep inspires insight. Nature 427: 352–355.

Wang X-J (1999) Synaptic basis of cortical persistent activity: The importance of NMDA receptors to working memory. Journal of Neuroscience 19: 9587–9603.

Watson RA (2006) Compositional Evolution: The Impact of Sex, Symbiosis, and Modularity on the Gradualist Framework of Evolution. Cambridge, MA: MIT Press.

Watson RA, Buckley CL, Mills R (2009) The Effect of Hebbian Learning on Optimisation in Hopfield Networks. Technical Report, ECS, University of Southampton.

Watson RA, Hornby GS, Pollack JB (1998) Modelling building-block interdependency. In Parallel Problem Solving from Nature, no. 1498 of LNCS. A Eiben, T Back, M Schoenauer, HP Schwefel, eds.: 97–106. Berlin: Springer.

Weir AAS, Chappell J, Kacelnik A (2002) Shaping of hooks in New Caledonian crows. Science 297: 981.

Willshaw D, Von der Malsburg C (1976) How patterned neural connections can be set up by self-organisation. Proceedings of the Royal Society of London B194: 431–445.

Young JM, Waleszczyk WJ, Wang C, Calford MB, Dreher B, Obermayer K (2007) Cortical reorganization consistent with spike timing—but not correlation-dependent plasticity. Nature Neuroscience 10: 887–895.

Young JZ (1979) Learning as a process of selection and amplification. Journal of the Royal Society of Medicine 72: 801–814.

Zipser D, Kehoe B, Littlewort G, Fuster GM (1993) A spiking network model of short-term active memory. Journal of Neuroscience 13: 3406–3420.

V EVOLUTIONARY DEVELOPMENTAL BIOLOGY

10 Facilitated Variation

Marc W. Kirschner and John C. Gerhart

Before the rediscovery of Mendel's laws at the turn of the twentieth century, there was confusion about the source of variation upon which natural selection was thought to act, and about the means by which this variation could be inherited. By the mid-twentieth century much of that confusion had dissipated and a consensus view of evolution, sometimes called the Modern Synthesis, had incorporated population genetics, selection, and chromosomal inheritance into a robust model of evolution. What was still missing were the cellular and molecular mechanisms underlying the generation of the phenotype, particularly the anatomy, physiology, and behavior of multicellular organisms. How relevant such mechanistic understandings were to the theory of evolution was then unclear. Though it is not impossible that molecular discoveries could be both intrinsically fascinating and irrelevant to understanding evolution, we come to the opposite conclusion, that incorporation of these recent cellular and developmental understandings is indispensable for a satisfactory theory of evolution. In our view, evolution as a theory would be incomplete without understanding the nature of phenotypic variation. Such a need was foreshadowed by Sewall Wright, who wrote in 1931:

The evolution of complex organisms rests on the attainment of gene combinations which determine a varied repertoire of adaptive cell responses in relation to external conditions. The older writers on evolution were often staggered by the seeming necessity of accounting for the evolution of fine details, for example, the fine structure of all the bones. From the view that structure is never inherited as such, but merely types of adaptive cell behavior which lead to particular types of structure under particular conditions, the difficulty to a considerable extent disappears. (Wright 1931: 147)

Although the nature of "adaptive cell behavior" could not have been anticipated by Sewall Wright, recent understandings have shown us a great deal about how cells and organisms produce variation both in response to the environment and in response to genetic change. The discoveries since the 1970s have revealed a view of the nature of phenotypic variation different from the one assumed by evolutionary biologists in the 1950s and 1960s (Mayr 1963). As we shall argue, these new understandings require a reappraisal of the contention that natural selection is the sole creative force in evolution. Novelty arises from an interplay between the properties of the organism and mutation under selection. The nature of the developmental and cellular circuits contributes a great deal to the kinds of variation that selection can act upon. We have summarized these ideas recently, in a theory of facilitated variation (Kirschner and Gerhart 2005, Gerhart and Kirschner, 2007). We shall explore both the origins of the new understandings and the understandings themselves, and their implications for evolutionary theory.

The Problem of Phenotypic Variation

Variation is indispensable to evolution: individuals in a population of organisms inevitably vary slightly in phenotype. Some of these variations are heritable. In the environmental conditions met by the population, some variant individuals reproduce better than others, due to their heritable phenotypic difference. They are selected, and their offspring carry the genetic, and hence the phenotypic, difference. In the broad view of evolution, all living organisms are related by descent from a first ancestor, and over the eons have diverged to diverse forms and adaptive functions by the gradual accumulation of selected phenotypic variations. This picture of evolution is widely accepted; most of these propositions go back to Darwin himself. The most important addition, contributed by the Modern Synthesis in the mid-twentieth century, was the insight that heritable phenotypic change requires genetic change, and that genetic change is random with respect to targets and the environment. Selection was established as central, and genetic change was established as the underlying cause of heritable variation.

Today, genetic change is well understood in light of our knowledge of DNA structure, replication, rearrangements, damage and repair, and the comparison of genome sequences. Genetic variation is interesting because it is at the start of the causal chain to new phenotypes, because it is random, because it is easily tracked by mating experiments and sequencing, and because it is easily expressible in mathematical terms. Understanding that genetic change generates phenotypic change, and that the latter is selected upon, favorably or unfavorably, gives a self-consistent and satisfying picture of evolution. Since the selected pheno-type always carries its specific genetic alteration, evolution can be viewed, in the terms of population genetics, as a change of allele frequency in populations over time. From equations with few variables the rate of spread of the alleles and the size of the selection coefficients can be derived or predicted.

Phenotypic variation, on the other hand, seems to defy such abstrac-tion. In the standard mathematical approaches to evolution, phenotypic variation is assumed to be well behaved in coupling genotype and selec-tion. But its role in evolution appears not to be limiting if indeed, as is often assumed, it is small in its extent of change, it is copious in its amount, and it is isotropic in direction—that is, of many, perhaps all, kinds, yet with no relation to the selective environment (Gould 2002). Selection by the environment, then, is the creative force. Genetic change is both the measure and the cause of phenotypic variation, which per se need not explicitly appear in the analysis. Yet the nature of phenotypic variation in response to genetic variation, also described by some as a product of the so-called genotype-phenotype map, has been the major preoccupation of developmental biologists, cell biologists, and biochem-ists for the last decades. Can we now say something general about the consequences of mutation for phenotypic variation?

We first confront why evolution explained solely by a quantitative model that omits the details of phenotypic variation is a limited view. Such a model does not account for how genetic change leads to pheno-typic change; it just assumes that it does. Yet, with this omission, nothing can be said about the kind and amount of phenotypic variation, just that there is more than enough variety of kinds of variation. From this abstraction comes no glimpse of how a hand or an eye, much less a horse or a zebra, was actually formed in evolution. When evolution is viewed across its broad scope, the individual events of selection and variation pale, whereas the epic story of descent with modification stands out, with its vast range of anatomical and physiological diversification. Then other questions may suggest themselves: What was modified, and by how much, for example, when animals first walked on land? In novel traits, are some aspects new and some old, or is everything changed a little? Are many genetic changes needed for a modification of the phenotype, or only a few? These questions beg concrete answers that cannot come from the existing mathematical analysis. It is of course possible that no generality will emerge about phenotypic variations; every case may be different. But it is also possible that powerful generalities about phenotypic variation can now be deduced from our modern understanding of organisms, especially their development and physiology, which will mean it is no longer necessary to relegate phenotypic variation to the background of evolutionary theory.

Recent molecular insights to be discussed here indicate that the organism itself plays a large role in creating the conditions for, and facilitates strongly, the generation of nonlethal and selectable phenotypic variation. The assumption of the Modern Synthesis shared by many evolutionary biologists today is that variation is nonlimiting, small in extent of change around the mean, copious in amount, and isotropic. This assumption has not been borne out by modern biological observations. Instead, over time, evolution seems to have crafted the kind of variation that has emerged from developmental processes. We present a set of concepts surrounding the generation of variation that we have unified in our theory of facilitated variation, intended to explain the connections between genotypic and phenotypic variation (Kirschner and Gerhart 2005). The theory has three major parts.

To begin with, we build on the unexpected deep conservation of processes at the cellular level. A group of "conserved core processes" is used over and over in the generation and operation of the phenotype. We argue that these conserved core processes have features that highly deconstrain the generation of phenotypic variation by regulatory changes, although changes in the processes themselves are severely selected against.

We then discuss some general properties of biological systems that deconstrain phenotypic change.

And finally, we argue that physiological variation, particularly the variation used in embryonic development, is an important substrate for evolutionary change.

Premolecular Thinking about Phenotypic Variation

There have been several attempts to understand phenotypic change, both within and in opposition to Darwin's ideas. To his lasting discredit, Lamarck is remembered as giving the organism an all-encompassing role in the generation of variation. The individual organism underwent anatomical and physiological changes in response to the environment, and these acquired characteristics were inherited. Darwin first considered variation to be random and small with respect to selective conditions, but later moved toward Lamarck in thinking the organism responds to conditions and furnishes the gametes with information enhancing the response. However, environment-specific variation would, in theory, render selection much less important. Weismann later refuted the inheritance of acquired characteristics by showing that the *germ line* conducted inheritance, whereas the *soma* developed the traits on which selection could act, leading to the genotype-phenotype distinction. Until recently there was no known mechanism by which physiological change could directly change the genome. However, the discovery that stable marks on chromatin can be inherited for some period of time has opened the door to direct effects of experience on the genome of germ cells (Jablonka and Lamb 2005). This mechanism was denied by biologists for a century, but the possibility has recently reemerged. Whether this could ever occur in a Lamarckian way, directing the environmental perturbation to a specific adaptive circuit, is still far from clear.

In a 30-year period of confusion after Darwin, various evolutionists made variation the major factor in evolutionary change, some even dismissing selection. Proponents of orthogenesis saw evolution as a slow playing out of an intrinsic development of the organism, with variations coming systematically from within (Whitman 1919). In the narrow form of Haeckel's biogenetic law, new phenotypes were added to the previous adult stage, as if the organism had an internally defined way of making changes (Haeckel 1866). Finally, some early geneticists thought that macromutations, those of large effect, underlay species change (De Vries 1909).

The Modern Synthesis dispelled many of these ideas by combining Darwin's original hypothesis with the new insights of transmission genetics, population genetics, and paleontology, eliminating macromutation, directed mutation, and acquired characteristics. Selection was restored to its central place. Developmental biology was not needed, and in all fairness, not much was then known about it. Within the Modern Synthesis, genetic change was required for a heritable phenotypic change, and selection over time would shape the chaotic profusion of small phenotypic variations into a phenotype favored by (adapted to) the selective conditions (Gould 2002).

Running parallel to the Modern Synthesis was a less well-known lineage of ideas about phenotypic variation, now very relevant to our molecular understandings. Baldwin (1896, 1902) reconciled aspects of Lamarck's and Darwin's ideas in what is now called the Baldwin effect. Animals, he noted, have a great capacity to make physiological and behavioral adjustments to short-term changes of the environment, and these reversible adjustments provide for marginal survival under long-term harsh conditions, allowing time for the emergence of new heritable traits. In this view, physiological adaptability can play a role in facilitating evolutionary change. Schmalhausen, writing in the 1940s, carried these ideas further (Schmalhausen 1986). He drew attention to the range of phenotypes an organism can generate when exposed to a range of environmental conditions. In his account, almost all that was new in a seemingly new trait was already available in the organism and evocable by the environment, even without genetic change, and was merely stabilized and enhanced by genetic change. Waddington, independently in the 1940s and 1950s, developed similar ideas, under the name of "genetic assimilation," for the stabilization of physiological adaptations by genetic change under selective conditions. He produced substantial phenotypic changes in *Drosophila* by environmental treatments (ether, heat, etc.), and then through selection obtained organisms that exhibited these phenotypes without further need of the stimulus (Waddington 1953). Rutherford and Lindquist (1998) have recently carried these demonstrations further by using molecular tools to uncover latent phenotypic variation. Mary Jane West-Eberhard (2003) has developed a thorough analysis of the means by which selected phenotypic plasticity can contribute to evolutionary change.

Before the recent breakthroughs in the study of development, Dawkins (1996) and a few others called attention to the possibility of developmental mechanisms that might change easily and spawn the radiations of animals. Key innovations had this quality; Liem (1990) drew attention to the special pharyngeal jaws of cichlids that have undergone a wide range of modifications related to the feeding specializations of the diverse species. Gould (1977) and others looked at heterochrony as evidence that development is organized in modules that can be moved in time relative to one another, giving different phenotypes. Though the modules would not change, the regulatory agents controlling timing would. And Raff (1996) and others noted that constraints on developmental mechanisms meant that organisms could not produce all varieties of phenotypic variation (Brakefield 2006).

The major conclusion about phenotypic variation that emerges from these studies is that when novelty is achieved in the course of variation and selection, the novel trait may contain rather little that is new. In most cases the components are largely unchanged, and the novelty rests on regulatory changes, such as in moving the expression of a genetic program in time, or stabilizing and enhancing what was already present as a physi-

processes and their molecular components have changed little. About two billion years ago the eukaryotes "invented" the compartmentalized cell with an extensive functioning cytoskeleton, chromosomes, a phosphorylation- and proteolysis-driven cell cycle, mitochondria, and many signaling systems. That suite of cell biological processes also remains unchanged today. Furthermore, waves of innovation in cell-cell interactions, signaling, and development accompanied the first metazoans perhaps a billion years ago. Recent analysis of the *Nematostella* (sea anemone) genome indicates that by the time of the Precambrian eumetazoan ancestor, probably 20% new genes had been added to the repertoire retained from eukaryotic single-celled ancestors (Putnam et al. 2007). New genes associated with phosphotyrosine-based signaling, along with cadherins, may have arisen as early as the ancestor shared by choanoflagellates and metazoa (King et al. 2008). Then, from Cambrian times forward, further waves of innovation accompanied the emergence of phylum-wide body plans among bilateral animals, followed by the first vertebrate fins and limbs or arthropod appendages more recently, depending on the lineage we choose to pursue. There are episodes of innovation, and then the long periods of conservation of these fundamental processes, their components, and the genes encoding them.

Conserved core processes represent the basic machinery of the multicellular organism, specified for its specific diversified functions by regulatory control. Thus the eukaryotic transcriptional machinery is a conserved core process with many substitutable components. The actin- based cytoskeleton is another, and along with actin is a suite of highly conserved actin-binding, -capping, and -severing proteins. The MAP kinase pathway, the splicing machinery, the vesicular trafficking machinery, and integrin-based signaling are all conserved core processes of eukaryotes. Conserved core processes can be grouped hierarchically in development to generate composite processes, which in certain lineages are also conserved as developmental processes. Thus aspects of the segmentation system of insects, made up of conserved core transcriptional and signaling components, are maintained in the entire arthropod lineage, and a similar assemblage of signaling and transcriptional circuits makes up the conserved core processes of the vertebrate limb bud. Laid on top of these are many specialized gene products which are neither conserved nor core, in that they are not used over and over for different functions. Roughly estimated, two-thirds of the genes of an animal's genome might encode components of the core processes (Waterston et al. 2002), and perhaps a few hundred processes operate in the animal as modules of function, each with many components. Their role is the formation and operation of the phenotype, the animal's development and physiology, and they are present in all extant animals. They can be considered to be the functional components responsible for generating the genotype-phenotype map. Studies of development have added a refinement to the appreciation of conservation, namely, different parts of an embryo use different combinations and amounts of the same processes to develop the different traits of the adult. While a few gene products may be unique to a trait, such as the specific forms of keratin in hair and skin, or ion channels in the heart, most aspects of a trait are produced by some combination of the shared core processes.

The stasis in the conserved core processes means that mutations in these components were generally deleterious and were removed by purifying selection. This might imply that evolutionary change should have been impeded by the accumulation of these conserved core processes. Paradoxically, they seem to have facilitated evolution in terms of morphological and physiological change, at least that is our contention. Otherwise, we would imagine that the accrual of a large number of genes whose deletion would be lethal, would have slowed further evolutionary change. In investigating this facilitation of morphological and physiological change, we asked what is the nature of these processes that might have enabled them to remain unchanged, and yet have promoted all the selected anatomical and physiological variation that flourished in all the evolution of all the animals since the Cambrian. We argue that the core processes are easily recombined in different groupings, in different amounts, and at different times by regulatory change involving limited genetic change. But what is it about these processes that makes them so prone to regulation? While a great deal of attention has been given to regulation, it is the nature of the processes themselves that sets the envelope of what can be generated by regulatory change. And this view reveals why there can be constraints on change in the core processes that can deconstrain regulatory change.

Toward a Theory of Facilitated Variation

The conserved core processes are called forth in various combinations at various times and places to construct the phenotype. They increase evolvability, which we have defined as "the capacity to generate heritable, selectable phenotypic variation" (Kirschner and Gerhart 2005). That definition may seem circular, but evolvability takes on real significance if we divide it into two components, the selection component and the variation component. The selection component simply refers to the concordance between an organism's characteristics and the nature of the environment. As such, there is nothing the organism contributes over time to the selection component. It may reflect, for example, a resistance to heat that fortuitously coincides with global warming. The variation component is of more interest to us. We have defined the variation component of evolvability as "capacity to generate phenotypic variation in response to genotypic variation," with special reference to the type of variability that is produced: (1) it maximizes the amount of phenotypic variation for a given amount of genotypic variation; (2) it minimizes the lethality of phenotypic variation; (3) it produces phenotypic variation that is most appropriate to environmental conditions, even conditions never before encountered in the lineage (this feature will be explained). The theory of facilitated variation is an attempt to look over the most important understandings in cell and developmental biology in order to extract some principles that would underlie the variation component of evolvability (Kirschner and Gerhart 2005; Gerhart and Kirschner 2007). Three of these will be described below as (1) exploratory processes, (2) weak linkage, and (3) compartmentation.

Exploratory Processes

The idea of variation and selection and its retention through heredity was a great intellectual leap by Darwin. Though hiding in plain sight in many other areas of biology, variation and selection was unappreciated until it appeared in Darwin's and Wallace's writings. It always seems to be an explanation of last resort. A classic story of variation and selection outside of evolution was the discovery of the mechanism of adaptive immunity by the Australian virologist Frank Macfarlane Burnet (1899–1985). The competing explanation for the immune response was formulated by the great chemist Linus Pauling, who thought that all antibodies were initially the same, but on encountering an antigen, they were induced to fold in a complementary way to neutralize the antigen by binding. But when MacFarlane Burnet made antibodies to staphylococcal enterotoxin in rabbits, he found that there was a slow response to the first injection but a much more rapid response to the second injection, and that response increased logarithmically for the next three days. This indicated to him that it could not be that antibodies were being converted one by one by antigens, which might be expected to occur at a nearly constant rate, but that something, presumably the antibody-producing cells, was multiplying. Other evidence that antibody-producing cells preexisted in the normal circulation suggested a very different model for generating specific variation in the immune system: that variation preexisted and the appropriate response was generated by selective multiplication of those antibody-producing cells that interacted with the antigen.

Preprogrammed variation followed by selection is widespread in cellular systems. Dynamic assembly and local stabilization of microtubules occurs in virtually all cells of every multicellular organism, as well as in unicellular eukaryotes. In this case an input of energy makes each microtubule polymer unstable in a very special way. Microtubules undergo excursions of growth and shrinkage in a seemingly pointless exploration of cytoplasmic space. Yet because of the fast turnover of structure, local stabilization at the end of a microtubule can appear to re-direct the entire array so that it is polarized toward the stabilizing signal. The localized stabilization agents select a particular microtubule sub-population in a certain region of the cell. The resulting polarized array of microtubules is not generated directly, but selected one microtubule at a time from a rapidly changing unpolarized array. As a result, cell morphology is highly adaptive. Although the microtubule protein is conserved, and hence internally constrained, microtubule formation generates a diversity of potential cellular morphologies each time it is employed. Like adaptive immunity, much of this diversity is wasted in each specific context; yet it can find new uses in evolution.

Exploratory processes as a core principle of facilitated variation contribute to all three attributes of the variation component of evolvability. They can generate phenotypic variation with a small amount of genetic variation because the physiological variation is engineered directly into the process of generating every phenotype. To vary the phenotype, one does not need to generate the variation, which is already built into the system, but merely to achieve a new selection. It also reduces the lethality of phenotypic variation because there is tolerance of imprecision. Finally, the same mechanism for assembling a mitotic spindle can serve with different cell types of different morphologies and, of course, cell types of new morphologies. Hence the cytoskeletal mechanism does not need to anticipate new morphologies; it can adapt to new states without any rewiring.

It is perhaps on the developmental level that exploratory mechanisms can best be seen to confer evolvability. Much of metazoan development concerns the generation of anatomical novelty, and often novelty poses special problems by requiring several simultaneous modifications to achieve any function. Darwin considered these issues under "organs of extreme perfection" (Darwin 1859) and gave the best arguments he could that this did not undermine his theory. Yet developmental mechanisms now provide a better understanding of how novelty is established (see chapter 12 in this volume). Behind every change of anatomical structure is a modification of development, which we now understand to a great level of precision. Exploratory processes and selection on a cellular and molecular level play an important part in this understanding.

If we consider the evolution of the vertebrate limb with all its impressive modifications, fins, and autopods, and their modifications into flippers, hands, wings, burrowing tools, and so on, each one seems to require several simultaneous innovations. For example, the limb is defined by the placement of cartilage and bone, but to function, it also requires the correct placement of muscles, nerves, and blood vessels. Yet, the vascular system, the nervous system, and the muscle system are all exploratory systems. The muscle precursors migrate from the trunk and take their positions relative to the bones, even if the limb bud is experimentally placed in an ectopic location. Limb buds can be transplanted to the head, and the resulting ectopic limbs will receive muscle cells. The motor neurons are generated in superfluous numbers, extend axons from the spinal cord, and innervate the muscles. If they don't find muscle, they die off, like unstabilized microtubules. Later synapse elimination and consolidation establishes proper connections. And the vascular system fills all space, biasing its random growth by its attraction to hypoxic environments, generally without need for a road map. Our demand-based vascular system is highly adaptive and even supports tumors, whose mass and location are not encoded in the germ line. Without any mutation the plasticity and adaptability of the core developmental processes responsible for muscle migration, nerve cell pathfinding, and angiogenesis can accommodate to major skeletal innovations driven by heritable changes of regulation. Later mutations may perfect or stabilize these circuits, but exploratory systems allay concern for the need of simultaneous modifications or an improbable sequence with zero or negative selection.

Weak Linkage

The molecular biology revolution that began in the 1950s was distinguishable by an interest in how biological systems transmitted information. Initially this concerned how information was transferred unidirectionally from the genome to the proteins or, roughly speaking, from the genotype to the phenotype. Yet there is much more to the phenotype than making a protein encoded by the gene, and in recent years molecular biology and now systems biology have been concerned with how information flows within cells, between cells, between tissues, and between organisms. With the realization that the number and types of genes, especially the large number of genes comprising the core processes, is limited while evolutionary innovation has continued, we are left with the conclusion that it is the employment of these conserved genes and processes in a wide range of combinations, times, and places that undergoes the most rapid and the most consequential change during the epochs between th e invention of new core processes. The details of regulation fill textbooks of cell and molecular biology, but here we are concerned less with regulation per se and more with what makes processes regulatable. Mere combinatoric juxtaposition of elements in a circuit does not assure variational evolvability. The theory of facilitated variation looks to those features within the core processes that facilitate new combinations, and not just any combination, but combinations that are nonlethal and have a good chance of being useful in new conditions.

Weak linkage has emerged as a central property of core processes that transfer information. Weak linkage refers to specific biochemical features of information systems in biology, where signals of low information content evoke complex, preprogrammed responses from the core process. The specifics of the signal have a weak relationship to the specifics of the outcome. The term "weak linkage" also alludes to the typically low energy and low specificity of the interactions between signals and responders. The first experiments to investigate the nature of gene regulatory control, by Jacob and Monod in the 1960s, established the principle of how a simple molecule such as lactose can mobilize the machinery for making a complex enzyme to degrade lactose. Further studies by Monod and his colleagues led to the concept of allostery, which provided critical details about how regulation can occur on the protein level. Their findings led to the notion that biological systems are composed of preprogrammed responses with built-in mechanisms of autoinhibition. These systems sit at the knife edge of on-ness and off-ness, and simple and/or weak signals can elicit very complex responses. To employ a complex response in a new situation, the response does not need to create an output from many fragments of molecular machinery, but simply to stabilize an existing state. In this way weak linkage has some similarities to exploratory processes that have already created a state that then needs to be selected. Signal transduction processes in cells are replete with switch-like molecules, such as GTP- binding proteins, that can flip between states driven by kinetic rather than thermodynamic means. This furthers the ease with which weak interactions can entrain large molecules, which in turn can bind to sites on other large molecules that are themselves often self-inhibited. The transcriptional machinery is rife with weak linkage. For example, binding of transcription factors to the genome can mobilize enzymes that modify chromatin. These factors do not themselves physically contact the core transcriptional machinery, whose modification is not required to create a new input-output relation.

Weak linkage probably exists on every level of biology, though of course our understanding of it may not be so complete and fundamental as it is on the molecular level. On a cellular level the basic biology of neuronal transmission is an example of weak linkage. The neuron connects many chemical inputs with chemical outputs often transmitted over great distances. To link these signals, the basic machinery of the neuron generates two states, resting and active, which differ in their membrane potential. The resting state blocks the secretion of neurotransmitters and has a negative membrane potential. The active state is permissive for secretion and has a less negative potential. The "weak" linkage provided by the membrane voltage insulates the complexity of the input side (receptors and ion channels) from the complexity of the output side (the secretory mechanism), so they do not have to co-evolve. With these features new input-output relations can be established in one step, since it is easy to install or remove receptors and ion channels without reconfiguring the process of secretion. The contrast to weak linkage would be strong linkage, where every input would make its unique chemical or physical contact with its output. Not only would it be hard to generate such input-output relations, it would be extremely difficult to multiplex inputs and outputs in the same cell.

On the even higher level of a developmental pathway, there may be no better example of weak linkage than embryonic induction, first described by Hans Spemann and Hilde Mangold (Spemann and Mangold 1924). They showed that a small group of cells, the "organizer," elicits the whole axial skeleton, musculature, and nervous system of a vertebrate animal. This experiment inspired many efforts to understand the chemical basis of the signaling and tissue organization. The surprising discovery of the 1990s is that the organizer accomplishes this reprogramming and complex morphogenesis by producing simple chemical signals, a few secreted antagonists of bone morphogenetic protein (De Robertis 2006). These antagonists of course cannot instruct the organism how to make the body plan, but they release a self-inhibited process to carry out its function. Thus the early vertebrate embryo is provided with an auto-inhibited process for generating its axial structure throughout the blastula. Simple signals, which can easily be moved, replaced, or modulated, can elicit even the most complex developmental responses, wherever they are located. The ease with which simple signals can entrain complex processes reflects the weak linkage and hierarchical nature of intracellular signal transduction and transcription.

Compartmentation

In multicelluar organisms a positive innovation in one region could have negative consequences in another, so-called "negative pleiotropy." Compartmentation of a very specific type vastly increases the power of regulatory mutations to generate nonlethal phenotypic changes. Compartmentation has been the preferred strategy for increasing anatomical diversity while maintaining both gene number and the limited set of conserved core processes. It has a very specific meaning in developmental biology, as a special kind of modularity which makes wide use of weak linkage. Each spatial compartment in an embryo is defined by a small set of unique selector genes, which encode transcription factors or signaling molecules that are expressed uniquely in that compartment. The selector gene can then "select" any other gene to be expressed or repressed in its compartment. This allows the core processes to be apportioned for usage in a specific compartment. For example, a process that causes cell proliferation in one compartment may cause differentiation or cell death, or not be expressed, in a different compartment, depending on what else has been selected to occur there. The amplitude or timing of any response can easily be modulated. Thus, different combinations and amounts of the core processes can be engaged in parallel in different regions of the animal (Gerhart and Kirschner 1997; Kirschner and Gerhart 2005). For example, just changing the degree of proliferation can account for much of the difference between the hind limbs and forelimbs of a kangaroo. Compartments arise in the embryo, not at the earliest stages but once there are several hundred to thousands of cells; subcompartments may also form. The insect embryo is divided into about 100 contiguous spatial compartments at the germ band stage. (Depending on the developmental strategy, the entire complement of compartments may not be visible at one time, though it is in *Drosophila*.) In vertebrates there are about 200 embryonic compartments.

One example of compartmentation is the difference between the bone-forming cells in the thoracic vertebrae, which form ribs, and the bone-forming cells in the cervical vertebrae, which do not. Both types of cells perform the basic bone-forming functions, but they produce different structures. This difference is inherent to the type of cell, as shown by transplantation experiments (Kieny et al. 1972). The differences between them are due solely to their having arisen in different compartments with different selector genes. These region-specific differences come close to Sewall Wright's example of the "adaptive cell behaviors" of bone-forming cells. Similarly, *Drosophila* has a single program for forming appendages: in the thorax imaginal disc this program produces a leg, but in the head it will produce a biting mouth part. Likewise, the forelimbs and hind limbs and fins of vertebrates differ because of compartment differences. Compartmentation allows easy respecification of function at spatially defined locations. It permits regulatory gene function, such as cis-regulatory binding sites for transcription factors, to evolve separately in different regions of an animal, thus facilitating phenotypic variation.

Many evolutionary changes, such as the specification of the vertebrae or the development of different functions of insect appendages, would be impossible to imagine without compartmentation. The alternative to compartmentation would be a completely different developmental mechanism for each anatomical feature. The three inventions of wings in vertebrates—pterosaurs, birds, and bats—each involved a different compartmental modification to achieve the common morphology of a wing, a morphology dictated by aerodynamics. Recently Niswander studied the development of the bat wing (Sears et al. 2006). The fossil evidence suggested that the bats may have arisen rapidly. perhaps involving few genetic changes. The developmental evidence indicates that a single protein, bone morphogenetic protein 2, is highly expressed in the bat forelimb as compared to the mouse, and may be the key event in bat wing evolution. The integration of these changes into the rest of the anatomy of the bat is a tribute to the exploratory behavior and physiological plasticity of the muscles, nerves, vasculature, and skin. However, the restriction of bone elongation to the forelimb requires compartmentation of the regulation. Recent studies by Cretekos et al. (2008) have identified a specific sequence in the DNA that upregulates the transcription of a gene that stimulates bone growth. This sequence activates expression only in the forelimb, effectively compartmentalizing the response.

We argue that weak linkage, exploratory behavior, and compartmentation are major underlying mechanisms that facilitate phenotypic variation. Another mechanism, which we have not discussed, is state variation in preexisting physiological systems, an idea close to that of Baldwin (1896) and West-Eberhard (2003). All of these mechanisms are conserved, but they have the property of deconstraining change around them. Though our analysis has focused on metazoans, we would expect both similar and different mechanisms that facilitate nonlethal variation in other branches of life. We now can begin to understand why the extensive regulatory tinkering with and recombination of transcriptional circuits and genetic structures so often produces successful and nonlethal outcomes, and why biology, based on core processes, is so regulatable.

Furnishing Evidence for Facilitated Variation

Evolution has been characterized as a historical process driven by chance and contingency. It has even been argued that replaying the tape of life would not be predictable and would generate vastly different outcomes (Gould 1989). There is, even among some, a kind of resignation that in a subject where chance plays such a big role, little can be said. But does this really mean that we cannot say anything useful about phenotypic variation, or maybe even about evolution itself? In reviewing our book on facilitated variation (Gerhart and Kirschner 2007), the evolutionary geneticist Brian Charlesworth wrote:

Until we have a predictive theory of developmental genetics, our understanding of the molecular basis of development—however fascinating and important in revealing the hidden history of what has happened in evolution—sheds little light on what variation is potentially available for the use of selection. As a result, it is currently impossible to evaluate the idea that developmental systems have special properties that facilitate variation useful for evolution. (Charlesworth 2005: 1619)

Charlesworth and others set a high bar. They say that without full understanding, we have nothing. They seem to forget that Darwin's theory was not a predictive theory, nor was the cell theory, nor was MacFarlane Burnet's clonal selection theory of immunity; yet by most accounts all of these were very useful. We do not have a completely predictive theory of the weather, but we understand the major features that contribute to it. We have a quantum mechanical theory that predicts the properties of the hydrogen atom, but not a similar predictive theory of a protein molecule or even an amino acid. Yet the knowledge gleaned from theories of the hydrogen atom has led to qualitative understandings in chemistry including the chemistry of proteins. Charlesworth's insistence on predictive theory is a red herring. Most theories in science are empirical and semiquantitative. These theories help explain phenomena and make rough predictions. That is our goal in understanding phenotypic variation. It is a developmental genetic theory insofar as it explains how heritable changes in gene regulation can be so effective in generating changes of anatomy and physiology. At this time, such considerations are of great interest to systems biologists asking the general question of what properties a complex system must have in order to be able to evolve.

There is a general epistemological question of what constitutes proof in a historical field, when events such as encounters with asteroids are absolutely important to the outcome; even in statistical physics and quantum mechanics there are limits on the precision of prediction. The standard test of redoing the experiment, impossible though it is for evolution, would not be guaranteed to replay the same scenes, even if everything were the same. Charlesworth demands a predictive theory, but we seek a more modest goal, a theory that helps us understand the role of the organism in generating variation available for selection. We wish to understand qualitatively whether the products of selection are tempered in very important ways by the nature of the evolved developmental processes. We wish to find out whether these processes, as they are elaborated in the metazoan lineages, serve to constrain evolutionary change or serve to deconstrain it. We wish to know whether among the profusion of cellular processes there are any rules as to what changes and what does not change in evolution, and whether change is facilitated or inhibited. In thinking about the core processes of cell and developmental biology, we wish to know the conditions under which they evolved. Finally, though these goals are qualitative, they should be falsifiable or extendable. How would that be achieved?

To verify facilitated variation, we may ask questions of the new data in comparative developmental biology. How are developmental processes altered to generate new phenotypes in evolution? Do the changes suggest that substantial innovation in anatomy and physiology is possible with easy-to-achieve genetic modification? Do they suggest that cell and developmental processes can reduce the lethality of change? Do they suggest that changes in certain dimensions are easier to achieve than in others, and that these dimensions carry with them a bias from successful previous selections? Do they suggest that evolutionary changes depend on the kinds of processes that we just highlighted: exploratory systems, weak linkage, and compartmentation? To answer these questions we have to turn to real examples.

Abzhanov et al. (2004) explored the nature of anatomical variation in the evolution of the Galapagos finch beaks. In comparing narrow, forceps-like beaks of one species with the broader, cutter-like beaks of another, they found a correlation between the higher expression of bone morphogenetic protein 4 (BMP4) and the greater width and depth of the finch beak. Similarly, in a subsequent paper they reported a strong correlation between the level of expression of the calcium transducer calmodulin and the length of beaks (Abzhanov et al. 2006). The importance of these changes beyond simple correlations was proven by the addition of BMP4 and calmodulin to chick embryos locally where the beak forms. Additional BMP4 increases the width and depth of the chick beak, and further expression of calmodulin increases the length. Thus simple changes in two factors can cause proportional changes in the upper and lower beak in a geometrically simple way covering much of the observed range of evolutionary change. It is easy for birds to accumulate mutations that would change the effective level of BMP4. For example, through regulatory mutations one could easily alter the transcription, translation, control of secretion, and processing of the BMP4, its receptor, and the secreted inhibitors of the receptor. A tolerance for large changes is clearly seen in the chick experiments. The beak not only enlarges, it also successfully articulates with the head. Head morphology does not need separate regulation; it accommodates to novelty in the beak, thus enabling large changes without lethality. Whereas global changes in BMP expression would be expected to alter every bone in the animal, it is likely that BMP changes in these examples are restricted to the region of the beak by compartmentation through selector genes, perhaps in the same way bone elongation was restricted to the forelimb of the bat. The exploratory nature of the neural crest, which gives rise to the beak and bone-forming cells, gives these cells wide susceptibility to the surrounding and unlimited proliferative power. Weak linkage most likely underlies the regulatory change. It also permeates the BMP pathway, so that it is readily perturbed by the appropriate ligand to change the phenotype while being robust and self-limiting, and therefore reducing lethality of genetic changes in this process.

In studies of changes in the armor of stickleback fish, genetics, genomics, and breeding experiments allow us not only to identify the changes that underlie morphological evolution but also to show how significant those changes are. Stickleback fish have undergone rapid geographical isolation and directional selection. One of the best examples is the monotonic reduction in armor (loss of dorsal spines and reduction in pelvis) in a 21,500-year fossil record in an ancient Nevada lake, where presumably the population was not open to predation as experienced in the ocean (Hunt et al. 2008). Yet the fossil record does not tell us how many mutations were required and how readily phenotypic variation was generated. Since the late 1970s, as the waters of Lake Washington in Seattle have become less turbid, there has been a steady increase in armor in its stickleback population, a reverse of the usual evolutionary loss of armor. In all known cases the loss or gain of armor is due to a hypomorphic allele or normal allele, respectively, of ectodysplasin, a membrane-bound signaling molecule that regulates cell interactions and differentiation. On one level, this change in gene frequency is just what we would expect from a simple evolutionary model, but the surprising thing is that the level of this single gene can cause such uniform, nonlethal changes in body armor all over the organism (Kitano et al. 2008). It is really an example of weak linkage, because the signals are information-poor, yet they elicit a complex response, such as the spines on the fish. Evolutionary modification is limited by compartmentation only to certain epithelia that give rise to the bony plates. We have further genetic information about the change in the large pelvic bones, which are reduced in the freshwater species, where predation is less severe. These species have been separated a short enough time that it is possible by genetic crosses to isolate the gene responsible for the pelvic elaborations. It is a transcription factor, known as Pitx1, widely employed in embryonic development, and the heritable regulatory change affects its expression (Shapiro et al. 2004). Compartmentation, presumably through Hox selector genes, likely limits the effects to the pelvic girdle, and thus facilitates the nonlethality of this variation.

Although today we have only a limited number of examples, it is becoming clearer that recurrent use of the same mechanism occurs in widely separated adaptations. The pelvic reduction that takes place in sticklebacks in the United States and Canadian lakes is probably a reflection of reduced predation, but the pelvic reduction in the Scottish lakes is thought to be a response to a shortage of calcium. Yet both involve localized suppression of Pitx1 expression by regulatory means (Coyle et al. 2007). The reuse of the same mechanism, even in widely divergent lineages, suggests that certain processes have the capacity to produce varied morphologies that are nonlethal and have been found useful in previous selections. Some may also be tied to selected forms of phenotypic plasticity. For example, regulation of another growth factor, c-Kit, reduces pigmentation in both sticklebacks and humans (Miller et al. 2007). All three examples are of major quantitative trait loci (QTL) and are present as standing variation in the populations.

Morphological genes of large effect challenge the neo-Darwinian assumption of small, gradual change throughout the genome. David Stern's laboratory has studied the morphological innovations in the larvae of different *Drosophila* species. The protrusions of the cuticle called denticles or trichomes provide frictional resistance for the larva to crawl. The difference in trichome distribution reflects the difference in trichome- producing cells, which develop in all related species through differences in the localized expression of one specific transcription factor, shavenbaby. Thus these important elements of morphology are never produced by convergence of different processes, as Ernst Mayr assumed, but by the repeated change in the compartmentation of shavenbaby expression (Sucena et al. 2003). Detailed molecular dissection of the transcriptional regulatory circuit for shavenbaby shows that it is complex and modular, containing common elements that limit expression to the epidermis and elements that further limit the domain of expression. Shavenbaby is a hot spot for changing trichome position in one step, but the complex trichome pattern that differs from species to species appears to have been generated by a concretion of small mutations that make discrete changes in the location of trichome-producing cells. The elaboration of the morphological features of fly larvae comes from weak linkage and compartmentation. Weak linkage provides the molecular process by which a single mutation can entrain an entire developmental program, trichome cell formation, in one step. Compartmentation tells us exactly how organisms avoid pleiotropic effects, that is, how they confine evolutionary novelty to specific regions while avoiding detrimental effects outside that region.

We can even begin to understand how extreme novelty such as antlers in deer and elk can arise. Antlers emerge from a special site on the frontal bone, called the pedicle, that is formed by the neural crest (Kierdorf 2007). The neural crest itself is an exploratory system par excellence. It is migratory, taking clues from the environment. It is multi-potential, giving rise to bone, nerve cells, dermis, dentine, and other tissues in the head, and it has nearly infinite reproductive capacity. The generation of antlers each season is controlled by hormones and most likely by local paracrine factors involved in wound healing. These signals, of low information content, initiate a complex process, as expected for weak linkage. The action of these simple factors suggests that the pedicle osteoblasts are self-inhibited, and await permissive signals. Finally, the limitation of pedicle formation to a small region indicates that compartmentation determines the location of these structures and assures that permissive signals do not initiate antler formation anywhere else in response to global hormonal signals or commonly available paracrine signals. Thus antler formation, growth, loss, and regrowth, one of the most impressive and bizarre forms of morphogenesis in mammals, is most likely a concretion of commonly available activities put together by core processes facilitating regulatory variation by way of their weak linkage, exploratory behavior, and compartmentation.

There are probably no extant organisms that lack the key features of facilitated variation. Hence, it would not be possible to establish the importance of facilitated variation by comparing the evolutionary history of organisms that have it with those that do not. Mathematical models may offer the opportunity to ask "what-if" questions, which are not possible in the natural world, such as "What would happen to the process of evolution if some of the features important for facilitated variation were not present?" There are three kinds of relevant mathematical models. The first are heuristic models; they do not aim to imitate biology closely, but instead to examine the rules for facilitated variation in abstraction or by analogy. In a heuristic model Kashtan, Noor, and Alon ask about the structure of logical circuits that are generated by variation and selection to produce a specific logical statement as an outcome (Kashtan et al. 2007). They probe how the initial circuit is set up and how it affects the rate of selection. They conclude that modularity and variation in selection conditions give the most rapid convergence on the goal. Thus logic circuits that have been trained to adapt to different goals have an increased ability to adapt to a novel goal, as if these circuits, like facilitated variation, have internalized their adaptive properties into their structure. Kashtan et al. interpret this in terms of weak linkage and reduced pleiotropy.

A second type of model addresses a real chemical or physical process to ask how the structure of the phenotype facilitates future evolutionary change. Here fitness landscapes have a chemical reality and the genotype is literally the sequence of nucleotides. Using RNA structure, where the phenotype is the three-dimensional fold and the genotype is the sequence of nucleotides, Ancel and Fontana (2000) showed that different sequences sharing the same 3-D fold (the same phenotype) differ in their ability to generate new structures (i.e., they have different evolvability). In a recent study Alon's group showed that the evolution of RNA structures incorporated many of the key elements of facilitated variation, particularly learning from previous adaptations (Kashtan et al. 2009).

A third type of model looks directly at some specific piece of biology to understand what about this biological system allows for adaptation and evolvability. For instance, the polymerization properties of microtubules were shown to optimize their ability to search space within a cell, thus optimizing their ability to find stabilizing activities anywhere (Holy and Leibler 1994). The evolvability of gene regulatory networks has similarly been probed in studies such as that of Hinman and Davidson (2007).

Finally, the relationship between evolutionary variation and physiological variation has been examined quantitatively for hemoglobin evolution in mammals. This study provided evidence for how previous selections for physiological robustness in hemoglobin might have paved the way for evolutionary change (Milo et al. 2007). In all of these theoretical studies the focus is not on genetic variation or on selection, but on the properties of biological systems that furnish a specific favorable kind of phenotypic variation. Although not all of them discuss facilitated variation as such, all indicate that the organism has special evolved properties that support or deconstrain phenotypic variation, given a certain amount of genotypic variation. This is a systems approach.

Facilitated Variation and Evolution

Ernst Mayr wrote that "the most fascinating aspect of Darwin's confusions and misconceptions concerning variation is that they did not prevent him from promoting a perfectly valid, indeed a brilliant theory of evolution" (Mayr 1982: 682). Yet Darwin's theory always needed to be grounded in mechanism, and many biologists, such as Ernst Haeckel, T. H. Morgan, and William Bateson, responded to that very call in the years after the publication of the *Origin of Species*. In the late 19th century the goal of many young scientists was to prove Darwin right or wrong. For many, the discovery of genetics offered a particularly important opportunity to address an important weakness in Darwin's theory. Today, 150 years after Darwin's magnum opus, our understanding of genotypic variation is very sophisticated, but our understanding of phenotypic variation has lagged. This represents a major incompleteness of evolutionary theory, not unlike the lack of understanding of heredity before the rediscovery of Mendel and before population genetics. That for many years we had little understanding of the nature or the frequency of phenotypic variation is not surprising, given the complexity of the cell and of embryonic development. But that has changed since the late 1980s. Formulating the theory of facilitated variation was in itself a test for us, as to whether the current understandings of developmental biology, cell biology, and genomics can now make important contributions to evolution. We think they can.

We should probably think of the theory of evolution as based on three subsidiary theories: a theory of natural selection, a theory of heredity, and a theory of phenotypic variation. Darwin succeeded in sketching the whole picture while understanding only one of these, selection. The Modern Synthesis combined that understanding with the new (at the time) concepts of genetics. But the Modern Synthesis did not and could not incorporate any understanding of how the phenotype is generated. Yet, many evolutionary biologists of that time were able to largely ignore that problem by what looks today like an unconvincing argument. They asserted that as long as phenotypic variation was plentiful, isotropic, and very small, such that significant phenotypic change required a summation of many small selected changes, then the process of phenotypic variation was never a limiting condition. Or, as Gould put it, phenotypic variation could not be the creative force in evolution. This was left for selection. But modern biology tells us that none of those assumptions turned out to be true. At one extreme, many mutations can elicit the same kind of nonlethal phenotypic variation, so that one might say that in some circumstances mutation is plentiful and not very creative. In others it will be just the opposite. As the initial efforts toward understanding facilitated variation suggest, the nature of phenotypic variation is a complex problem, but not one lacking general conceptual principles. We are proceeding rapidly to identify the molecular mechanisms of development, and we are finding that there are only a few core processes, and these rest on even fewer rules of operation.

There is no question that biology has made major inroads in answering the most important questions about how the phenotype is generated. These accomplishments have greatly enlarged our understanding of evolution, but they have not undermined the previous achievements of evolutionary theory. With facilitated variation, genetic change is still required, and mutations leading to regulatory change are the most important kind, but the entire burden of creativity in evolution does not have to rest on selection alone, nor on mutation alone. The complex existing phenotype determines the kind and amount of phenotypic variation. This variation will be based on new combinations, times, and places of use of the unchanging core processes. The biological system has modes of responding physiologically, developmentally, and genetically, and these responses are elicited in many ways by mutation, acting through regulatory modification. This biases the kinds of phenotypic variation that is generated, and some of those biases reflect the selected adaptability of the organism. The evolution of new forms of weak linkage, through mutation, is facilitated by constrained core systems that are already poised to act. Signals can have very low information content, yet elicit complex and useful outcomes. Conservation has not created more constraint, as some have argued. Rather, these systems—though internally constrained to change—provide more than compensating deconstraint for regulatory change. These views are not at all Lamarckian, nor are they arguments of selection for future good.

There is no conceptual problem in imagining the evolution of facilitated variation. Physiological adaptation and developmental programs are themselves selected properties that depend on previous mutations and selections. Poised systems are selected for their physiological responsiveness, and they contribute to robustness on the level of the individual. Evolutionary deconstraint by way of facilitated variation is simply a by-product. It is not a gratuitous by-product because the same molecular features that allow for versatility and robustness in an organism's lifetime can easily be seen to provide versatility and robustness over the evolutionary long run when genetically encoded. Furthermore, the systems of robust physiology lead inevitably to the buffering of the effects of genetic variation, which in turn leads to the accumulation of more genetic variation in populations. In this way, the capacity to maintain nonlethal phenotypic variation results in increased accumulated genetic variation, which has long been argued to increase the rate of evolutionary change (Schmalhausen 1986).

A subtext of the Altenberg meeting was to consider the Modern Synthesis in light of recent discoveries. The Modern Synthesis was a great intellectual accomplishment in an important era for evolutionary biology. Viewed today it is neither modern nor much of a synthesis. For the three-legged stool upon which evolutionary theory is based—genetic variation, phenotypic variation, and selection—the Modern Synthesis could in its fullest realization provide only two legs, leaving an unstable scaffold. Fortunately modern biology, particularly cell and developmental biology, has contributed a lot to the third leg. It may not look exactly the way evolutionary biologists predicted in the 1940s, but it supports the same edifice. Furthermore, the three legs are supporting more than the theory of evolution. Theodosius Dobzhansky's famous epigraph, "Nothing in biology makes sense except in the light of evolution" (1973), has for almost four decades been honored more in the breach than in the observance. Despite the appealing sentiment, virtually none of the great mechanistic, cellular, and developmental advances in biology since Dobzhansky's comment have depended at all on knowledge of the theory of evolution (or, even more specifically, on the Modern Synthesis). Yet Dobzhansky's epigraph is more relevant today. What other basis do we have to understand the strange admixture of conserved processes and divergent phenotypes that constitute life in general and the human phenotype and its pathologies in particular? If one reads current papers in modern genomics, cellular biology, and developmental biology, one sees a "shifting balance" from asking the question "What can molecular biology do for evolution?" to the question "What can evolution do for molecular biology?" For this to be complete, evolutionary theory must expand to incorporate our modern understandings of phenotypic variation.

References

Abzhanov A, Kuo WP, Hartmann C, Grant BR, Grant PR, Tabin CJ (2006) The calmodulin pathway and evolution of elongated beak morphology in Darwin's finches. Nature 442: 563–567.

Abzhanov A, Protas M, Grant BR, Grant PR, Tabin CJ (2004) Bmp4 and morphological variation of beaks in Darwin's finches. Science 305: 1462–1465.

Ancel LW, Fontana W (2000) Plasticity, evolvability, and modularity in RNA. Journal of Experimental Zoology: Molecular and Developmental Evolution 288: 242–283.

Baldwin JM (1896) A new factor in evolution. American Naturalist 30: 441–451.

Baldwin JM (1902) Development and Evolution. New York: Macmillan.

Brakefield PM (2006) Evo-devo and constraints on selection. Trends in Ecology and Evolution 21: 362–368.

Charlesworth B (2005) Evolution: On the Origins of Novelty and Variation. Science 310: 1619–1620.

Coyle SM, Huntingford FA, Peichel CL (2007) Parallel evolution of Pitx1 underlies pelvic reduction in Scottish threespine stickleback (*Gasterosteus aculeatus*). Journal of Heredity 98: 581–586.

Cretekos CJ, Wang Y, Green ED, Martin JF, Rasweiler JJ 4th, Behringer RR (2008) Regulatory divergence modifies limb length between mammals. Genes and Development 22: 141–151.

Darwin C (1859) On the Origin of Species by Means of Natural Selection, or the Preservation of Favoured Races in the Struggle for Life. London: John Murray.

Dawkins R (1996) The Blind Watchmaker: Why the Evidence of Evolution Reveals a Universe Without Design. Updated ed. New York: Norton.

De Robertis EM (2006) Spemann's organizer and self-regulation in amphibian embryos. Nature Reviews Molecular and Cell Biology 7: 296–302.

De Vries H (1909) The Mutation Theory: Experiments and Observations on the Origin of Species in the Vegetable Kingdom. JB Farmer, AD Darbishire, trans. London: Kegan Paul.

Dobzhansky T (1973) Nothing in biology makes sense except in the light of evolution. American Biology Teacher 35: 125–129.

Gerhart J, Kirschner M (2007) The theory of facilitated variation. Proceedings of the National Academy of Sciences of the USA 104(Suppl 1): 8582–8589.

Gerhart JG, Kirschner MW (1997) Cells, Embryos and Evolution: Toward a Cellular and Developmental Understanding of Phenotypic Variation and Evolutionary Adaptability. Boston: Blackwell Science.

Gould SJ (1977) Ontogeny and Phylogeny. Cambridge, MA: Harvard University Press.

Gould SJ (1989) Wonderful Life: The Burgess Shale and the Nature of History. New York: Norton.

Gould SJ (2002) The Structure of Evolutionary Theory. Cambridge, MA: Belknap Press of Harvard University Press.

Haeckel E (1866) Generelle Morphologie der Organismen. 2 vols. Berlin: Georg Reimer.

Hinman VF, Davidson EH (2007) Evolutionary plasticity of developmental gene regulatory network architecture. Proceedings of the National Academy of Sciences of the USA 104: 19404–19409.

Holy TE, Leibler S (1994) Dynamic instability of microtubules as an efficient way to search in space. Proceedings of the National Academy of Sciences of the USA 91: 5682–5685.

Hunt G, Bell MA, Travis MP (2008) Evolution toward a new adaptive optimum: Phenotypic evolution in a fossil stickleback lineage. Evolution 62: 700–710.

Jablonka E, Lamb MJ (2005) Evolution in Four Dimensions: Genetic, Epigenetic, Behavioral, and Symbolic Variation in the History of Life. Cambridge, MA: MIT Press.

Kashtan N, Noor E, Alon U (2007) Varying environments can speed up evolution. Proceedings of the National Academy of Sciences of the USA 104: 13711–13716.

Kashtan N, Mayo AE, Kalisky T, Alon U (2009) An analytically solvable model for rapid evolution of novel structure. PLoS of Computational Biology 5: e1000355.

Kieny M, Mauger A, Sengel P (1972) Early regionalization of somitic mesoderm as studied by the development of axial skeleton of the chick embryo. Developmental Biology 28: 142–161.

Kierdorf U, Kierdorf H, Szuwart T (2007) Deer antler regeneration: Cells, concepts, and controversies. Journal of Morphology 268: 726–738.

King N, Westbrook MJ, Young SL, Kuo A, Abedin M, Chapman J, Fairclough S, Hellsten U, Isogai Y, Letunic I, Marr M, Pincus D, Putnam N, Rokas A, Wright KJ, Zuzow R, Dirks W, Good M, Goodstein D, Lemons D, Li W, Lyons JB, Morris A, Nichols S, Richter DJ, Salamov A, Sequencing JGI, Bork P, Lim WA, Manning G, Miller WT, McGinnis W, Shapiro H, Tjian R, Grigoriev IV, Rokhsar D (2008) The genome of the choanoflagellate *Monosiga brevicolis* and the origin of metazoans. Nature 451: 783–788.

Kirschner MW, Gerhart JC (2005) The Plausibility of Life: Resolving Darwin's Dilemma. New Haven, CT: Yale University Press.

Kitano J, Bolnick DI, Beauchamp DA, Mazur MM, Mori S, Nakano T, Peichel CL (2008) Reverse evolution of armor plates in the threespine stickleback. Current Biology 18: 769–774.

Liem KF (1990) Key evolutionary innovations, differential diversity, and symecomorphosis. In Evolutionary Innovations. M Nitecki, ed.: 147–170. Chicago: University of Chicago Press.

Mayr E (1963) Animal Species and Evolution. Cambridge, MA: Belknap Press of Harvard University Press.

Mayr, E (1982) The Growth of Biological Thought: Diversity, Evolution, and Inheritance. Cambridge, MA: Belknap Press of Harvard University Press.

Miller CT, Beleza S, Pollen AA, Schluter D, Kittles RA, Shriver MD, Kingsley DM (2007) Supplemetary data: cis-regulatory changes in Kit ligand expression and parallel evolution of pigmentation in sticklebacks and humans. Cell 131: 1179–1189.

Milo R, Hou JH, Springer M, Brenner MP, Kirschner MW (2007) The relationship between evolutionary and physiological variation in hemoglobin. Proceedings of the National Academy of Sciences of the USA 104: 16998–17003.

Putnam NH, Srivastava M, Hellsten U, Dirks B, Chapman J, Salamov A, Terry A, Shapiro H, Lindquist E, Kapitonov VV, Jurka J, Genikhovich G, Grigoriev IV, Lucas SM, Steele RE, Finnerty JR, Technau U, Martindale MQ, Rokhsar DS (2007) Sea anemone genome reveals ancestral eumetazoan gene repertoire and genomic organization. Science 317: 86–94.

Raff RA (1966) The Shape of Life. Chicago: University of Chicago Press.

Rutherford SL, Lindquist S (1998) Hsp90 as a capacitor for morphological evolution. Nature 396: 336–342.

Schmalhausen II (1986) Factors in Evolution: The Theory of Stabilizing Selection. T Dobzhansky, ed. Chicago: University of Chicago Press.

Sears KE, Behringer RR, Rasweiler JJ 4th, Niswander LA (2006) Development of bat flight: Morphologic and molecular evolution of bat wing digits. Proceedings of the National Academy of Sciences of the USA 103: 6581–6586.

Shapiro MD, Bell MA, Kingsley DM (2006) Parallel genetic origins of pelvic reduction in vertebrates. Proceedings of the National Academy of Sciences of the USA 103: 13753–13758.

Shapiro MD, Marks ME, Peichel CL, Blackman BK, Nereng KS, Jónsson B, Schluter D, Kingsley DM (2004) Genetic and developmental basis of evolutionary pelvic reduction in threespine sticklebacks. Nature 428: 717–723.

Spemann H, Mangold H (1924) Über Induktion von Embryonanlagen durch Implantation artfremder Organisatoren. Roux' Archiv für Entwicklungsmechanik 100: 599–638.

Sucena E, Delon I, Jones I, Payre F, Stern DL (2003) Regulatory evolution of shavenbaby/ovo underlies multiple cases of morphological parallelism. Nature 424: 935–938.

Waddington CH (1953) Genetic assimilation of an acquired character. Evolution 7: 118–126.

Waterston RH et al. (2002) Initial sequencing and comparative analysis of the mouse genome. Nature 420: 520–562.

West-Eberhard MJ (2003) Developmental Plasticity and Evolution. Oxford: Oxford University Press.

Whitman CO (1919) Orthogenetic Evolution in Pigeons. Carnegie Institute of Washington Publication no. 257.

Wright S (1931) Evolution in Mendelian populations. Genetics 16: 97–159.

11 Dynamical Patterning Modules

Stuart A. Newman

Multicellular organisms employ a variety of means to attain their definitive forms. Different cell types are generated and deployed into specific patterns, and the tissues they constitute are molded into three-dimensional shapes. The use of conserved developmental mechanisms and the products of variant genes to achieve species-specific outcomes reflects the common ancestry and particular evolutionary trajectories of modern organisms.

Recognizing this, however, provides no actual insights into how such systems emerge or evolve. The intracellular genetic mechanisms that arose and were refined over several billion years in the single-celled ancestors of the modern animals are not obvious bases for developmental processes, particularly ones involving pattern formation and morphogenesis. This provides an enormous challenge to the Modern Synthesis and, to a certain extent, to more broadly conceived Darwinian accounts for early phases in the evolution of multicellular life, since such explanations are by default incremental and sequentially adaptive.

One aspect of multicellular evolution that does have a discernible basis in unicellular life is cell differentiation. Although we have only modern forms to examine, all unicellular organisms, be they bacteria, fungi, protists, or algae, exhibit alternative states of differentiation, both reversible and irreversible, under different conditions (Pan and Snell 2000; Kaiser 2001; Ryals et al. 2002; Blankenship and Mitchell 2006; Süel et al. 2007; Vlamakis et al. 2008). This must also have been the case for the single-celled ancestors of the Metazoa, that is, the ancient and modern animals.

The biosynthetic states of all cells are determined by the dynamics of transcription factor-mediated gene regulatory networks (GRNs) (Davidson 2006). Such networks, containing feedback and feed-forward loops by which the transcription factors promote and suppress their own and each other's synthesis, exhibit multistability (Forgacs and Newman 2005). The systems can thus switch among discrete states, the number of states always being much smaller than the total number of genes in the organism's genome. Since the genes that specify nontranscription factor proteins and regulatory RNAs are themselves subject to transcriptional control, the alternative stable states of the GRNs specify cell types distinguished by extensive biosynthetic differences.

Any plausible model of complex regulatory networks shows that multistable behavior is inevitable, and therefore not necessarily adaptive (Kauffman 1969; Keller 1995; Laurent and Kellershohn 1999; Kaneko 2006; Kaufman et al. 2007). Indeed, embryonic cells can be coaxed into differentiated fates that are uncharacteristic of the organ they are normally destined to form, and even of the taxonomic class from which they are derived (Mezentseva et al. 2008). One effect of evolution is to tame the inherent propensity of complex systems to switch between alternative cell states, so that the resulting cell types contribute to the organism's functioning, and disruptive or superfluous and costly types are suppressed.

But apart from generating new cell types, development must arrange them in appropriately coherent spatiotemporal patterns (Salazar-Ciudad et al. 2003; Gilbert et al. 2006). Unlike cell type switching mechanisms, however, mechanisms of developmental pattern formation and tissue morphogenesis cannot have existed before multicellularity: the processes that generate spatial organization on the multicellular level are entirely different from those which operate in individual cells.

In this chapter I will describe a plausible basis for the emergence of key mechanisms of developmental pattern formation from ingredients that existed in the unicellular world (Newman et al. 2006; Newman and Bhat 2008). This scenario, like the mechanisms of cell type switching described above, involves roles for certain molecular components of the single-celled ancestors of multicellular forms. But unlike cell type switching, it also involves certain formative physical processes that came into play only with the appearance of the multicellular state. The relevant physical determinants (as will become clear) were not new to the physical world, but rather became newly relevant to living systems in conjunction with a change in their spatial scale and cell-cell proximity.

I suggest that the ancient and continuing role of certain physical mechanisms in the molding and patterning of multicellular aggregates has provided a fount of complex forms that could be selected and refined over the course of evolution. The all but inevitable emergence, in this view, of organismal motifs that were not products of natural selection, but rather served as its raw material, raises questions concerning both the necessity and sufficiency of the mechanisms of the neo-Darwinian Modern Synthesis for the origination of ancient multicellular forms.

Although several taxonomic groups, such as prokaryotes, protists, fungi, and plants, have multicellular members, I focus here on the Metazoa. In the first place, the evolutionary history of the metazoans is relatively well described, and was initiated with remarkable rapidity during the late Precambrian and early Cambrian periods (Rokas et al. 2005; Larroux et al. 2008). Second, development in all the metazoan phyla is mediated by a group of only a few dozen cell signaling and transcription factors, termed the "developmental–genetic tool kit" (Carroll et al. 2005), the members of which have been conserved in many of their patterning functions for more than half a billion years. Finally, the genome of a representative of unicellular sister clade of the Metazoa, the choanoflagellate *Monosiga brevicollis*, has recently been sequenced (King et al. 2008), permitting comparisons with a hypothetical metazoan molecular "ground state." All these features make the Metazoa ideal for presenting the concepts mentioned, although by their nature the general principles should be equally applicable to the other multicellular taxonomic groups.

The Metazoa and Their Antecedents

The extant Metazoa have classically been divided into the Eumetazoa, organisms which exhibit true tissues, epithelia with polarized cells, cell–cell junctions, a well-defined basement membrane, and neurons and muscle cells; and the sponges (Porifera) and Placozoa, which lack all of the aforementioned features. The single known type of placozoan, *Trichoplax adhaerens*, contains several cell types and layers. But unlike the sponges (which exhibit gastrulation-like movements during development and complex labyrinthine morphologies) (Larroux et al. 2006), it has a simple, flat body without internal cavities. Recent sequencing of the genome of *T. adhaerens* (possibly one of several species; Miller and Ball 2005) has complicated this categorization, however, showing Placozoa to have greater genetic affinity to the Eumetazoa than the earlier-diverging sponges (Srivastava et al. 2008).

The Eumetazoa fall into two main classes. The diploblasts, consisting of the Cnidaria (e.g., hydroids and corals) and, traditionally, the Ctenophora (e.g., comb jellies), have two epithelial body layers and true lumens. The triploblasts (chordates, echinoderms, arthropods, mollusks, etc.), in contrast, have a third, mesenchymal, body layer. Recent work has partly revised this scheme, suggesting on genetic grounds that the Ctenophora are not in the main line of the animals, constituting instead a sister clade of the Metazoa (Dunn et al. 2008).

Essentially all the triploblastic metazoan body plans emerged within the space of no more than 20 million years, beginning about 535 million years ago (Conway Morris 2006), during the well-known Cambrian explosion. Simpler sheetlike and hollow spherical forms (Yin et al. 2007), and budding and segmented tubes (Droser and Gehling 2008), are seen beginning about 630 million years ago in fossil beds of the Precambrian Ediacaran period. The full range of body plans of these enigmatic organisms also seems to have emerged rapidly (within 10 million years), during the newly designated "Avalon explosion" (Shen et al. 2008). Although the affinities of the Ediacaran biota and metazoan fauna have not been resolved, Erwin (2008) has conjectured that the first Cnidaria may have been holdovers from the earlier evolutionary episode. Modern animals, and perhaps some of the Ediacaran forms, have a common ancestry in the Precambrian with the choanoflagellates, some of whose extant members are transiently colonial (Wainright et al. 1993; Lang et al. 2002; King et al. 2003; Philippe et al. 2004).

A striking aspect of the metazoan radiation is the fact that a common set of highly conserved gene products, the developmental–genetic tool kit (which includes determinants of both cell type switching and cell pattern formation), has been used to a nearly exclusive extent to generate animal body plans and organ forms for the more than half a billion years since the inception of this taxonomic kingdom (Wilkins 2002; Carroll et al. 2005). A surprising number of the tool kit genes, including some that have key roles in morphogenesis and pattern formation, are found in the genome of *M. brevicollis*, an exclusively unicellular choanoflagellate (King et al. 2008). As will be described below, a few additional genes appear in the tool kit concomitant with the emergence of the sponges, and a few more are found in the simplest eumetazoans, the cnidarians. The Cambrian explosion followed with no more significant additions to the tool kit.

How was metazoan complexity achieved in such a rapid fashion with an essentially unchanging set of ingredients? An early evolutionary step that was absolutely essential to multicellularity was the acquisition by unicellular antecedents of the capacity to remain attached to one another after dividing. Standard evolutionary scenarios would envision the emergence of new genes and gene products to mediate this function. However, the genome of *M. brevicollis* contains 23 putative cadherin genes, as well as 12 C-type lectins (Abedin and King 2008; King et al. 2008). Both classes of molecules mediate cell attachment and aggregation in metazoan organisms, although they require sufficient levels of extracellular calcium ion to do so. This may have been supplied by rising oceanic Ca^{2+} levels during the establishment of multicellularity (Kazmierczak and Kempe 2004).

The acquisition of new functions for previously evolved genes or other features has variously been termed "gene sharing" (Piatigorsky and Wistow 1989), "moonlighting" (Tompa et al. 2005), or, more generally, "exaptation" (Gould and Vrba 1982). By mobilizing the physical force of adhesion, the homophilic cadherins and sugar-binding C-type lectins of the single-celled ancestors of the Metazoa simultaneously acquired a novel cell–cell function which mediated the emergence of multicellularity. The multicellular state then set the stage for additional physical processes to come into play, specifically those that pertain to matter on the meso (or middle) scale: >100 μm in linear dimension (Newman and Comper 1990; Newman et al. 2006). We have referred to functional modules in which one or more of the tool kit gene products mobilize physical processes on this scale so as to mediate the formation of new patterns and forms, "dynamical patterning modules" (DPMs) (Newman and Bhat 2008, 2009).

As will be described in the following section, the DPMs, in conjunction with cell type-defining and switching networks, transformed simple, spherical, topologically solid cell clusters into hollow, multilayered, elongated, segmented, folded, and appendage-bearing structures. They thus founded the pathways that evolved into the developmental programs of modern animals.

The Molecules, Physics, and Outcomes of the Major DPMs

Metazoan embryos employ a variety of patterning and shaping processes (Salazar-Ciudad et al. 2003), some of which are used in all of the kingdom's taxonomic groups and others of which are used in most of those groups. As noted above, the first DPM, designated ADH (table 11.1), results in the formation of a multicellular cluster. Within such a cluster, any or all of the following can occur: the local coexistence of cells of more than one epigenetic state or type, the formation of distinct cell layers, the formation of an internal space or lumen, the elongation of

Table 11.1
Names, components and roles in evolution and development of major metazoan dynamical patterning modules (DPMs)

DPM	Molecules	Physical Effect	Role
ADH	cadherins	adhesion	multicellularity
DAD	cadherins	differential adhesion	multilayering
LAT	Notch	lateral inhibition	coexistence of alternative cell types
POL_a	Wnt	cell surface anisotropy	lumen formation
POL_p	Wnt	cell shape anisotropy	tissue elongation
MOR	TGF-β/BMP; Hh, FGFs	diffusion	pattern formation, induction
TUR	MOR + Wnt + Notch	chemical waves	periodic patterning
OSC	Wnt + Notch	chemical oscillation and synchronization	segmentation
ECM	collagen; chitin; fibronectin	stiffness; dispersal and cohesion	epithelial elasticity; skeletogenesis; epithelial-mesenchymal transformation

the cluster, the formation of repeated metameres or segments, the change in state of one region of the cell cluster due to local or long-range signals from another region, the change in stiffness or elasticity of a cell layer, and the dispersal of cells while they continue to remain part of an integral tissue (reviewed in Forgacs and Newman 2005). At the origin of the Metazoa the DPMs implemented all of the above transformations by mobilizing physical forces and processes characteristic of viscoelastic, chemically active materials on the spatial scale of cell aggregates and tissues. In physical terms, this is "soft matter" (De Gennes 1992) which is simultaneously an "excitable medium" (Mikhailov 1990).

In the next subsections I briefly summarize the properties of the major DPMs, focusing on how preexisting molecules served to bring to bear one or another basic mesoscale physical force, effect, or process on cells and cell clusters. I also describe how DPMs can combine spatiotemporally so as to embody more complex physical phenomena (e.g., biochemical oscillation, reaction-diffusion patterning instabilities) that also play developmental roles. Each DPM is given a three-letter designation (which is also used in table 11.1). Additional details of the molecular and physical aspects of the DPMs are provided in Newman and Bhat (2008, 2009).

Adhesion and Differential Adhesion

As mentioned above, the emergence of multicellularity at the origin of the Metazoa depended on cadherins and lectins of single-cell ancestors taking on the new function of cell–cell adhesion. Once this happened, a second DPM arose in consequence: differential adhesion (DAD; table 11.1). If subsets of cells within an aggregate contain sufficiently different levels of cell adhesion molecules on their surfaces, there will be a sorting into islands of more adhesive cells within lakes of less adhesive ones (Steinberg and Takeichi 1994). Random cell movement will cause the islands to coalesce and an interface to be established, across which cells will not intermix (Steinberg 2003), an effect with the same physical basis as phase separation of two immiscible liquids, such as oil and water (reviewed in Forgacs and Newman 2005). Whether cell adhesion differences arise randomly, or in a controlled fashion (see below), nonmixing layers of tissue will inevitably form.

Lateral Inhibition and Choice Between Alternative Cell Fates

Development of morphologically complex organisms always employs lateral inhibition (LAT; table 11.1), whereby early differentiating cells signal to cells adjacent to them to take on a different fate (Rose 1958; Meinhardt and Gierer 2000). In the Metazoa, lateral inhibition is mediated by the Notch signal transduction pathway, specifically, interaction of the cell surface receptor Notch with members of a class of other integral membrane proteins (Delta, Serrate/Jagged, and Lag2: the DSL proteins) that act as ligands for the receptor and mediators of Notch activity (Ehebauer et al. 2006). This mechanism does not determine the particular fate of any cell, but only enforces the coexistence of alternative fates in adjacent cells in the same cluster or aggregate.

The Notch pathway involves the translocation of an intracellular fragment of the Notch receptor to the cell nucleus as a consequence of Notch's binding a DSL ligand on an adjacent cell. The fragment converts a transcriptional repressor of the CBF-1-Su(H)-Lag-1 (CSL) category into a transcriptional activator (Lai 2002; Ehebauer et al. 2006). CSL factors (but not Notch) are present in fungi (Převorovský et al. 2007), including yeast, and proteins other than the Notch fragment can modulate the effect of CSL (Koelzer and Klein 2003). CSL factors are thus likely to have operated as dual-function mediators of cell state switching in the single-celled antecedents of the Metazoa. This capability, joined to juxtacrine (i.e., direct cell–cell) signaling in multicellular aggregates, is the basis of lateral inhibition in animal embryos.

The choanoflagellate *M. brevicollis* contains protein modules of Notch receptors, though not all are encoded in the same genes (King et al. 2008). The morphologically simplest metazoans to contain the Notch receptor (plausibly evolved by gene shuffling in a choanoflagellate-like ancestor) are sponges (Nichols et al. 2006). Lateral inhibition would have enabled basic cell pattern formation in these organisms. The placozoan *T. adhaerens* lacks a Notch receptor, though it has Notch ligands and intracellular components (Srivastava et al. 2008). This may account for its being morphologically much simpler than the sponges, despite having many more bilaterian-associated genes than the latter.

Induction of Apical-Basal and Planar Cell Polarity

Since cell aggregates behave like viscoelastic liquid droplets (reviewed in Forgacs and Newman 2005), their default morphology is solid in the topological sense (i.e., having no lumen), and spherical. Metazoans contain cells that can be polarized in one of two ways, leading the tissues composed of such cells to form lumens or become elongated. Lumens can arise in aggregates of cells that are anisotropic along their surfaces (referred to as apical–basal (A/B) polarization; Karner et al. 2006a). Specifically, when this polarization leads cells to have lowered adhesiveness on one portion of their surface, they will preferentially attach to their neighbors on their more adhesive (lateral) portions, leaving the less adhesive (basal) portions adjoining an interior space (Newman 1998). Apical–basal polarity is also important in layered tissue arrangements where the affinity of one surface of a sheet of cells to a cellular or acellular substratum needs to be different from the affinity of its opposite surface.

Tissue elongation may occur when cells individually polarize in shape (instead of surface properties), a phenomenon called planar cell polarity (PCP; Karner et al. 2006b). Planar-polarized cells can intercalate along their long axes, causing the tissue mass to narrow in the direction parallel to the cell's long axis, and consequently elongate in the orthogonal direction. This tissue reshaping is known as convergent extension (Keller et al. 2000; Keller 2002).

Both A/B polarity and PCP are mediated by secreted factors of the Wnt family, which interact in a paracrine fashion with receptors of the Frizzled family. Intracellularly, they depend on a polarization mechanism that extends as far back as the common ancestor of metazoans and fungi (Mendoza et al. 2005). Which type of polarization occurs depends on the presence of different accessory proteins with the A/B- and PCP-inducing pathways referred to, respectively, as the canonical and noncanonical Wnt pathways. In each case, the structural alterations of individual cells have novel consequences in a multicellular context, permitting multicellular aggregates to overcome the morphological defaults of solidity and sphericity. We designate the DPMs involving the Wnt pathway operating in a multicellular context as POL_a and POL_p (table 11.1).

Although Wnt genes are not present in choanoflagellates (King et al. 2008), they are found, along with genes for their Frizzled receptors, in Porifera (Nichols et al. 2006). These components are also present in Placozoa (Srivastava et al. 2008). Sponges, of course, have many interior spaces, while the placozoan *Trichoplax*, despite containing only four cell types, has them arranged in three distinct layers, which is possible only if the cells are polarized.

Putative "embryos"—small, hollow, cell clusters identified in the Precambrian Doushantuo Formation in China (Chen et al. 2004; Hagadorn et al. 2006; Yin 2007)—may actually have been the definitive forms of the earliest metazoans and metazoan-like organisms (Newman et al. 2006). It is plausible that the origination of these hollow forms at the transition between the Ediacaran biota and those of the Cambrian explosion was based on the presence of POL_a, which would have caused cell aggregates to develop interior spaces.

Significantly, genes specifying components of the noncanonical Wnt pathway have not been reported in sponges and placozoans, but are present in the morphologically more complex cnidarians (Guder et al. 2006). It is reasonable to speculate that the Ctenophora, which despite indications of their early divergence from the Metazoa (Dunn et al. 2008) are morphologically more complex than the sponges or Placozoa, may utilize the POL_p DPM, via components of the Wnt noncanonical pathway or equivalent mediators of PCP.

Morphogen Gradients and Activator–Inhibitor Systems

While single-celled organisms have the ability to change their physiological state in response to molecules secreted into the microenvironment by other such cells (Luporini et al. 2006), this effect has novel developmental consequences when it occurs in a multicellular context. Secreted molecules that act as patterning signals in metazoan embryos by mediating concentration-dependent responses are termed morphogens (MOR; table 11.1). Wnt, discussed above, is a very locally acting morphogen, but Hedgehog acts over longer distances (Zhu and Scott, 2004). The genome of the marine sponge *Oscarella carmela* contains genes specifying both these categories of morphogens and their receptors (Nichols et al. 2006), as well as receptor tyrosine kinases (Sudhop et al. 2004; Nichols et al. 2006), which in Eumetazoa transduce the effects of morphogens such as FGF and EGF. Morphogens of the FGF class exist in arthropods, chordates, cnidarians (Rentzsch et al., 2008), and echinoderms (Röttinger et al., 2008). Although some components of the intracellular pathway that mediate signaling by the TGF-β class of morphogens in Eumetazoa are present in *O. carmela*, morphogens or receptors of this type are not seen until the appearance of the Placozoa (Srivastava et al. 2008) and Cnidaria (Holstein et al. 2003).

The ability of one or a small group of cells to influence other cells via morphogens enables the generation of nonuniform cellular patterns. The function of morphogens is likely to have originally been tied to the physical principle of diffusion. Based on the time–distance–concentration relationships inherent in macromolecular diffusion (Crick 1970), such patterns would form over tens of hours on a spatial scale of 100 μm– 1 mm. Although this is realistic for many developmental systems, evolution has often produced transport processes that are formally equivalent to diffusion, but which, by using additional cell-dependent modalities, are faster or slower than the simple physical process (Lander 2007).

When morphogens are positively autoregulatory, that is, directly or indirectly stimulatory of their own synthesis in target cells, they tend not to be maintained as gradients, since all cells eventually become morphogen sources. This tendency can be held in check, however, if the positively autoregulatory morphogen elicits a mechanism of lateral inhibition (such as the LAT DPM associated with Notch signaling). In this case, a zone will be induced around any peak of morphogen activity within which activation will not spread (Gierer and Meinhardt 1972; Meinhardt and Gierer 2000). Peaks of activation in such systems can form only at distances from one another at which the effects of the inhibitor are attenuated. This arrangement, termed local autoactivation–lateral inhibition (LALI) (Meinhardt and Gierer 2000; Nijhout 2003; Newman and Bhat 2007), which includes the chemical pattern-forming systems described by Turing (1952), can produce regularly spaced spots or stripes of morphogen concentration (TUR; table 11.1). In contemporary metazoans the TUR DPM has been proposed to underlie pattern formation of the vertebrate limb skeleton (Newman and Frisch 1979; Hentschel et al. 2004), the dentition (Salazar-Ciudad and Jernvall 2002), the feather germs (Jiang et al. 2004), and the hair follicles (Sick et al. 2006).

Oscillations in Cell State

As noted in the introduction, cell states, including their "types" (i.e., differentiated states), are determined by intracellular transcription factor- and cis-regulatory sequence module-based gene regulatory networks (GRNs). Just as sufficiently complicated GRNs are inevitably multi-stable, so an appropriate but fairly generic balance of positive and negative feedbacks will cause such systems to exhibit temporal oscillations in concentration of gene products (Goldbeter 1996; Reinke and Gatfield 2006). When coordinated across cell boundaries by juxtacrine (e.g., Notch pathway) and short-range paracrine (e.g., Wnt pathway) signaling, such oscillations have the potential to drive morphogenetic change. Indeed, periodic alteration of adhesion in a growing system is, by itself, sufficient to cause the formation of segmented or partly segmented forms (Newman 1993), which may explain the apparently independent emergence of segmentation in distinct metazoan lineages. We designate the oscillation-associated DPM as OSC (table 11.1).

The OSC DPM is used in conjunction with the MOR DPM in vertebrate somitogenesis. Somitogenesis is the process by which blocks of tissue, the primordia of vertebrae and associated muscles, form in a progressive spatiotemporal order along the central axis of vertebrate embryos. In the presomitic mesoderm of chicken and other vertebrate embryos, the expression of certain genes (particularly that specifying the Notch pathway mediator Hes1) undergoes temporal oscillation with a period similar to the formation of the somites (Dequéant et al. 2008). These oscillations then become synchronized by Notch-mediated juxtacrine signaling (Giudicelli et al. 2007; Kageyama et al. 2007; Riedel-Kruse et al. 2007). In conjunction with an FGF8 morphogen gradient with its source at one end of the extended embryo, the Hes1 and associated oscillations provide the basis for the generation of somites in vertebrate embryos (Dequéant et al. 2008). It has been suggested on theoretical grounds that the OSC DPM has an analogous role in the segmentation of some arthropods (Salazar-Ciudad et al. 2001), and there is some experimental evidence for this in spiders (Damen et al. 2005) and cockroaches (Pueyo et al. 2008).

The synchronization of oscillations will coordinate cell state across broad tissue domains even in the absence of other factors required for segmentation, permitting concerted responses to a variety of developmental signals. For this reason, we have proposed (Newman and Bhat 2009) that the OSC DPM is at the basis of the ubiquious but mechanistically elusive phenomenon of the "morphogenetic field" (Gilbert 2006).

Extracellular Matrices

The DPMs ADH and DAD mediate the formation of "epithelioid" tissues and tissue layers, defined as composed of cells that are directly attached to each other. The physical property of viscosity in epithelioid cell aggregates and tissues depends on the ease with which cells slip past one another while maintaining their often transient attachments. Epithelioid tissue elasticity is primarily a function of the cytoskeleton, since this makes up the bulk of the tissue, and its cohesiveness is determined by the force (dependent on cell adhesion molecules) required to separate the cells.

The other major cell aggregate or tissue type, "mesenchyme," is composed of cells that are embedded in a secreted macromolecular microenvironment, the extracellular matrix (ECM; Comper 1996; see table 11.1). In mesenchymal tissues viscosity, elasticity, and cohesiveness are largely determined by the ECM, making them subject to a range of physical processes not seen in epithelioid tissues. The ECM molecules and the physics they mobilize thus constitute a novel DPM.

Most metazoans produce the glycosaminoglycan hyaluronan and fibrillar (e.g., type I) collagen, which occupy the interstitium between mesenchymal cells and the cells of more mature connective tissues. They also produce network (i.e., type IV) collagen and laminin, which are components of the basement membrane that attaches epithelial sheets to mesenchymal and connective tissues. ECM proteins of all types bind to cell surfaces via transmembrane proteins known as integrins, a subunit of which is specified in the *M. brevicollis* genome (King et al. 2008). Although the Cnidaria, a eumetazoan group, appear to lack interstitial ECMs, genes specifying these components are found in the choanoflagellate *M. brevicollis* (King et al. 2008), where their function is obscure, and in sponges (Nichols et al. 2006), where they serve a role similar to that in the eumetazoans. *M. brevicollis*, like Porifera, Cnidaria, and Placozoa, also has genes for basement membrane components, including subunits of type IV collagen and laminin. Again, it is unclear what function these molecules perform in the single-celled organism, or would have served in its common ancestor with the Metazoa, but they have clearly been recruited to new roles in the multicellular context.

Most sponge cells reside within an ECM called the "mesohyl," to which they bind via integrins (Wimmer et al. 1999). Sponges actively remodel their branched skeletal structures by the continuous movement of their cells (Bond 1992), and their morphology exhibits environment-dependent plasticity (Uriz et al. 2003).

In the Eumetazoa ECM is employed in a more limited fashion, and in the most primitive of these, the Cnidaria, as well as in the Ctenophora, it is used primarily to cement together the epithelial sheets of which these organisms are constructed. A basement membrane endows an epithelial sheet with bending elasticity (Mittenthal and Mazo 1983; Newman 1998), permitting it to exhibit a range of folding, buckling, and wrinkling effects (Gierer 1977; Forgacs and Newman 2005). The other DPMs mentioned above (ADH, DAD, LAT, POL, etc.) also have freer play in the epithelial context, making the ctenophores and the cnidarians morphologically more elaborate than the sponges.

Although organisms that are purely epithelial (cf. Ctenophora, Placozoa, Cnidaria) are structurally feasible, purely mesenchymal organisms would lack structural integrity, and are therefore not likely to persist. It is in the triploblasts, which are simultaneously epithelial and mesenchymal, that all the DPMs (including several not discussed here; Newman and Bhat 2008, 2009) can operate. For example, epithelia can disperse by producing interstitial ECM, a developmentally important effect known as "epithelial–mesenchymal transformation" (Hay 2005) that is used in gastrulation of most triploblasts and in the formation of the neural crest of vertebrates. Triploblastic animals, which include the arthropods, annelids, echinoderms, mollusks, and chordates, thus exhibit all of the morphological motifs potentially generated by what we have termed the metazoan "pattern language" (Newman and Bhat 2009) (figure 11.1).

New Forms, New Niches

The scenarios described above, whereby novel forms emerged relatively abruptly by the mobilization of previously irrelevant physical processes in the multicellular state (itself brought into existence by the newly employed force of cell–cell adhesion), raise the perennial specter of the "hopeful monster" (Goldschmidt 1940) and its presumed incompatibility with modern evolutionary theory (Gould 2002). Unlike Goldschmidt's hypothesis that systemic mutations or macromutations were the driving force of evolutionary innovation, the DPM-based scenarios are more like exaptations (Gould and Vrba 1982) in that all or most of the genetic evolution needed for the mobilization of new physical processes has already occurred by the time the DPMs are brought into play. Since the resulting pattern or form would potentially self-organize in a significant portion of the founding population, there

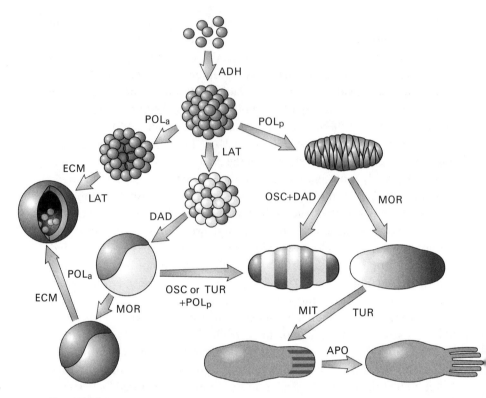

Figure 11.1
Schematic representation of metazoan forms potentially generated by single and combinatorial action of dynamical patterning modules (DPMs). Cells are represented individually in the upper tiers; the middle and lower tiers are shown at the scale of tissues. Beginning at the top, single cells form cell aggregates by the action of the ADH (adhesion; e.g., cadherins) module. The POL (polarity; Wnt pathway) module has two versions, apical–basal (POL_a) and planar (POL_p). POL_a causes cells to have different surface properties at their opposite ends, leading to structurally polarized epithelial sheets and lumens within cell aggregates. POL_p, in contrast, causes cells to elongate and intercalate in the plane, which leads to convergent extension and elongation of the cell mass. The LAT (lateral inhibition; Notch pathway) module transforms an aggregate of homotypic cells into one in which two or more cell types coexist in the same aggregate, while the expression of ADH molecules to different quantitative extents leads to sorting out by the action of the differential adhesion (DAD) module. Production of diffusible molecules leads to morphogen (MOR; e.g., TGF–β/BMP, Hh, FGF) gradients which can affect the same or different cells. Biochemical oscillation (OSC) of key components of the Notch and Wnt pathways, for example, in conjunction with the DAD module, can generate segments. Appropriate feedback relationships among activating and inhibitory morphogens can lead to patterns with repetitive elements by Turing-type or similar LALI processes (TUR). The action of mitogenesis (MIT) and apoptosis (APO) DPMs (Newman and Bhat 2009), which respectively add and remove material (i.e., cells) from the system, reshapes developing primordia. The secretion of extracellular matrix (ECM; e.g., collagen, fibronectin) between cells or into tissue spaces creates new microenvironments for cell translocation or novel mechanical properties in cell sheets or masses. See Newman and Bhat 2008, 2009 for additional details.

would be no question of a single, isolated individual needing to become established on its own.

No doubt new, morphologically aberrant, subpopulations brought into being by DPMs would be poorly adapted to the ecological niches inhabited by the originating species. But niches are not preexisting slots in the natural environment passively occupied by organisms that have the right set of characters. They are explored, selected by, and in many cases constructed by their inhabitants (Levins and Lewontin 1985; Odling-Smee et al. 2003). When novelties arise (see chapter 12 in this volume), particularly (as could be the case with DPM-based innovations) in multiple members of a population, there is no requirement for the new forms to stay put. The phenomenon of "transgressive segregation" in plants, whereby hybrids exhibit phenotypes that are extreme or novel relative to the parental lines, leading to niche divergence (Rieseberg et al. 1999, 2003), provides an analogous mode (albeit with different underlying causation) of function following form in the invention of new ways of life.

Since DPMs by their nature are capable of being elicited by microenvironmental change, the niche to which the new form would initially be best adapted is the one that provided the conditions for its existence. Environmental conditions are generally labile, however, and the most effective way a phenotypically plastic organism can found a morphotype independent of externalities is via consolidating genetic or epigenetic change. Selection for persistence of an environmentally induced phenotype, variously termed genetic assimilation (Waddington 1961) or genetic accommodation (West-Eberhard 2003), can convert condition-dependent morphological characters into products of evolutionary lineage-specific developmental programs (Newman 1994; Newman and Müller 2000) (figure 11.2).

Discussion

The view described here of the origination of metazoan morphological and pattern motifs, and ultimately of body plans and organ forms, both diverges from and complements the Modern Synthesis in several ways. According to Levins and Lewontin (1985: 32), Darwin's mechanism of evolution can be summarized in the following three propositions:

1. Individuals within a species vary in physiology, morphology, and behavior: the principle of variation.

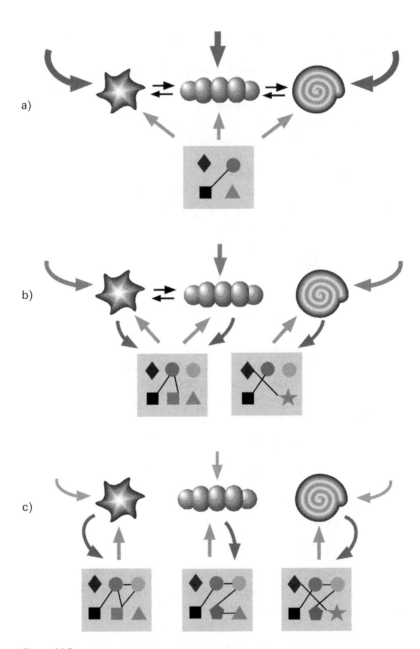

Figure 11.2
Schematic representation of evolutionary partitioning of a morphologically plastic ancestral organism into distinct morphotypes associated with unique genotypes. (A) A hypothetical primitive metazoan of the Precambrian is shown with a schematic representation of
its genome in the box below it. Developmental–genetic tool-kit genes, both those involved

2. Offspring resemble their parents on the average more than they resemble unrelated individuals: the principle of heredity.

3. Different variants leave different numbers of offspring: the principle of natural selection.

There is nothing in the DPM-based scenarios for origination and innovation of form that is inconsistent with the first proposition, the principle of variation, if "variation" is not confined to those phenotypic differences between individuals that are independent of external conditions. But Levins and Lewontin intended their characterization of Darwin's mechanism to include only the more limited definition of variation, as becomes clear in a later passage in the same work:

Before Darwin, theories of historical change were all transformational. That is, systems were seen as undergoing change in time because each element in the system underwent a transformation during its life history. Lamarck's theory of evolution was transformational in regarding species as changing because each individual organism underwent an individual transformation during its life history.... In contrast, Darwin proposed a variational principle, that individual members of the ensemble differ from each other in some properties and the system evolves by changes in the proportions of the different types. There is a sorting-out process in which some variant types persist while others disappear, so the nature of the ensemble changes without any successive changes in the individual members. (Levins and Lewontin 1985: 86)

in cell-type-determining GRNs and those involved in form-and-pattern-determining DPMs, are shown as shaded geometric objects; interactions between them, by lines. Determinants of the organism's form include the products of expression of its genes (arrows extending from genomes to forms) and the physicochemical external environment (broad arrows pointing to forms from top) acting on its inherent physical properties. At this stage of evolution the organism is highly plastic, exhibiting several condition-dependent forms that are mutually interconvertible (black horizontal arrows). (B) Descendants of organism in A after some stabilizing evolution. Gene duplication, mutation, and so on have led to genetic integration, development, and assimilation of some outcomes that were previously more dependent on the environment, as well as some subpopulations being biased toward subsets of the original morphological phenotypes. Determinants of form are still gene products and the physical environment, but the effect of the latter has become attenuated (smaller, fainter arrows from the top) as development has become more programmatic. There is also causal influence of the form on the genotype (arrows from forms to genomes), exerted over evolutionary time, as niche construction filters out those variant genotypes that are not compatible with the established form. Some morphotypes remain interconvertible at this stage of evolution, but others are not. (C) Modern organisms descended from those in B. Further stabilizing evolution has now led to each morphotype being uniquely associated with its own genotype. Physical causation is even more attenuated. In this idealized example the forms have remained unchanged while the genes and mechanisms for generating the forms have undergone extensive evolution. This is a highly schematic representation which is not meant to suggest that the Precambrian forms were identical to modern ones. (Adapted, with changes, from Newman et al. 2006).

Strictly speaking, DPMs could act in a variational fashion: they could be evoked in a subset of a population's members as a result of environment-independent variation in genes for the associated molecular components. But physical processes are also components of DPMs: a temperature shift could cause a previously nonoscillatory genetic network to oscillate, a pressure change could affect the assembly and rheological properties of ECM molecules, a concentration change in ambient cations could render a cell surface protein sticky. In such cases DPMs would act in a transformational fashion.

The DPM framework also extends evolutionary change beyond the second Darwinian proposition, the principle of heredity. Insofar as parents and offspring are not distinguished by phenotypic novelties (e.g., segmentation) due to the action of DPMs in members of only one of the generations, they will resemble each other more than they do less related individuals. But if offspring of unrelated, unsegmented forms undergo DPM-related transformations that render them segmented (due either to genetic or to environmental means), they will resemble each other in this important respect more than either group resembles their parents. As this example demonstrates, the evolutionary perspective described here includes genetics, but is not genetic-determinist.

Regarding the third proposition, the principle of natural selection, the DPM framework assigns it a lesser role than does the Modern Synthesis. While it is not needed to produce morphological and, more generally, phenotypic novelties, it will always be available to act in a phenotype-consolidating or -stabilizing capacity. Because genetic variability is an inevitable feature of populations of living organisms, genotypes associated with increased reliability of developmental outcome will often be selected, particularly if there is a premium on breeding true, leading to what has been termed genetic assimilation or accommodation (Waddington 1961; West-Eberhard 2003; see also chapter 14 in this volume).

As other contributions to this volume attest, the Modern Synthesis is a flexible and open framework (see also Pigliucci 2007). It is capable of incorporating such concepts as multilevel selection, variations in tempo and mode, niche construction, and evolvability. But it is a framework, and is therefore not infinitely malleable. The hallmark of Darwin's theory of evolution and its lineal successors is incremental change via selection of genetically variant individuals or groups. A theory that featured saltational changes via collective transformation, followed by genetic consolidation via drift or stabilizing selection, would not be

Darwinian by the criteria of Levins and Lewontin or most other descriptions (see above).

One thing that the Modern Synthesis was never claimed to contain is a theory of the origination and innovation of form. The new focus on this missing element by the emerging field of evolutionary–developmental biology (Robert 2004; Müller and Newman 2005; Müller 2007; Callebaut et al. 2007; Moczek 2008; Minelli and Fusco 2008) has brought to center stage a number of properties of developmental systems that were previously marginal to evolutionary theory. These include developmental and phenotypic plasticity and genotype–phenotype discordances (Newman 1994; Trut et al. 2009; Pigliucci 2001; West-Eberhard 2003; Badyaev 2005; Badyaev et al. 2005; Salazar-Ciudad 2006; Goodman 2008), determination of form by physical and epigenetic factors (Müller and Streicher 1989; Newman and Comper 1990; Newman and Müller 2000), and inheritance systems that extend beyond the gene (Jablonka and Lamb 1995, 2005).

The DPM hypothesis for the emergence of the Metazoa incorporates these various conceptual strains in a natural fashion by indicating how the development–genetic tool kit that preexisted morphologically complex animals mobilized mesoscale physical processes once the first of the DPMs, cell–cell adhesion (ADH), brought about multicellularity. The rapidity by which this would have occurred, considering the relatively small degree of genetic change required to produce many disparate morphotypes from a primordial metazoan, provides a plausible scenario for a "radiation compressed in time" (Rokas et al. 2005), notwithstanding the striking conservation of the developmental–genetic tool kit (Newman 2006).

Since adaptive selection is not the engine of morphological innovation in this new view (Müller 2007; see also chapter 12 in this volume), the problem of enhanced fitness of incrementally different intermediate forms (a difficulty of strict Darwinian accounts) is no longer an issue. But since potential saltational change is an element of this scenario, the relation of novel forms to their originating niches raises new issues of its own. Here the concept of niche construction enters the picture (see chapter 8 in this volume), providing a means for novel forms to establish new ways of life, with isolation eventually leading to genetic divergence from the originating population and loss of the capacity to revert to the initial phenotype (figure 11.2).

Finally, it will be noted that the view presented here differs from the classic Darwinian paradigm in one more important sense: the role of uniformitarianism. This is a tenet that Darwin adopted from the geologists Hutton and Lyell, which holds that the natural processes operating in the past are the same as those that can be observed to operate in the present (Gould 1987). Clearly, physical laws and basic genetic mechanisms have remained constant throughout the evolutionary history of all taxonomic lineages, including the Metazoa. However, the degree of physically based plasticity of the evolving forms will decrease over time due to the progressive consolidation of generative pathways by canalizing (Waddington 1942) or stabilizing (Schmalhausen 1949) selection (figure 11.2). As has been frequently noted (e.g., Conway Morris 2006), once the metazoan phyla were established more than half a billion years ago, no additional groups of similar grade emerged. Darwinian uniformitarianism would have predicted otherwise.

The view presented here provides an integrated account of macro- and microevolutionary change for the early phases of multicellular life. At more advanced stages of evolution, however, when plasticity will be less prevalent and stricter genotype–phenotype associations become the norm, the mode of incremental modification by adaptive selection featured in the Modern Synthesis, all of whose tenets are entirely consistent with the DPM framework, will come to assume a governing role in organismal change.

Acknowledgments

I thank Massimo Pigliucci and Gerd Müller for providing a forum and an incubator for these ideas at the Konrad Lorenz Institute, and Ramray Bhat, Gerd Müller, and Isaac Salazar-Ciudad for helpful discussions. This work was supported by the National Science Foundation program in Frontiers of Integrative Biological Research.

References

Abedin M, King N (2008) The premetazoan ancestry of cadherins. Science 319: 946–948.

Badyaev AV (2005) Stress-induced variation in evolution: From behavioural plasticity to genetic assimilation. Proceedings of the Royal Society of London B272: 877–886.

Badyaev AV, Foresman KR, Young RL (2005) Evolution of morphological integration: Developmental accommodation of stress-induced variation. American Naturalist 166: 382–395.

Blankenship JR, Mitchell AP (2006) How to build a biofilm: A fungal perspective. Current Opinion in Microbiology 9: 588–594.

Bond C (1992) Continuous cell movements rearrange anatomical structures in intact sponges. Journal of Experimental Zoology 263: 284–302.

Callebaut W, Müller GB, Newman SA (2007) The organismic systems approach: Streamlining the naturalistic agenda. In Integrating Evolution and Development: From Theory to Practice. R Sansom and RN Brandon, eds.: 25–92. Cambridge, MA: MIT Press.

Carroll SB, Grenier JK, Weatherbee SD (2005) From DNA to Diversity: Molecular Genetics and the Evolution of Animal Design. 2nd ed. Malden, MA: Wiley-Blackwell.

Chen J-Y, Bottjer DJ, Oliveri P, Dornbos SQ, Gao F, Ruffins S, Chi H, Li C-W, Davidson EH (2004) Small bilaterian fossils from 40 to 55 million years before the Cambrian. Science 305: 218–222.

Comper WD, ed. (1996) Extracellular Matrix. Vol I, Tissue Function; vol. II, Molecular Components and Interactions. Amsterdam: Harwood Academic Publishers.

Conway Morris S (2006) Darwin's dilemma: The realities of the Cambrian "explosion." Philosophical Transactions of the Royal Society of London B361: 1069–1083.

Crick FHC (1970) Diffusion in embryogenesis. Nature 225: 420–422.

Damen WGM, Janssen R, Prpic N-M (2005) Pair rule gene orthologs in spider segmentation Evolution & Development 7: 618–628.

Davidson EH (2006) The Regulatory Genome: Gene Regulatory Networks in Development and Evolution. 2nd rev. ed. Amsterdam: Elsevier; London: Academic Press.

De Gennes PG (1992) Soft matter. Science 256: 495–497.

Dequéant ML, Pourquié O (2008) Segmental patterning of the vertebrate embryonic axis. Nat Rev Genet 9: 370–382.

Droser ML, Gehling JG (2008) Synchronous aggregate growth in an abundant new Ediacaran tubular organism. Science 319: 1660–1662.

Dunn CW, Hejnol A, Matus DQ, Pang K, Browne WE, Smith SA, Seaver E, Rouse GW, Obst M, Edgecombe GD, Sørensen MV, Haddock SHD, Schmidt-Rhaesa A, Okusu A, Kristensen RM, Wheeler WC, Martindale MQ, Giribet G (2008) Broad phylogenomic sampling improves resolution of the animal tree of life. Nature 452: 745–749.

Ehebauer M, Hayward P, Arias AM (2006) Notch, a universal arbiter of cell fate decisions. Science 314: 1414–1415.

Erwin DH (2008) Wonderful Ediacarans, wonderful cnidarians? Evolution and Development 10: 263–264.

Forgacs G, Newman SA (2005) Biological Physics of the Developing Embryo. Cambridge: Cambridge University Press.

Gierer A (1977) Physical aspects of tissue evagination and biological form. Quarterly Reviews of Biophysics 10: 529–593.

Gierer A, Meinhardt H (1972) A theory of biological pattern formation. Kybernetik 12: 30–39.

Gilbert SF (2006) Developmental Biology. 8th ed. Sunderland, MA: Sinauer.

Giudicelli F, Ozbudak EM, Wright GJ, Lewis J (2007) Setting the tempo in development: An investigation of the zebrafish somite clock mechanism. PLoS Biology 5: e150.

Goldbeter A (1996) Biochemical Oscillations and Cellular Rhythms: The Molecular Bases of Periodic and Chaotic Behaviour. Cambridge: Cambridge University Press.

Goldschmidt RB (1940) The Material Basis of Evolution. New Haven, CT: Yale University Press.

Goodman RM (2008) Latent effects of egg incubation temperature on growth in the lizard Anolis carolinensis. Journal of Experimental Zoology A309: 1–9.

Gould SJ (1987) Time's Arrow, Time's Cycle: Myth and Metaphor in the Discovery of Geological Time. Cambridge, MA: Harvard University Press.

Gould SJ (2002) The Structure of Evolutionary Theory. Cambridge, MA: Belknap Press of Harvard University Press.

Gould SJ, Vrba E (1982) Exaptation: A missing term in the science of form. Paleobiology 8: 4–15.

Guder C, Philipp I, Lengfeld T, Watanabe H, Hobmayer B, Holstein TW (2006) The Wnt code: Cnidarians signal the way. Oncogene 25: 7450–7460.

Hagadorn JW, Xiao S, Donoghue PCJ, Bengtson S, Gostling NJ, Pawlowska M, Raff EC, Raff RA, Turner FR, Chongyu Y, Zhou C, Yuan X, McFeely MB, Stampanoni M, Nealson KH (2006) Cellular and subcellular structure of neoproterozoic animal embryos. Science 314: 291–294.

Hay ED (2005) The mesenchymal cell, its role in the embryo, and the remarkable signaling mechanisms that create it. Developmental Dynamics 233: 706–720.

Hentschel HGE, Glimm T, Glazier JA, Newman SA (2004) Dynamical mechanisms for skeletal pattern formation in the vertebrate limb. Proceedings of the Royal Society of London B271: 1713–1722.

Holstein TW, Hobmayer E, Technau U (2003) Cnidarians: An evolutionarily conserved model system for regeneration? Developmental Dynamics 226: 257–267.

Jablonka E, Lamb MJ (1995) Epigenetic Inheritance and Evolution: The Lamarckian Dimension. Oxford: Oxford University Press.

Jablonka E, Lamb MJ (2005) Evolution in Four Dimensions: Genetic, Epigenetic, Behavioral, and Symbolic Variation in the History of Life. Cambridge, MA: MIT Press.

Jiang T-X, Widelitz RB, Shen W-M, Will P, Wu D-Y, Lin C-M, Jung H-S, Chuong C-M (2004) Integument pattern formation involves genetic and epigenetic controls: Feather arrays simulated by digital hormone models. International Journal of Developmental Biology 48: 117–135.

Kageyama R, Masamizu Y, Niwa Y (2007) Oscillator mechanism of Notch pathway in the segmentation clock. Developmental Dynamics 236: 1403–1409.

Kaiser D (2001) Building a multicellular organism. Annual Review of Genetics 35: 103–123.

Kaneko K (2006) Life: An Introduction to Complex Systems Biology. Berlin and New York: Springer.

Karner C, Wharton KA, Carroll TJ (2006a) Apical–basal polarity, Wnt signaling and vertebrate organogenesis. Seminars in Cell and Developmental Biology 17: 214–222.

Karner C, Wharton KA Jr., Carroll TJ (2006b) Planar cell polarity and vertebrate organogenesis. Seminars in Cell and Developmental Biology 17: 194–203.

Kauffman SA (1969) Metabolic stability and epigenesis in randomly constructed genetic nets. Journal of Theoretical Biology 22: 437–467.

Kaufman M, Soulé C, Thomas R (2007) A new necessary condition on interaction graphs for multistationarity. Journal of Theoretical Biology 248: 675–685.

Kazmierczak J, Kempe S (2004) Calcium build-up in the Precambrian sea: A major promoter in the evolution of eukaryotic life. In Origins: Genesis, Evolution and Diversity of Life. J Seckbach, ed.: 329–345. Dordrecht: Kluwer.

Keller AD (1995) Model genetic circuits encoding autoregulatory transcription factors. Journal of Theoretical Biology 172: 169–185.

Keller R (2002) Shaping the vertebrate body plan by polarized embryonic cell movements. Science 298: 1950–1954.

Keller R, Davidson L, Edlund A, Elul T, Ezin M, Shook D, Skoglund P (2000) Mechanisms of convergence and extension by cell intercalation. Philosophical Transactions of the Royal Society of London B355: 897–922.

King N, Hittinger CT, Carroll SB (2003) Evolution of key cell signaling and adhesion protein families predates animal origins. Science 301: 361–363.

King N, Westbrook MJ, Young SL, Kuo A, Abedin M, Chapman J, Fairclough S, Hellsten U, Isogai Y, Letunic I, Marr M, Pincus D, Putnam N, Rokas A, Wright KJ, Zuzow R, Dirks W, Good M, Goodstein D, Lemons D, Li W, Lyons JB, Morris A, Nichols S, Richter DJ, Salamov A, Sequencing JGI, Bork P, Lim WA, Manning G, Miller WT, McGinnis W, Shapiro H, Tjian R, Grigoriev IV, Rokhsar D (2008) The genome of the choanoflagellate *Monosiga brevicollis* and the origin of metazoans. Nature 451: 783–788.

Koelzer S, Klein T (2003) A Notch-independent function of suppressor of Hairless during the development of the bristle sensory organ precursor cell of *Drosophila*. Development 130: 1973–1988.

Lai EC (2002) Keeping a good pathway down: Transcriptional repression of Notch pathway target genes by CSL proteins. EMBO Reports 3: 840–845.

Lander AD (2007) Morpheus unbound: Reimagining the morphogen gradient. Cell 128: 245–256.

Lang BF, O'Kelly C, Nerad T, Gray MW, Burger G (2002) The closest unicellular relatives of animals. Current Biology 12: 1773–1778.

Larroux C, Fahey B, Liubicich D, Hinman VF, Gauthier M, Gongora M, Green K, Wörheide G, Leys SP, Degnan BM (2006) Developmental expression of transcription factor genes in a demosponge: Insights into the origin of metazoan multicellularity. Evolution and Development 8: 150–173.

Larroux C, Luke GN, Koopman P, Rokhsar DS, Shimeld SM, Degnan BM (2008) Genesis and expansion of metazoan transcription factor gene classes. Molecular Biology and Evolution 25: 980–996.

Laurent M, Kellershohn N (1999) Multistability: A major means of differentiation and evolution in biological systems. Trends in Biochemical Sciences 24: 418–422.

Levins R, Lewontin RC (1985) The Dialectical Biologist. Cambridge, MA: Harvard University Press.

Luporini P, Vallesi A, Alimenti C, Ortenzi C (2006) The cell type-specific signal proteins (pheromones) of protozoan ciliates. Current Pharmaceutical Design 12: 3015–3024.

Meinhardt H, Gierer A (2000) Pattern formation by local self-activation and lateral inhibition. BioEssays 22: 753–760.

Mendoza M, Redemann S, Brunner D (2005) The fission yeast MO25 protein functions in polar growth and cell separation. European Journal of Cell Biology 84: 915–926.

Mezentseva NV, Kumaratilake JS, Newman SA (2008) The brown adipocyte differentiation pathway in birds: An evolutionary road not taken. BMC Biology 6: 17.

Mikhailov AS (1990) Foundations of Synergetics I: Distributed Active Systems. Berlin: Springer.

Miller DJ, Ball EE (2005) Animal evolution: The enigmatic phylum Placozoa revisited. Current Biology 15: 26–28.

Minelli A, Fusco G, eds. (2008) Evolving Pathways: Key Themes in Evolutionary Developmental Biology. Cambridge: Cambridge University Press.

Mittenthal JE, Mazo RM (1983) A model for shape generation by strain and cell–cell adhesion in the epithelium of an arthropod leg segment. Journal of Theoretical Biology 100: 443–483.

Moczek AP (2008) On the origins of novelty in development and evolution. BioEssays 30: 432–447.

Müller GB (2007) Evo-devo: Extending the evolutionary synthesis. Nature Reviews Genetics 8: 943–949.

Müller GB, Newman SA, eds. (2003) Origination of Organismal Form: Beyond the Gene in Developmental and Evolutionary Biology. Cambridge, MA: MIT Press.

Müller GB, Newman SA (2005) The innovation triad: An EvoDevo agenda. Journal of Experimental Zoology B304: 487–503.

Müller GB, Streicher J (1989) Ontogeny of the syndesmosis tibiofibularis and the evolution of the bird hindlimb: A caenogenetic feature triggers phenotypic novelty. Anatomy and Embryology 179: 327–339.

Newman SA (1993) Is segmentation generic? BioEssays 15: 277–283.

Newman SA (1994) Generic physical mechanisms of tissue morphogenesis: A common basis for development and evolution. Journal of Evolutionary Biology 7: 467–488.

Newman SA (1998) Epithelial morphogenesis: A physico-evolutionary interpretation. In Molecular Basis of Epithelial Appendage Morphogenesis. C-M Chuong, ed.: 341–358. Austin, TX: R.G. Landes.

Newman SA (2006) The developmental–genetic toolkit and the molecular homology–analogy paradox. Biological Theory 1: 12–16.

Newman SA, Bhat R (2007) Activator–inhibitor dynamics of vertebrate limb pattern formation. Birth Defects Research C Embryo Today 81: 305–319.

Newman SA, Bhat R (2008) Dynamical patterning modules: Physico-genetic determinants of morphological development and evolution. Physical Biology 5: 015008.

Newman SA, Bhat R (2009) Dynamical patterning modules: A "pattern language" for development and evolution of multicellular form. International Journal of Developmental Biology (in press).

Newman SA, Comper WD (1990) "Generic" physical mechanisms of morphogenesis and pattern formation. Development 110: 1–18.

Newman SA, Forgacs G, Müller GB (2006) Before programs: The physical origination of multicellular forms. International Journal of Developmental Biology 50: 289–299.

Newman SA, Frisch HL (1979) Dynamics of skeletal pattern formation in developing chick limb. Science 205: 662–668.

Newman SA, Müller GB (2000) Epigenetic mechanisms of character origination. Journal of Experimental Zoology B288: 304–317.

Nichols SA, Dirks W, Pearse JS, King N (2006) Early evolution of animal cell signaling and adhesion genes. Proceedings of the National Academy of Sciences of the USA 103: 12451–12456.

Nijhout HF (2003) Gradients, diffusion and genes in pattern formation. In Origination of Organismal Form: Beyond the Gene in Developmental and Evolutionary Biology. GB Müller, SA Newman, eds.: 165–181. Cambridge, MA: MIT Press.

Odling-Smee FJ, Laland KN, Feldman MW (2003) Niche Construction: The Neglected Process in Evolution. Princeton, NJ: Princeton University Press.

Pan J, Snell WJ (2000) Signal transduction during fertilization in the unicellular green alga, Chlamydomonas. Current Opinion in Microbiology 3: 596–602.

Philippe H, Snell EA, Bapteste E, Lopez P, Holland PWH, Casane D (2004) Phylogenomics of eukaryotes: Impact of missing data on large alignments. Molecular Biology and Evolution 21: 1740–1752.

Piatigorsky J, Wistow GJ (1989) Enzyme/crystallins: Gene sharing as an evolutionary strategy. Cell 57: 197–199.

Pigliucci M (2001) Phenotypic Plasticity: Beyond Nature and Nurture. Baltimore: Johns Hopkins University Press.

Pigliucci M (2007) Do we need an extended evolutionary synthesis? Evolution 61: 2743–2749.

Převorovský M, Půta F, Folk P (2007) Fungal CSL transcription factors. BMC Genomics 8: 233.

Pueyo JI, Lanfear R, Couso JP (2008) Ancestral Notch-mediated segmentation revealed in the cockroach Periplaneta americana. Proceedings of the National Academy of Sciences of the USA 105: 16614–16619.

Reinke H, Gatfield D (2006) Genome-wide oscillation of transcription in yeast. Trends in Biochemical Sciences 31: 189–191.

Rentzsch F, Fritzenwanker JH, Scholz CB, Technau U (2008) FGF signalling controls formation of the apical sensory organ in the cnidarian *Nematostella vectensis*. Development 135: 1761–1769.

Riedel-Kruse IH, Muller C, Oates AC (2007) Synchrony dynamics during initiation, failure, and rescue of the segmentation clock. Science 317: 1911–1915.

Rieseberg LH, Archer MA, Wayne RK (1999) Transgressive segregation, adaptation and speciation. Heredity 83: 363–372.

Rieseberg LH, Widmer A, Arntz AM, Burke JM (2003) The genetic architecture necessary for transgressive segregation is common in both natural and domesticated populations. Philosophical Transactions of the Royal Society of London B358: 1141–1147.

Robert JS (2004) Embryology, Epigenesis, and Evolution: Taking Development Seriously. Cambridge and New York: Cambridge University Press.

Rokas A, Kruger D, Carroll SB (2005) Animal evolution and the molecular signature of radiations compressed in time. Science 310: 1933–1938.

Rose SM (1958) Feedback in the differentiation of cells. Scientific American 199: 36–41.

Röttinger E, Saudemont A, Duboc V, Besnardeau L, McClay D, Lepage T (2008) FGF signals guide migration of mesenchymal cells, control skeletal morphogenesis and regulate gastrulation during sea urchin development. Development 135: 353–365.

Ryals PE, Smith-Somerville HE, Buhse HE, Jr. (2002) Phenotype switching in polymorphic *Tetrahymena*: A single-cell Jekyll and Hyde. International Review of Cytology 212: 209–238.

Salazar-Ciudad I (2006) On the origins of morphological disparity and its diverse developmental bases. BioEssays 28: 1112–1122.

Salazar-Ciudad I, Jernvall J (2002) A gene network model accounting for development and evolution of mammalian teeth. Proceedings of the National Academy of Sciences of the USA 99: 8116–8120.

Salazar-Ciudad I, Jernvall J, Newman SA (2003) Mechanisms of pattern formation in development and evolution. Development 130: 2027–2037.

Salazar-Ciudad I, Solé RV, Newman SA (2001) Phenotypic and dynamical transitions in model genetic networks. II. Application to the evolution of segmentation mechanisms. Evolution and Development 3: 95–103.

Schmalhausen II (1949) Factors of Evolution. T Dobzhansky, ed. Philadelphia: Blakiston.

Shen B, Dong L, Xiao S, Kowalewski M (2008) The Avalon explosion: Evolution of Ediacara morphospace. Science 319: 81–84.

Sick S, Reinker S, Timmer J, Schlake T (2006) WNT and DKK determine hair follicle spacing through a reaction–diffusion mechanism. Science 314: 1447–1450.

Srivastava M, Begovic E, Chapman J, Putnam NH, Hellsten U, Kawashima T, Kuo A, Mitros T, Salamov A, Carpenter ML, Signorovitch AY, Moreno MA, Kamm K, Grimwood J, Schmutz J, Shapiro H, Grigoriev IV, Buss LW, Schierwater B, Dellaporta SL, Rokhsar DS (2008) The *Trichoplax* genome and the nature of placozoans. Nature 454: 955–960.

Steinberg MS (2003) Cell adhesive interactions and tissue self-organization. In Origination of Organismal Form: Beyond the Gene in Developmental and Evolutionary Biology. GB Müller and SA Newman, eds.: 137–163. Cambridge, MA: MIT Press.

Steinberg MS, Takeichi M (1994) Experimental specification of cell sorting, tissue spreading, and specific spatial patterning by quantitative differences in cadherin expression. Proceedings of the National Academy of Sciences of the USA 91: 206–209.

Sudhop S, Coulier F, Bieller A, Vogt A, Hotz T, Hassel M (2004) Signalling by the FGFR-like tyrosine kinase, Kringelchen, is essential for bud detachment in *Hydra vulgaris*. Development 131: 4001–4011.

Süel GM, Kulkarni RP, Dworkin J, Garcia-Ojalvo J, Elowitz MB (2007) Tunability and noise dependence in differentiation dynamics. Science 315: 1716–1719.

Tompa P, Szász C, Buday L (2005) Structural disorder throws new light on moonlighting. Trends in Biochemical Sciences 30: 484–489.

Trut L, Oskina I, Kharlamova A (2009) Animal evolution during domestication: the domesticated fox as a model. BioEssays 31: 349–360.

Turing AM (1952) The chemical basis of morphogenesis. Philosophical Transactions of the Royal Society of London B237: 37–72.

Uriz M-J, Turon X, Becerro MA, Agell G (2003) Siliceous spicules and skeleton frameworks in sponges: Origin, diversity, ultrastructural patterns, and biological functions. Microscopy Research and Technique 62: 279–299.

Vlamakis H, Aguilar C, Losick R, Kolter R (2008) Control of cell fate by the formation of an architecturally complex bacterial community. Genes & Development 22: 945–953.

Waddington CH (1942) Canalization of development and the inheritance of acquired characters. Nature 150: 563–565.

Waddington CH (1961) Genetic assimilation. Advances in Genetics 10: 257–293.

Wainright PO, Hinkle G, Sogin ML, Stickel SK (1993) Monophyletic origins of the metazoa: An evolutionary link with fungi. Science 260: 340–342.

West-Eberhard MJ (2003) Developmental Plasticity and Evolution. New York: Oxford University Press.

Wilkins AS (2002) The Evolution of Developmental Pathways. Sunderland, MA: Sinauer.

Wimmer W, Perovic S, Kruse M, Schröder HC, Krasko A, Batel R, Müller WE (1999) Origin of the integrin-mediated signal transduction: Functional studies with cell cultures from the sponge *Suberites domuncula*. European Journal of Biochemistry 260: 156–165.

Yin L, Zhu M, Knoll AH, Yuan X, Zhang J, Hu J (2007) Doushantuo embryos preserved inside diapause egg cysts. Nature 446: 661–663.

Zhu AJ, Scott MP (2004) Incredible journey: How do developmental signals travel through tissue? Genes & Development 18: 2985–2997.

12 Epigenetic Innovation

Gerd B. Müller

The evolution of organismal forms consists of the generation, fixation, and variation of phenotypic characters. The Modern Synthesis concentrated on the last, adaptive variation, essentially avoiding the problem of how complex morphological traits originate and how specific combinations of traits become stabilized as body plans. This variational bias of the Modern Synthesis theory derives from three assumptions necessary for its population genetic formalism to work: (1) all evolution is continuous and gradual; (2) its basis is random and heritable genetic variation; and (3) genetic variation relates directly to phenotypic variation. The resulting abstract scenario accounts for the behavior of gene variation in ideal populations, and natural selection is the sole factor that causes specific phenotypic solutions, which, by the same token, are all adaptive. Therefore, since all its starting points are variational, the Modern Synthesis is confined to explaining, in principle, how existing traits diversify.

Despite the concentration of standard evolutionary theory on variation, the problem of innovation has not escaped attention. Darwin (1859), responding to some extent to Mivart's criticism, had mentioned on several occasions that "characters may have originated from quite secondary causes, independently from natural selection," and one of the key architects of the Modern Synthesis called innovation "a neglected problem ... in spite of its importance in a theory of evolution" (Mayr 1960). Schmalhausen, among others (e.g., Rensch 1959), acknowledged the problem of "new differentiations" and suggested considering developmental "tissue reactivity" as its source (Schmalhausen 1949), but the overwhelming success of the population genetic approach during the decades following the Modern Synthesis all but sidelined the issue of innovation. It was more rewarding to calculate the variation of the existing rather than to puzzle over the origination of the unprecedented.

During the past few decades, and through the rise of EvoDevo in particular, the situation has changed. Some of the central tenets of the Modern Synthesis have come under scrutiny, and new insights have gained acceptance: phenotypic evolution is not always gradual (Eldredge and Gould 1972; Pagel et al. 2006); not all traits are necessarily adaptive (Gould and Lewontin 1979; Alberch and Gale 1985); and the relationship between genetic variation and the phenotype is far from being simple or direct (Altenberg 2005). Rather, a complex apparatus of developmental transformation intervenes between genotype and phenotype, and the science of EvoDevo has begun to elucidate the evolutionary roles of this apparatus. This has made it possible to begin to address a suite of problems at the phenotype level of evolution that were excluded by the Modern Synthesis approach, such as the origin of structural complexity, biased variation, rapid change of form, and others (Müller and Newman 2005a), the problem of innovation figuring prominently among them. Since the early 1990s a rising number of research papers (overviews in Müller and Newman 2005b; Mozek 2008), books (Nitecki 1990; Margulis and Fester 1991; Schwartz 1999; Müller and Newman 2003; Reid 2007), special issues of journals (Müller and Newman 2005b), and doctoral dissertations (Love 2005) have focused on evolutionary innovation, addressing both empirical and conceptual themes.

Several essential questions arise from the treatment of innovation as a special subject in evolutionary theory. Foremost, do phenotypic novelties exist at all, and if so, how are they distinguished from ordinary forms of variation? This is of central importance, because if novelties merely represented a special case of variation, no major consequences for evolutionary theory would ensue. Moreover, if a distinction between phenotypic variation and innovation is possible, are the underlying mechanisms in the realization of innovations also distinct, and what is the role of development in this process? And, finally, does innovation have specific consequences for the patterns and dynamics of organismal evolution? The present chapter explores these questions from an epigenetic perspective (see chapter 7 in this volume for a characterization of the main types of epigenetic research) and summarizes the conceptual contributions of the innovation approach to an extended evolutionary synthesis.

The Variation–Innovation Distinction

In the standard scheme of evolutionary theory, the problem of phenotypic innovation, if addressed at all, was treated as part of the variation issue by calling all changes of form "variants," irrespective of whether or not they were actually based on the variation of a character already present in the primitive condition. Accordingly, morphological novelties were seen as "major variants," without any need for further criteria that would allow one to distinguish innovation from variation. This was permissible as long as the theory treated evolution as a statistical relationship between genetic variation and phenotypic change, but the turn toward the mechanistic explanation of phenotypic change introduced by EvoDevo meant that innovation and novelty could not just be treated in the same way as variation (Müller and Wagner 1991; Müller and Newman 2005b; Salazar-Ciudad and Jernvall 2005). Two major problems of phenotypic innovation were identified: (1) the first origin of clade-specific combinations of structural building elements (body plans) and (2) the insertion of new elements into existing body plans during further evolution.

The origin of metazoan body plans is a unique evolutionary occurrence near the Precambrian–Cambrian boundary, a period during which the majority of clade-specific anatomical architectures appeared in what looks, geologically, like a burst of forms. Even though only around 35 body plans (corresponding to major phyla) exist today, estimates are that more than 100 may actually have originated before and during the Cambrian explosion (Valentine 2004), indicating that a greater generative potential had existed than can be deduced from present forms. Still, the number of body plans ever produced is minute compared with the millions of species that are differentiated elaborations of the basic set. This points to the peculiarity of the evolutionary problem of body plan origination, not covered by the variational approach to evolution. Whence this limited array of basic anatomical themes?

Although the pre-metazoan organisms from which complex forms arose already possessed genomes, the anatomical assemblies that constituted first metazoan body plans were not built directly from genes, but from cells that combined as a consequence of their physical properties, which were, in part, mediated by gene products. Hence these structures were not immediate results of genetic evolution, but represented an emergent consequence of cell and tissue organization. The recent attention given to comparative gene expression in the embryonic development of basic body architectures (Carroll et al. 2005; Davidson 2006; Wilkins 2002) addresses the problem at the level of extant gene regulation. But what is needed for an understanding of primordial body plans is a vocabulary of cell and tissue assemblies that can physically result

Table 12.1
Classification of morphological novelty

type I novelty	Primary anatomical architecture of a metazoan body plan
type II novelty	Discrete new element added to an existing body plan
type III novelty	Major change of an existing body plan character

from differential cell properties. Only the consideration of the rules that determine the generation of macroscopic tissue components can lead to a mechanistic explanation of body plan evolution. The molecular and cellular mechanisms underlying this aspect of innovation, also referred to as the "origination problem" (Müller and Newman 2005b), is the topic of chapter 11 in this volume. Hereafter the origin of the primary architectures of metazoan body plans will be called "type I novelty" (table 12.1).

In this chapter we concentrate on a second problem of innovation, the occurrence of morphological novelty in established metazoan lineages. Although they were previously regarded as rare events, compared with the multitude of phenotypic changes realized through the variation of existing parts, recent literature indicates that innovations are more frequent than usually perceived (Erwin 1993; Jernvall et al. 1996; see also chapter 13 in this volume). Higher taxa originations are often associated with newly introduced features that are added to, or individualized from, the established body plan of a lineage. Again the question is not one of variation but one of innovation, that is, given the existing basic architecture, how could unprecedented elements be added if variation were the only mechanistic possibility?

In order to approach this question in a meaningful way, it is necessary to determine the actual phenomenon that requires explanation, namely, what constitutes a phenotypic innovation or a morphological novelty. Are any definitions at hand? In the vein of the Modern Synthesis this problem was usually avoided by defining novelty at a functional level, as in Mayr (1960), who proposed that "tentatively, one might restrict the designation 'evolutionary novelty' to any newly acquired structure or property which permits the assumption of a new function." This kind of definition recognizes the scientific problem but makes it impossible to distinguish between quantitative variation and qualitative innovation, since both can permit new functions.

To grasp the phenomenon of novelty in phenotypic evolution, major quantitative change that is based on a progressive transformation of a

A) type II novelties   B) type III novelties

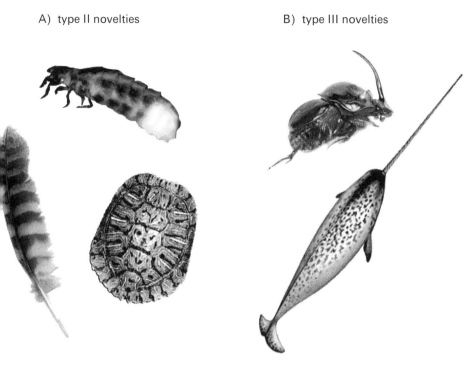

Figure 12.1
Examples of (A) type II novelties (discrete new elements, such as avian feathers, the turtle carapace, or the light organ of fireflies) and (B) type III novelties (progressive individualizations, such as the narwhal tusk or beetle horns).

feature needs to be discerned from the appearance of discrete new elements that are not continuous with an individualized precursor element. Whereas the former can represent a major morphological or functional change, which may even amount to a "key innovation" in phylogeny, or to a taxonomical novelty (apomorphy), a precursor organ always exists, such as a tooth or a horn, which may be progressively modified until it appears as a distinct character (figure 12.1b). In such cases, termed "type III novelty" (table 12.1), even though a distinct novel feature has formed, no individualized element or structural unit is added to the general body plan, and standard variation–selection mechanisms suffice to account for this kind of quantitative change. Accordingly, later definitions held that significant novelties represent deviations from quantitative variation and constitute qualitative differences in development (Müller 1990; West-Eberhard 2003), albeit without establishing an operational criterion for the designation of novelty. This is also not accomplished by the term "key innovation," defined as a novel trait that permits the exploitation of a new adaptive zone or facilitates species diversification (Liem 1990; Galis and Drucker 1996).

A criterion excluding the type III variational characters, or characters that deviate only quantitatively from the ancestral condition, was introduced by a definition based on the comparative character concept of homology. The definition states that a morphological novelty is "a new constructional element in a bodyplan that neither has a homologous counterpart in the ancestral species nor in the same organism" (Müller and Wagner 1991). While this definition excludes the cases of mere quantitative change of a preexisting structure, it applies to all cases in which no individualized precursor element had existed, as well as to those cases in which a new character arose through combinations or subdivisions of previously existing elements to form a new unit or element of the body plan, that is, a new homologue or type II novelty (table 12.1). Any further consequences associated with the new character, whether functional, phylogenetic, or taxonomic, do not play a role in this definition of novelty. Although it can be argued that the definition is narrow (e.g., Moczek 2008), it permits the identification of unambiguous cases of type II novelty, not only in morphology but also in physiology or behavior (if the term "constructional" is adapted correspondingly). More detailed discussions of novelty definitions and the associated problem of homology can be found in Müller (2003b), Müller and Newman (2005b), Minelli and Fusco (2005), Love and Raff (2006), and Moczek (2008).

Among the range of cases of phenotypic novelty that satisfy a homology-based definition (figure 12.1a), and whose development is known in some detail, are the turtle carapace (Gilbert et al. 2001; Moustakas 2008), avian (and theropod) feathers (Prum and Brush 2002), the corpus callosum of mammalian brains (Mihrshahi 2006), external cheek pouches in rodents (Brylski and Hall 1988), arthropod limbs (Williams 1999; Minelli and Fusco 2005), insect wings (Averof and Cohen 1997), butterfly wing patterns (Nijhout 2001), the lantern organ of fireflies (Oertel et al. 1975), cephalopod brain ganglia and arms (Lee et al. 2003), and floral organs in plants (Kramer and Jaramillo 2005). Many more examples, though usually less understood in their developmental detail, can be found in paleontological and comparative surveys (Nitecki 1990; Erwin 1993; see also chapter 13 in this volume).

A particularly rich source of cases of novelty is the vertebrate skeleton. Vertebrate evolution can be described as a series of additions of elements to the axial skeleton (which in itself represented a vertebrate innovation), such as the additions of upper and middle limb, wrist, and digit bones, of pelvic and shoulder girdle elements, and of the various bones contributing to the skull. Even in higher systematic clades, additions of elements to the basic skeletal organization appear, such as the phalangeal bones in cetaceans, extra "digits" in pandas, the naviculare in horses, the patella in birds and mammals, the calcar in bats, the falciform in moles, and additional skull roof elements in tree frogs. No doubt the genetic and cellular innovations that permitted the formation of mineralized tissues enabled the developmental formation of these structures, but the question of phenotypic novelty is why and how these processes were initiated in specific patterns and at specific locations of the vertebrate body.

We may conclude that the individualization of existing phenotypic traits can lead to type III novelties, which can include major functional innovation and may amount to taxonomical apomorphy, and which are explained by the standard variational paradigm. But for the kind of qualitative phenotypic novelty that implies the emergence of new units of construction, standard variation cannot be considered their source. These type II novelties require a distinct explanation.

Causes of Type II Novelty

Innovation theory argues that the mechanisms by which a new homologue is introduced into an existing character assembly (body plan) are rooted in the emergent effects of interacting levels of biological organization (molecules, cells, tissues) that are mobilized during evolutionary change. Since evolution acts upon highly interconnected, dynamical systems of development, the reactions to change in one of their components by necessity will affect other components that were not the target of the initial change. In this scenario, novelties emerge as side effects of evolutionary modifications to developmental systems (Müller 1990; Newman and Müller 2000). This is not to suggest that there exists a unique mechanism for evolutionary innovation, but rather that type II novelty generation is a systems reaction that does not belong within the incremental variation paradigm. Below we will examine some of the developmental systems properties responsible for novelty generation. Addressing the issue of innovation, West-Eberhard (2003) made the important distinction between "initiation" and "sources" of novel traits, roughly corresponding to what had earlier been called ultimate and proximate causation (Mayr 1961). We follow this argument in differentiating general "initiating conditions" from specific "realizing conditions" of innovation.

Initiating Conditions

In spite of their focus on variation, Modern Synthesis adherents speculated early on about whether special initiating conditions applied to the origin of novelties, even though no explicit distinctions were drawn between variation and innovation. It was intuitively clear, though, that natural selection could not act on characters that were not yet in existence and, hence, could not by itself account for the appearance of novelties. Without naming them as such, additional factors were taken into account, always intended, however, to remain in keeping with the Modern Synthesis paradigm.

Behavioral change and change of function, for instance, were considered as initiating factors in early discussions. Sewertzoff (1931) had given special attention to the principle of intensification of function, whereas Mayr (1958, 1960) regarded change of function as "by far the most important principle" in the origin of novelty, arguing for a "behavioral change comes first" mode of innovation. He reached the conclusion, though, that a new structure resulting from shifts of function "is merely a modification of a preceding structure," thereby resituating the novelty problem squarely into the realm of variation. More recent evaluations of functional shift and functional decoupling (Galis 1996; Ganfornina and Sánchez 1999) reemphasize such modes of novelty initiation, and the "behavioral change comes first" position also gained new support from developmental psychology. Behavioral flexibility based on developmental plasticity is argued to result in behavioral neophenotypes, which in turn cause morphological innovation followed by genetic integration (Johnston and Gottlieb 1990). The term "exaptation" (Gould and Vrba 1982), often brought up in this context, does not refer to a particular mode of novelty initiation, but designates the evolutionary principle of function shift.

Mutation pressure was another lingering idea to explain novelty initiation. The Modern Synthesis replaced Goldschmidt's (1940) macromutational ideas with "proper" micromutational concepts, as advocated by Mayr (1960), who thought that "the problem of the emergence of evolutionary novelties [then] consists in having to explain how a sufficient number of small gene mutations can be accumulated until the new structure has become sufficiently large to have selective value." Although, somewhat enigmatically, Mayr eventually reached the conclusion that "mutation pressure plays a negligible role in the emergence of evolutionary novelties" (Mayr 1960), it remained a standard Modern Synthesis view to regard molecular evolution as primary, and phenotypic evolution as secondary. More recently, the classical notion of independent, incremental, single-locus mutations has been replaced by gene duplications and coordinated gene regulatory change (e.g., through promoter mutations), but the basic idea that genetic evolution propels phenotypic evolution forward, and hence may also be responsible for major phenotypic novelty, is still alive (Arthur 2001; Davidson 2006). By contrast, recent work indicates that gene duplications increase mutational robustness and a developmental system's ability to withstand mutations, and thereby facilitate the evolution of innovations (A. Wagner 2008). Hence the evolution of robustness may represent an important background condition for the generation of novelty.

EvoDevo and phenotypic plasticity theory have recently focused attention on another potent initiating factor: the environment. According to this approach, environmental factors can elicit innovation not via natural selection but through their direct influence on developmental systems. The immediate influence of external physical and chemical factors on developmental processes is well known (Gilbert 2001; Hall et al. 2003; Müller 2003a; West-Eberhard 2003), but because of the seemingly Lamarckian connotations, its evolutionary role was considered out of bounds with respect to the Modern Synthesis paradigm. In fact, "environmental induction," the "most important initiator of evolutionary novelties" (West-Eberhard 2003), does not imply a Lamarckian mechanism of individual transmission of acquired traits, but represents a response to an external perturbation of the developmental systems of several members of a population at once. Since this stimulus–response reaction will likely be repeated over many generations, and may be phenotypically integrated through plastic properties of development, variably termed "phenotypic accommodation" (West-Eberhard 2003), "epigenetic integration" (Müller 2003b), or "exploratory processes" (see chapter 10 in this volume), no immediate genetic fixation is necessary. The environmentally induced change may eventually become genetically consolidated by subsequent natural selection via processes akin to assimilation (Waddington 1956; West-Eberhard 2003) discussed later in this chapter.

Environmental induction must be considered a realistic initiating condition for innovation. Although phenotypic plasticity of modern forms is extensive (Pigliucci 2001), it is likely that ancient forms were even more plastic (Newman and Müller 2000). The fact that organismal development has inherent morphogenetic properties that govern its response to external effects (see chapter 11 in this volume) indicates that accounts based on environmental conditions and those based on epigenetic factors are closely interdependent. Equally, the correspondences between macroevolutionary innovation rates and large-scale ecological patterns (see chapter 13 in this volume) support the initiating role of environmental conditions, even though not all environmental change would have its evolutionary effects through direct induction on development.

Standard natural selection cannot be disregarded as an initiating condition either, as long as it is clear that it cannot "aim" for novelty. But, as with environmental induction, yet more indirectly, natural selection may lead to innovation through the mobilization of developmental reactions that were not the immediate target of selection. The modes of natural selection need not be further elaborated here, since they are a principal subject of all classical treatments. But whether initiation modes are behavioral, genetic, selectional, or environmental, the specificity of the phenotypic outcome will depend in all cases on the "realizing conditions" that are present in the populations undergoing evolutionary change.

Realizing Conditions

The substrates on which all initiating factors act in organismal evolution are developmental systems. These form-generating processes consist of highly integrated, genetically routinized systems of interactions between gene products, cells, and tissues that exhibit dynamical feedback between their components (figure 12.2). Because these systems include autonomous properties of their components (cell behaviors, tissue geometries, etc.) as well as the capacities for self-organization, and respond to local and global external conditions, they are not mere executors of deterministic genetic programs. Hence they are termed "epigenetic" (from "epigenesis," not to be confounded with "epigenetics," the non-DNA–based forms of gene modulation, although the latter represents one aspect of the former; see also chapter 13 in this volume). It is a well-known property of self-regulatory, dynamical systems that they react in nonlinear fashions to changes in the initiating conditions. In embryonic systems such nonlinear reactions are based on two characteristic properties of development: the threshold behavior of regulatory processes, and the self-organizational behavior of cells and tissues.

Threshold behaviors are a widespread phenomenon at all levels of development. A physiological example is the MAP–kinase-dependent

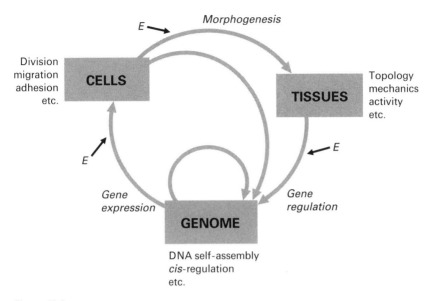

Figure 12.2
Dynamical interactions in development. Examples of the autonomous properties of each level of organization are given next to the boxes. E, environmental influences (after Müller 1991).

activation of the Yan protein in *Drosophila* development (Melen et al. 2005). The Yan protein is a transcriptional repressor of pointed (pnt) target genes involved in creating sharp boundaries between adjacent territories in the ventral ectoderm of the embryo. Experiments show that the level of the active form of Yan rises suddenly as a reaction to the gradual decrease of the MAP–kinase activity, depending on the distance from the midline in the embryo (figure 12.3). Any change that would bring the kinase activity below or above a critical value would affect tissue boundary formation in a nonlinear fashion. Such bi-stable switches are recognized as an important principle of developmental patterning (Goldbeter et al. 2007).

The modeling of complex gene regulatory networks highlights similar threshold behaviors. Eukaryote promoters yield ultrasensitive responses to changes in transcription factor quantities. The balance between opposing positive feedback loops in histone modification, for instance, is strongly affected by small changes in the levels of a transcription factor binding to a single site, producing a large change in gene expression (Sneppen et al. 2008). In modes of gene regulation mediated by RNAs, the sensitivity and abruptness of response are even greater. Gene expression is ultrasensitive

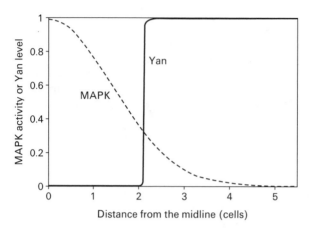

Figure 12.3
Yan protein levels rise suddenly as a reaction to the gradual decrease of the MAP–kinase (MAPK) activity, depending on the distance from the midline in the *Drosophila* embryo (from Goldbeter et al. 2007).

to small changes in sRNA, including threshold-linear responses in which the threshold is tunable (Levine et al. 2007). Interestingly, the targets of sRNA regulation are often themselves regulatory genes. Ultrasensitive threshold responses may have already acted in early metazoan evolution, since such modes exist even in viral and phage regulation (Kobiler et al. 2005). In metazoans, threshold responses affect morphological patterns via their cell behavioral consequences.

How developmental threshold effects mobilize cell and tissue self-organization is seen in the skeletal patterning of vertebrate limbs. The skeletogenic system involves the interaction of a number of biochemical–genetic core mechanisms that give rise to spatially discrete prechondrogenic cell condensations, which are the precursor structures of the embryonic cartilage elements. The basic system involves positive autoregulation of TGF-β and TGF-β–based induction of fibronectin, which in turn induces mesenchymal cell aggregation, leading to cartilage matrix production inside the condensations, and to an FGF-dependent elicitation of a lateral inhibitor of cartilage formation from the initial condensations themselves (Newman and Bhat 2007). In combination with the geometric confinements of the limb bud and its directional growth, establishing different zones of activity, regularly spaced cartilage elements will result. Computational models of this mechanism, based on "reaction" (i.e., gene regulation and biosynthesis of key products) and "diffusion" (i.e., local spread of secreted morphogens), indicate how skeletal pat-

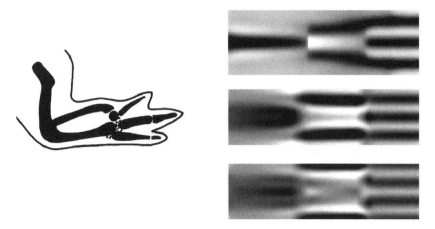

Figure 12.4
Computational model of limb skeletal patterning showing emergent patterns from three different parameter settings (after Kiskowski et al. 2004). Left: chick limb pattern.

terns are an emergent consequence of activation–inhibition thresholds in geometrically confined spaces (figure 12.4). The model demonstrates that the network is capable of producing skeletal patterns in a spatio-temporally regulated fashion without the need for programmed control (Hentschel et al. 2004). Besides the local concentrations of condensation-inducing molecules, further threshold parameters are represented by the cell number and cell density inside the prechondrogenic condensations, and in the biomechanical stimulation of cartilage matrix production (Newman and Müller 2005).

The case of the vertebrate limb also illustrates how epigenetic innovation generates type II novelty. Any changes in the initiating conditions that affect a general parameter of the limb system, such as the timing, size, or geometric shape of bud development, automatically give rise to different numbers and spatial arrangements of skeletal condensations. Evolutionary modifications of limb bud size or proportions, for instance, can result in new condensations or loss of condensations, based solely on the thresholds implicit in the autoregulatory chondrogenic mesenchyme. Experimental perturbations of the limb system expose such evolutionarily relevant thresholds in amphibian (Alberch and Gale 1985) and reptilian digital reduction studies (figure 12.5). This kind of experiment demonstrates how continuous reductions of cell numbers in developing vertebrate limbs can result not in a gradual reduction of limb size, but in a threshold-based deletion of complete digits, in the opposite

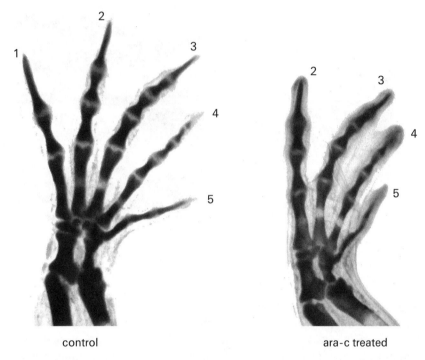

control   ara-c treated

Figure 12.5
One-step reduction of digit 1 in a forelimb of *Alligator mississippiensis* in response to a reduction of cell number induced through a temporary inhibition of cell proliferation by cytosin-arabinoside (ara-c) in early embryonic development (Müller unpublished).

sequence of their embryonic formation. The reverse effect, additions of complete digits to the basic pentadactyl pattern, appear, for instance, in polydactylous mutants in which the spatial proportions of the limb system are affected (Chan et al. 1995; Litingtung et al. 2002).

The patterns of digit reductions and digit additions found in natural populations match the experimental patterns, and show that developmental thresholds have specific evolutionary consequences. In the Australian skinks of the genus *Lerista*, for instance, trends toward body elongation are often combined with limb reduction. Different degrees of limb reduction, represented by different species, have resulted in the stepwise loss of complete digits (figure 12.6a), not in their gradual diminishment, as the standard variational theory would argue. Moreover, the sequence of digit loss can be predicted on the basis of the sequence of their developmental formation in reptilians (Müller and Alberch 1990). Complete digits also appear as natural multiples in spontaneous or

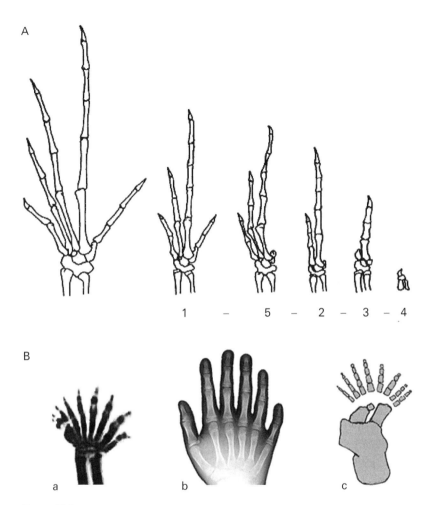

Figure 12.6
(A) Loss of complete digits associated with limb size reduction in skinks. 1–5 indicates the number of the digit lost (after Greer 1990). (B) Appearance of multiples of complete digits in (a) mutant mice (after Litingtung et al. 2002), (b) humans (Wikipedia), and (c) the primitive tetrapod *Acanthostega* (after Coates and Clack 1990).

induced mutations and in primitive vertebrate limbs, as in the case of *Acanthostega* (figure 12.6b), a late Devonian stem tetrapod with eight digits on its forelimbs and hind limbs.

Other kinds of developmental mechanisms with nonlinear response characteristics and evolutionary consequences include inductive tissue interactions (e.g., Brylski and Hall 1988), as well as the kinetics (Bolouri and Davidson 2003), dynamics (Cinqun and Demongeot 2005; Salazar-Ciudad and Jernvall 2005), and modularity (Von Dassow et al. 2000) of gene regulatory networks. The modeling of cell behavior, based on parameters of cell division, cell differentiation, and cell–cell signaling interactions, highlights the relationships between developmental complexity and the generation of innovation (Salazar-Ciudad and Jernvall 2005). In summary, emergent morphogenetic responses from altered developmental systems are a ubiquitous occurrence in metazoan evolution.

Epigenetic Innovation Theory

The arguments presented here (see also Müller 1990; Müller and Wagner 1991; Newman and Müller 2000; chapter 11 in this volume; and Newman and Bhat 2008), as well as aspects of the phenotypic plasticity concept (West-Eberhard 2003; see also chapter 14 in this volume) combine in a theory of epigenetic innovation. "Epigenetic" is here used in its traditional meaning of "contextual development" not in the sense of "gene silencing" or "epigenetic inheritance" (see chapter 7 in this volume). This approach does not ignore the genetic components of innovation, such as gene variation or gene regulatory evolution, but it takes these as background conditions that are always at work, while concentrating on the mechanisms underlying macroscopic novelty generation. The key components of epigenetic innovation theory can be summarized in the following points:

1. The realizing conditions for phenotypic evolution are embodied in developmental systems that are characterized by cellular self-organization, feedback regulation, and environment dependence.

2. Chemically and mechanically excitable cell aggregates provided emergent morphological motifs of the first multicellular organisms, which then became morphogenetic templates for further phenotypic evolution.

3. Complex developmental systems can react in nonlinear fashions to changing initiating conditions.

4. Any disturbance in one, or several, of a system's parameters mobilizes its self-organizing qualities. Minor disturbances will be buffered by the plastic dynamics of the system, but more substantial disturbances can take it to its limits (e.g., in cell number, blastema size, inductive capacity, etc.), and threshold responses ensue.

5. Gradual or nongradual regimes of evolutionary change can be developmentally transformed into discontinuous phenotypic outcomes. These may appear as losses of traits, or as combinations of previously independent traits, or as the kernels of new traits not present in the ancestral condition.

6. These morphogenetic products will be exposed to natural selection, and can be elaborated and refined through standard variational mechanisms.

7. Genetic evolution, while facilitating innovation, serves a consolidating role rather than a generative one, capturing and routinizing morphogenetic templates.

Epigenetic innovation theory represents a systems-oriented approach. It argues that genetic variation, natural selection, and environmental induction affect integrated developmental systems that generate specific phenotypic reactions when the canalized plasticity has reached its limits. In other words, a developmental system falls into a new steady-state interaction among its components. Hence the kernels of type II phenotypic novelty originally appear as by-products, or side effects (Müller 1990), of a change of parameters in the epigenetic organization of development. In such cases, natural selection is involved only in an indirect manner, changing initiating conditions, but the specificity of the morphological outcome will be dictated by the responsiveness of the developmental system.

A number of further questions arise from such a perspective. When novelty origination can be an event that is not based on the continuous variation of a preexisting character, but appears de novo from developmental systems reactions, then we need to ask how a new character can be maintained and integrated into the constructional, functional, and genetic systems, and how it can spread through a population. The first question is easily answered, because if a novelty arises from a developmental response to initiating conditions, the same response will be elicited in every new generation, provided the initiating condition persists. As pointed out by Kirschner and Gerhart (see chapter 10 in this volume), many other tissue systems of the embryo will automatically follow any change, and a novel trait can remain epigenetically integrated for extended periods of time before genetic integration takes place. In fact, it will commonly be the case that epigenetic integration precedes genetic integration (Newman and Müller 2000). Similar concepts that emphasize epigenetic integration have been called "generative entrenchment" (Wimsatt 1986) or "epigenetic traps" (G. P. Wagner 1989).

Genetic integration of type II novelty may turn out to be equally unproblematic, because any novel character necessarily includes new heritable variation in the expression of its phenotype. Waddington (1956) showed how genetic integration may follow from selection acting on the genetic variation that arises with a novel character. Today it is known that integration includes the recruitment of orthologous and paralogous regulatory circuits that acquire new developmental roles over the course of evolution (Wray 1999; Wray and Lowe 2000). Evolving structure–function interrelationships (Galis 1996) integrate novel characters at the phenotypic level but will also contribute to genetic integration, with selection favoring the genetic linkage of functionally coupled characters and evolvability (G. P. Wagner 1984; see also chapter 15 in this volume). The evolving genome will thus gain control over the epigenetic conditions that prevail during the origination of novelties, in a fashion that corresponds to what West-Eberhard (2003) terms "accommodation." Since epigenetic integration will usually come first, it can provide the templates for both phenotypic and genetic integration (Newman and Müller 2000; Müller 2003b). Genetic integration over-determines the generative processes, resulting in a stabilization of the phenotype.

It is noteworthy that the mechanisms that confer stabilization and robustness against perturbation are the very same conditions that are implicated in innovation. Developmental regulatory networks, such as the arthropod segmentation network, are generally overdetermined and highly robust, but will yield new phenotypes when certain thresholds of perturbation are exceeded (Von Dassow et al. 2000). It has been proposed that the substantial buffering capacities of such systems may allow the accumulation of genetic variation, which can be released in a threshold event, so that a resulting novel phenotype would be endowed with heritable variation from its very inception (Moczek 2007). Rutherford and Lindquist's (1998) findings on heat shock proteins indicate exactly this type of behavior. The role of robustness in innovation is also supported by the study of gene duplications. Gene duplication generally increases mutational robustness, which may counteract innovation in the short term, but may also allow molecular diversification and phenotypic innovation in the long term (A. Wagner 2008). Hence the frequent association of gene duplications with phenotypic innovation does not seem to reflect an immediate causal role, but a facilitating one.

Computer simulations of the evolution of genotype–phenotype relationships indicate that network robustness includes a tendency to progress from emergent to hierarchical control (Salazar-Ciudad et al. 2001a). Whereas in the emergent phase the resulting patterns are a consequence, predominantly, of the self-organizing capacities of a developmental system, such as cells responding stochastically to a reaction-diffusion mechanism, a more reproducible and fine-tuned outcome is achieved through hierarchical control networks. The evolution of insect segmentation is a case in support of this argument (Salazar-Ciudad et al. 2001b). This corresponds to the observation that emergent networks are characteristic of later stages of development, which can tolerate greater amounts of variation, whereas early developmental processes, which have many downstream consequences, tend to be replaced by hierarchical, more program-like networks.

The above considerations regarding the origin and stabilization of type II novelty, in which the threshold qualities of evolved gene regulatory and cell interaction systems have a central role, contrast with the main factors that governed type I novelty, the origination of primary body assemblies, in which biophysical properties of early cell aggregates would have had a dominant role in the absence of any tight genetic programming (see chapter 11 in this volume). This, and the observation of the plastic, environment-dependent relationships between genotype and phenotype (Gilbert 2001), as well as a paleontological record indicating high disparity rates in early phylogenetic radiations and a decrease of phenotypic innovations in later phases (Erwin 1993; Foote 1997; Salazar-Ciudad and Jernvall 2005; see also chapter 13 in this volume), have led to the proposal that organismal evolution progressed from a pre-Mendelian world to the Mendelian one we study today (Newman and Müller 2000; Newman 2005). Modern organisms are Mendelian, in the sense that genotype and phenotype are inherited in a close correlation, and development is under program-like genetic control. But this is likely to be a derived condition, enforced by natural selection, whereas in the pre-Mendelian world a much looser connection between genotype and phenotype would have prevailed, which permitted the generation of multiple forms from single genotypes, depending on environmental influences. Here the generic physical properties of cells, cell aggregates, and tissues would have been the decisive determinants of biological form, which would only later be harnessed by genetic routinization and overdetermination. In both worlds epigenetic factors have an important role in causing innovation, but whereas the type I novelties in the pre-Mendelian world would result predominantly from the physical properties of simple cell aggregates, type II and type III innovations in the Mendelian world would be based on the dynamical properties of integrated gene, cell, and tissue interactions. Type II and type III innovations would be further distinguished by the former being based on the discontinuous responses of developmental systems to genetic and environmental change, and the latter on the continuous responses.

Consequences for an Extended Evolutionary Synthesis

The origin of novelty and its role in the evolution of phenotypic complexity represent an evolutionary problem that had been sidelined by the Modern Synthesis, due to its focus on variation and population dynamics. Evolutionary developmental biology reemphasizes the phenotype in addressing the mechanistic factors of organismal change. The bulk of present-day EvoDevo is in the empirical study of the evolving molecular tool kit and its regulatory interactions (Carroll et al. 2005; Davidson 2006). But the major theoretical consequences of EvoDevo lie in the manifestations of developmental systems properties in the evolutionary process. Evolvability theory, with its associated issues of plasticity, modularity, and facilitated variation, represents one area of EvoDevo-related extensions (see chapters 10 and 15 in this volume), and the generative principles of phenotypic complexity, such as dynamical patterning modules (see chapter 11 in this volume) and epigenetic innovation discussed in the present chapter, are another kind of expansion. The theory extensions, then, consist, on the one hand, of the incorporation of additional factors into the evolutionary framework (cellular self-organization, nonlinear effects, etc.), adding principles of emergence to the principles of variation. On the other hand, through these inclusions, the explanatory capacity of evolutionary theory is extended to nonadaptive and nongradual phenomena of phenotypic evolution.

The inclusion of EvoDevo concepts in general, and of epigenetic innovation theory in particular, entails not only a conceptual widening but also a shift of emphasis in some of the traditional elements of the Modern Synthesis. One is the reduction of the overreliance on facile genetic explanations of organismal form and diversity. Whereas the population dynamic and molecular descriptions of genetic evolution address important phenomena in their own right, and form a backbone of evolutionary theory, they do not by themselves provide causal explanations of organismal form. EvoDevo demonstrates that knowledge of the rules of the processes that intervene between genotype and phenotype is necessary for understanding the causalities of biological form. It is this shift from a predominantly statistical and correlational approach to a causal–mechanistic approach that is one of the main characteristics of the Extended Synthesis. Whereas the explanation of adaptive change as a population dynamic event was the central goal of the Modern Synthesis, EvoDevo seeks to explain phenotypic change through the understanding of developmental mechanisms, the physical interactions among genes, cells, and tissue architecture in particular. Furthermore, EvoDevo aims at explaining how developmental systems evolve and how developmental processes are controlled by the interplay between genetic, epigenetic, and environmental factors. Various directions of empirical research have begun to address these issues (West-Eberhard 2003; Müller and Newman 2005a; Deacon 2006).

Another shift of emphasis, induced through innovation theory and EvoDevo, concerns the role of natural selection in phenotypic evolution. Whereas in the Modern Synthesis framework the burden of explanation of the phenotype rests on the action of natural selection, with genetic variation representing the necessary boundary condition that continuously provides a substrate for selection, the EvoDevo framework assigns more explanatory weight to the generative properties of developmental processes, with natural selection representing a boundary condition. In this interpretation, natural selection functions to release inherent developmental potential and to explore the residual plasticity of developmental systems, but the specificity of the phenotypic outcome is determined by the capacities of the developmental systems undergoing change. Thus, EvoDevo shifts the weight of explanation from the external and contingent to the internal and inherent (Newman and Müller 2006; Callebaut et al. 2007). Selection, then, in the case of innovation, functions as an initiating cause and becomes a factor of secondary stabilization, whereas the actual morphological solutions result from the specific properties of development.

Evolutionary theory, through the modifications outlined above, and through a multitude of other conceptual expansions, partially explored in this volume, becomes a much more pluralistic and systemic theory than it was under the Modern Synthesis paradigm (see also chapters 16 and 17 in this volume). Additional levels of analysis and explanation are included, and more factors and feedback interactions among these factors are taken into account. As a tangible consequence, the limitations of the Modern Synthesis with regard to the explanation of higher levels of organization are overcome by an Extended Synthesis.

Acknowledgments

I am grateful for the feedback from the workshop participants, and to Massimo Pigliucci and Stuart Newman for their valuable comments on the manuscript.

References

Alberch P, Gale EA (1985) A developmental analysis of an evolutionary trend: Digital reduction in amphibians. Evolution 39: 8–23.

Altenberg L (2005) Modularity in evolution: Some low-level questions. In Modularity: Understanding the Development and Evolution of Natural Complex Systems. W Callebaut, D Rasskin-Gutman, eds.: 99–128. Cambridge MA: MIT Press.

Arthur W (2001) Developmental drive: An important determinant of the direction of phenotypic evolution. Evolution and Development 3: 271–278.

Averof M, Cohen SM (1997) Evolutionary origin of insect wings from ancestral gills. Nature 385: 627–630.

Bolouri H, Davidson EH (2003) Transcriptional regulatory cascades in development: Initial rates, not steady state, determine network kinetics. Proceedings of the National Academy of Sciences of the USA 100: 9371–9376.

Brylski P, Hall BK (1988) Epithelial behaviors and threshold effects in the development and evolution of internal and external cheek pouches in rodents. Zeitschrift für zoologische Systematik und Evolutionsforschung 26: 144–154.

Callebaut W, Müller GB, Newman S (2007) The Organismic Systems Approach: Evo-Dvo and the streamlining of the naturalistic agenda. In Integrating Evolution and Development: From Theory to Practice. R Sansom, RN Brandon, eds.: 25–92. Cambridge, MA: MIT Press.

Carroll SB, Grenier JK, Weatherbee SD (2005) From DNA to Diversity: Molecular Genetics and the Evolution of Animal Design. 2nd ed. Malden, MA: Wiley-Blackwell.

Chan DC, Laufer E, Tabin C, Leder P (1995) Polydactylous limbs in Strong's Luxoid mice result from ectopic polarizing activity. Development 121: 1971–1978.

Cinquin O, Demongeot J (2005) High-dimensional switches and the modeling of cellular differentiation. Journal of Theoretical Biology 233: 391–411.

Coates MI, Clack JA (1990) Polydactyly in the earliest known tetrapod limbs. Nature 347: 66–69.

Darwin C (1859) On the Origin of Species. London: John Murray.

Davidson EH (2006) The Regulatory Genome: Gene Regulatory Networks in Development and Evolution. 2nd rev. ed. San Diego: Academic Press.

Deacon TW (2006) Reciprocal linkage between self-organized processes is sufficient for self-reproduction and evolvability. Biological Theory 1: 136–149.

Eldredge N, Gould SJ (1972) Punctuated equilibria: An alternative to phyletic gradualism. In Models in Paleobiology. TJM Schopf, ed.: 82–115. San Francisco: Freeman, Cooper.

Erwin DH (1993) Early introduction of major morphological innovations. Acta Palaeontologica Polonica 38: 281–294.

Foote M (1997) The evolution of morphological diversity. Annual Reviews of Ecology and Systematics 28: 129–152.

Galis F (1996) The application of functional morphology to evolutionary studies. Trends in Ecology and Evolution 11: 124–129.

Galis F, Drucker EG (1996) Pharyngeal biting mechanics in centrarchid and cichlid fishes: Insights into a key evolutionary innovation. Journal of Evolutionary Biology 9: 641–670.

Ganfornina MD, Sánchez D (1999) Generation of evolutionary novelty by functional shift. BioEssays 21: 432–439.

Gilbert SF (2001) Ecological developmental biology: Developmental biology meets the real world. Developmental Biology 233: 1–12.

Gilbert SF, Loredo GA, Brukman A, Burke AC (2001) Morphogenesis of the turtle shell: The development of a novel structure in tetrapod evolution. Evolution and Development 3: 47–58.

Goldbeter A, Gonze D, Pourquié O (2007) Sharp developmental thresholds defined through bistability by antagonistic gradients of retinoic acid and FGF signaling. Developmental Dynamics 236: 1495–1508.

Goldschmidt RB (1940) The Material Basis of Evolution. New Haven, CT: Yale University Press.

Gould SJ, Lewontin RC (1979) The spandrels of San Marco and the Panglossian paradigm: A critique of the adaptationist programme. Proceedings of the Royal Society of London B205: 581–598.

Gould SJ, Vrba ES (1982) Exaptation: A missing term in the science of form. Paleobiology 8: 4–15.

Greer AE (1990) Limb reduction in the Scincid lizard genus *Lerista*. 2. Variation in the bone complements of the front and rear limbs and the number of postsacral vertebrae. Journal of Herpetology 24: 142–150.

Hall BK, Pearson RD, Müller GB, eds. (2003) Environment, Development, and Evolution: Toward a Synthesis. Cambridge, MA: MIT Press.

Hentschel HGE, Glimm T, Glazier JA, Newman SA (2004) Dynamical mechanisms for skeletal pattern formation in the vertebrate limb. Proceedings of the Royal Society of London B271: 1713–1722.

Jernvall J, Hunter JP, Fortelius M (1996) Molar tooth diversity, disparity, and ecology in Cenozoic ungulate radiations. Science 274: 1489–1492.

Johnston TD, Gottlieb G (1990) Neophenogenesis: A developmental theory of phenotypic evolution. Journal of Theoretical Biology 147: 471–495.

Kiskowski MA, Alber MS, Thomas GL, Glazier JA, Bronstein NB, Pud J, Newman SA (2004) Interplay between activator–inhibitor coupling and cell-matrix adhesion in a cellular automaton model for chondrogenic patterning. Developmental Biology 271: 372–387.

Kobiler O, Rokney A, Friedman N, Court DL, Stavans J, Oppenheim AB (2005) Quantitative kinetic analysis of the bacteriophage lambda genetic network. Proceedings of the National Academy of Sciences of the USA 102: 4470–4475.

Kramer EM, Jaramillo MA (2005) Genetic basis for innovations in floral organ identity. Journal of Experimental Zoology B304: 526–535.

Lee PN, Callaerts P, De Couet HG, Martindale MQ (2003) Cephalopod Hox genes and the origin of morphological novelties. Nature 424: 1061–1065.

Levine E, Zhang Z, Kuhlman T, Hwa T (2007) Quantitative characteristics of gene regulation by small RNA. PLoS Biology 5: e229.

Liem KF (1990) Key evolutionary innovations, differential diversity, and symecomorphosis. In Evolutionary Innovations. MH Nitecki, ed.: 147–170. Chicago: University of Chicago Press.

Litingtung Y, Dahn RD, Li Y, Fallon JF, Chiang C (2002) Shh and Gli3 are dispensable for limb skeleton formation but regulate digit number and identity. Nature 418: 979–983.

Love AC (2005) Explaining evolutionary innovation and novelty: A historical and philosophical study of biological concepts. Doctoral dissertation, University of Pittsburgh.

Love AC, Raff RA (2006) Larval ectoderm, organizational homology, and the origins of evolutionary novelty. Journal of Experimental Zoology B306: 18–34.

Margulis L, Fester R, eds. (1991) Symbiosis as a Source of Evolutionary Innovation: Speciation and Morphogenesis. Cambridge, MA: MIT Press.

Mayr E (1958) Behavior and systematics. In Behavior and Evolution. A Roe, GG Simpson, eds.: 341–362. New Haven, CT: Yale University Press.

Mayr E (1960) The emergence of evolutionary novelties. In Evolution After Darwin. S Tax, ed.: 349–380. Cambridge, MA: Harvard University Press.

Mayr E (1961) Cause and effect in biology. Science 134: 1501–1506.

Melen GJ, Levy S, Barkai N, Shilo B-Z (2005) Threshold responses to morphogen gradients by zero-order ultrasensitivity. Molecular Systems Biology doi:10.1038/msb4100036.

Mihrshahi R (2006) The corpus callosum as an evolutionary innovation. Journal of Experimental Zoology B306: 8–17.

Minelli A, Fusco G (2005) Conserved versus innovative features in animal body organization. Journal of Experimental Zoology B304: 520–525.

Moczek AP (2007) Developmental capacitance, genetic accommodation, and adaptive evolution. Evolution and Development 9: 299–305.

Moczek AP (2008) On the origins of novelty in development and evolution. BioEssays 30: 432–447.

Moustakas JE (2008) Development of the carapacial ridge: Implications for the evolution of genetic networks in turtle shell development. Evolution and Development 10: 29–36.

Müller GB (1990) Developmental mechanisms at the origin of morphological novelty: A side-effect hypothesis. In Evolutionary Innovations. H Nitecki, ed.: 99–130. Chicago: University of Chicago Press.

Müller GB (1991) Experimental strategies in evolutionary embryology. American Zoologist 31: 605–615.

Müller GB (2003a) Embryonic motility: Environmental influences and evolutionary innovation. Evolution and Development 5: 56–60.

Müller GB (2003b) Homology: The evolution of morphological organization. In Origination of Organismal Form: Beyond the Gene in Developmental and Evolutionary Biology. GB Müller, SA Newman, eds.: 51–69. Cambridge, MA: MIT Press.

Müller GB (2007) Evo-Devo: Extending the evolutionary synthesis. Nature Reviews Genetics 8: 943–949.

Müller GB, Alberch P (1990) Ontogeny of the limb skeleton in *Alligator mississippiensis*: Developmental invariance and change in the evolution of archosaur limbs. Journal of Morphology 203: 151–164.

Müller GB, Newman SA, eds. (2003) Origination of Organismal Form: Beyond the Gene in Developmental and Evolutionary Biology. Cambridge, MA: MIT Press.

Müller GB, Newman SA, eds. (2005a) Evolutionary Innovation and Morphological Novelty. Special issue of Journal of Experimental Zoology B304.

Müller GB, Newman SA (2005b) The innovation triad: An EvoDevo agenda. Journal of Experimental Zoology B304: 487–503.

Müller GB, Wagner GP (1991) Novelty in evolution: Restructuring the concept. Annual Reviews of Ecology and Systematics 22: 229–256.

Newman SA (2005) The pre-Mendelian, pre-Darwinian world: Shifting relations between genetic and epigenetic mechanisms in early multicellular evolution. Journal of Biosciences 30: 75–85.

Newman SA, Bhat R (2007) Activator–inhibitor dynamics of vertebrate limb pattern formation. Birth Defects Research C81: 305–319.

Newman SA, Bhat R (2008) Dynamical patterning modules: Physico-genetic determinants of morphological development and evolution. Physical Biology 5: 15008.

Newman SA, Müller GB (2000) Epigenetic mechanisms of character origination. Journal of Experimental Zoology B288: 304–317.

Newman SA, Müller GB (2005) Origination and innovation in the vertebrate limb skeleton: An epigenetic perspective. Journal of Experimental Zoology B304: 593–609.

Newman SA, Müller GB (2006) Genes and form: Inherency in the evolution of developmental mechanisms. In Genes in Development: Rereading the Molecular Paradigm. E Neumann-Held, C Rehmann-Sutter, eds.: 38–73. Durham: Duke University Press.

Nijhout HF (1991) The Development and Evolution of Butterfly Wing Patterns. Washington, DC: Smithsonian Institution Press.

Nitecki MH, ed. (1990) Evolutionary Innovations. Chicago: University of Chicago Press.

Oertel D, Linberg KA, Case JF (1975) Ultrastructure of the larval firefly light organ as related to control of light emission. Cell and Tissue Research 164: 27–44.

Pagel M, Venditti C, Meade A (2006) Large punctuational contribution of speciation to evolutionary divergence at the molecular level. Science 314: 119–121.

Pigliucci M (2001) Phenotypic Plasticity: Beyond Nature and Nurture. Baltimore: Johns Hopkins University Press.

Prum RO, Brush AH (2002) The evolutionary origin and diversification of feathers. Quarterly Review of Biology 77: 261–295.

Reid RGB (2007) Biological Emergences: Evolution by Natural Experiment. Cambridge, MA: MIT Press.

Rensch B (1959) Evolution Above the Species Level. New York: Columbia University Press.

Rutherford SL, Lindquist S (1998) Hsp90 as a capacitor for morphological evolution. Nature 396: 336–342.

Salazar-Ciudad I, Jernvall J (2005) Graduality and innovation in the evolution of complex phenotypes: Insights from development. Journal of Experimental Zoology B304: 619–631.

Salazar-Ciudad I, Newman SA, Solé RV (2001a) Phenotypic and dynamical transitions in model genetic networks. I. Emergence of patterns and genotype–phenotype relationships. Evolution and Development 3: 84–94.

Salazar-Ciudad I, Solé RV, Newman SA (2001b) Phenotypic and dynamical transitions in model genetic networks. II. Application to the evolution of segmentation mechanisms. Evolution and Development 3: 95–103.

Schmalhausen II (1949) Factors of Evolution. Chicago: University of Chicago Press.

Schwartz JH (1999) Sudden Origins: Fossils, Genes, and the Emergence of Species. New York: Wiley.

Sewertzoff AN (1931) Morphologische Gesetzmäßigkeiten der Evolution. Jena: Gustav Fischer.

Sneppen K, Micheelsen MA, Dodd IB (2008) Ultrasensitive gene regulation by positive feedback loops in nucleosome modification. Molecular and Systems Biology 4: 182.

Valentine JW (2004) On the Origin of Phyla. Chicago: University of Chicago Press.

Von Dassow G, Meir E, Munro EM, Odell GM (2000) The segment polarity network is a robust developmental module. Nature 406: 188–192.

Waddington CH (1956) Genetic assimilation. Advances in Genetics 10: 257–290.

Wagner A (2008) Gene duplications, robustness and evolutionary innovations. BioEssays 30: 367–373.

Wagner GP (1984) Coevolution of functionally constrained characters: Prerequisites for adaptive versatility. BioSystems 17: 51–55.

Wagner GP (1989) The biological homology concept. Annual Reviews of Ecology and Systematics 20: 51–69.

West-Eberhard MJ (2003) Developmental Plasticity and Evolution. New York: Oxford University Press.

Wilkins A (2002) The Evolution of Developmental Pathways. Sunderland, MA: Sinauer.

Williams TA (1999) Morphogenesis and homology in arthropod limbs. American Zoologist 39: 664–675.

Wimsatt WC (1986) Developmental constraints, generative entrenchment, and the innate–acquired distinction. In Integrating Scientific Disciplines. W Bechtel, ed.: 185–208. Dordrecht: Martinus Nijhoff.

Wray GA (1999) Evolutionary dissociations between homologous genes and homologous structures. In Homology (Novartis Foundation Symposium 222). GR Bock, G Cardew, eds.: 189–203. Chichester, UK: Wiley.

Wray GA, Lowe CJ (2000) Developmental regulatory genes and echinoderm evolution. Systematic Biology 49: 151–174.

VI MACROEVOLUTION AND EVOLVABILITY

13 Origination Patterns and Multilevel Processes in Macroevolution

David Jablonski

The expansion of the Evolutionary Synthesis began in the 1970s, and considerable progress has been made. Many of the necessary conceptual elements, including history, scale, and hierarchy, as well as new perspectives on the evolutionary roles of intrinsic factors such as development and extrinsic factors such as mass extinctions and their aftermath, have been put on the table. New tools have also become available, from a spectacular array of molecular advances applicable to such areas as phylogeography and developmental biology, to phylogenetic comparative methods and novel approaches to factoring out biases in the fossil record.

Despite these successes, we have much to do. With some important exceptions, the subdisciplines of what Van Valen (1982 and elsewhere) termed "the evolutionary half of biology" are still somewhat isolated from one another, and many promising concepts remain theoretical constructs rather than part of the working methodology of evolutionary biology. Here I will address these issues in terms of scale and hierarchy in macroevolution, with an emphasis on how theory can guide, and be challenged by, empirical analyses of the spatial and temporal patterns of taxa and phenotypes. I will start by discussing the origin of evolutionary novelties, specifically their nonrandom origins in time and space, which demands novel approaches to the evolution of form. From there I will address the fates of novelties and the clades they define, including the failure of simple extrapolation from local, short-term observation to account for the long-term stability of species, the differential production and survival of species (species selection in broad and strict senses), the weak correlation between short-term biotic interactions and long-term clade dynamics, and the strong potential for organismic features to hitchhike on the array of intrinsic and extrinsic factors that govern clade dynamics. This is a book chapter and not a treatise, so I can only briefly touch on these phenomena, but hope to underscore that hierarchical approaches are needed to understand such large-scale, long-term evolutionary processes.

Nonrandom Origins of Evolutionary Novelties in Time and Space

The geologically sudden burst of animal diversity during the Cambrian radiation 530 million years (Myr) ago is one of the most dramatic events in the history of life. With one exception (bryozoans), all of the phyla that could reasonably be expected to have a fossil record, and a few that could not, such as priapulids, are documented in the first 10–15 Myr of the Phanerozoic (Valentine 2004). However molecular data are interpreted on the timing of phylogenetic branch points (although later rather than earlier branchings seem more likely; see, e.g., Peterson et al. 2008), the explosion of phenotypic diversity is striking. Later, if less extreme, bursts of evolutionary creativity do occur, of course, including those accompanying the invasion of land by plants and animals, the recovery of marine clades from the end–Paleozoic mass extinction, and the diversification of mammals in the early Cenozoic.

The major diversifications are increasingly well documented, but remain poorly understood in terms of underlying mechanisms. Comparative approaches should be useful here (even if all diversifications are not alike; e.g., Jablonski 1980 and in prep.; Erwin 1992). Thus, the Cambrian explosion is unique for its breadth and inventiveness, but it shares key features with certain other major radiations, such as the diversification of land plants (see Jablonski 2000, 2007; and Erwin 2007 for references; and Shen et al. 2008 for a similar view of the latest Precambrian, Ediacaran biota). For example, (1) there is a visible evolutionary fuse to the explosion, that is, small, ecologically marginal and phenotypically unimpressive stem groups can be detected in the fossil record before the main event. (2) Comparative developmental work in a phylogenetic context suggests that many of the essential regulatory gene networks were in place well before the event. (3) Diversity analyses show that morphological disparity (and/or the first occurrences of higher taxa) tends to outpace species richness in the early phases. (4) Phenotypic characters seem to show high degrees of variability and polymorphism (see Webster 2007 for the first robust demonstration in the Cambrian), and mosaic combinations of traits are common. And (5) The events are relatively short-lived: the radiations are not sustained at the fever pitch seen at the start, and it's back to evolutionary business as usual after

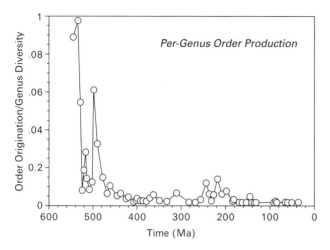

Figure 13.1
Time- and diversity-dependence in the origin of marine invertebrate orders through the Phanerozoic. Horizontal axis is millions of years before present (Ma); dip between 550 and 500 Ma is likely to be a sampling artifact, but might reflect a lull between the Cambrian Explosion and the Ordovician Diversification. (Courtesy of Gunther Eble, Université de Bourgogne.)

some tens of Myr, so that major evolutionary novelties appear to exhibit *negative* diversity dependence. Morphological and functional variety continue to accrue, as shown by fruit flies and orchids (e.g., Bambach et al. 2007 on marine invertebrates; Adamowicz et al. 2008 on arthropods), but the later process tends to operate as a slower diffusion through morphospace. Thus, the per-taxon rate of novelty production actually declines over the course of the diversification. This is hardly a new observation (e.g., Wright 1949; Valentine 1969), but it is a telling one, and moving down to a lower taxonomic level to increase statistical power, this diversity dependence prevails through the Phanerozoic at the ordinal level in marine invertebrates (Eble 1998, 1999) (see figure 13.1). Positive feedbacks, which intuitively should play a role at any level—diversity begets diversity, so that epiphytes, and the insects that exploit them, could not diversify until trees did—are evidently transient or weak here (see also Jablonski 2008a).

Several nonexclusive hypotheses can account for these features at the intersection of development and ecology, but we cannot yet evaluate them fully. One set of hypotheses might be termed intrinsic—the patterns in the fossil record reflect rules for the generation of form, with some aspect of the path from genotype to phenotype changing in a way that briefly promoted, then damped novelty production over time. However, while we have made great strides in understanding the broad developmental basis of important features such as heads and limbs, more information is needed on how the molecular underpinnings of those features were assembled, or how the evolving organization of the genome, or the genetic architecture of traits, might inhibit or promote particular modes or directions of evolutionary change. For example, does the complexity of regulatory networks eventually impose a cost in the form of reduced flexibility? Robustness and evolvability may be antagonistic features at the molecular level, but this may not be true for phenotypes (Wagner 2008); interaction of the two levels has not been much explored from this perspective. An overarching decline in evolvability at high levels might seem surprising, given the positive contribution that enhanced evolvability, in the sense of an ability to generate viable phenotypes, might be expected to make to clade persistence and proliferation (e.g., Pigliucci 2007; but see Lynch 2007, who questions the general advantage of evolvability, although definitions are crucial here).

Alternatively or additionally, the temporal pattern of novelty production may derive from extrinsic factors, for example, by ecological incumbency on a grand scale, in which the establishment of some functional or ecological group inhibits further diversification. This crowding scenario is attractive (e.g., Valentine 1980, 1995, Gavrilets and Losos 2009, Reznick and Ricklefs 2009), but has not been tested critically. Other models even suggest that increasing the number of antagonistic selection pressures, as might be expected as biotas diversify, should promote diversification, rather than slow it down, by rendering multiple, suboptimal adaptive peaks effectively equivalent (Niklas 1994, 2004; Marks and Lechowitz 2006; Marshall 2006). In any event, large-scale phenotypic diversifications often tend to flag even as species-level diversification persists or accelerates. Positive ecological feedbacks (which must have been present at some point) were evidently manifest at this scale for only a short time, or were only locally important, at least with regard to morphological and functional variety. As already noted, these aspects of biodiversity continued to accrue at a much slower rate and to less striking effect.

Perhaps both of these paradigms are true, and the Cambrian explosion is so dramatic because conditions were right for *both* intrinsic and extrinsic factors to promote the diversification of form. Then the great post–Paleozoic marine diversification—impressive, but no Cambrian explosion—after the huge end–Permian mass extinction removed about 95% of species, may show what the evolutionary system can do with plentiful ecological opportunities and a "mature" set of genomes. The post–Mesozoic diversification of birds and mammals might also be analyzed in this light. This is not to say that recoveries are so deterministic that all of the survivors of an extinction event participate equally in the recovery (Jablonski 2002), or that significant shifts in the fortunes of clades do not occur after mass extinctions (as seen for the large flightless birds that were top carnivores in the early Cenozoic as the mammals diversified; Witmer and Rose 1991).

Much less appreciated is the nonrandom origin of evolutionary novelties in space. Marine invertebrate orders in post–Paleozoic seas, representing body plans as disparate as heart urchins, sand dollars, feather stars, and scleractinian corals, show a strong tendency to appear in onshore, shallow-water environments, even when their present-day distributions extend entirely across the continental shelf or are exclusively deep-water (Jablonski 2005a) (see figure 13.2). This onshore locus of evolutionary creativity can be seen in the distribution of key derived characters (Jablonski and Bottjer 1990) and in the divergence of species from ancestral groups in morphospace (Eble 2000), and so is not an artifact of taxon definition or rank. Comparable patterns are also reported, in considerably less detail, for the Cambrian explosion, the Ordovician radiations of marine invertebrates, Paleozoic land plants, and flowering plants (references in Jablonski 2005a). In contrast, at least for the post–Paleozoic marine clades, lower-level novelties, as indicated by genus-level origination and derived characters within orders, lack an onshore origination bias and tend to originate according to their clade's bathymetric diversity gradient—at this lower level, diversity does seem to beget diversity (Jablonski and Bottjer 1990; Jablonski et al. 1997; Jablonski 2005a).

This discordance in origination across hierarchical levels is another unexpected dynamic lying at the intersection of development and ecology, and again, nonexclusive hypotheses fall broadly into intrinsic and extrinsic categories (see Jablonski and Bottjer 1990; Jablonski 2005a). Hypothesized mechanisms involving intrinsic factors that might promote greater novelty production onshore include short-term selection for plasticity or evolvability in more disturbed and variable onshore environments (see also chapter 15 in this volume; and Hollander 2008, who finds greater plasticity in marine species from more heterogeneous environments); the high rates of gene flow and chaotic

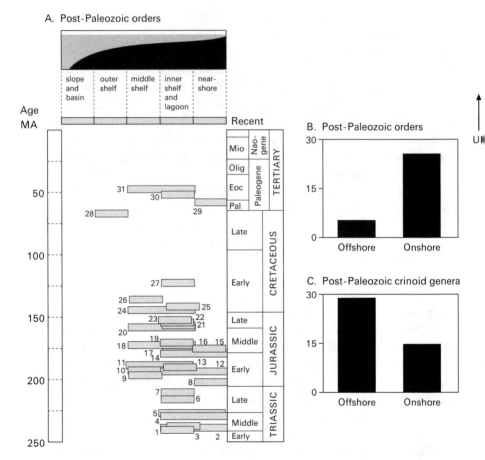

Figure 13.2
Environment-dependence in the origin of marine invertebrate orders through the Mesozoic and Cenozoic. (A, B) Of the 31 well-preserved orders whose first occurrences are mapped here, 80% first appear in onshore environments (first 2 right-hand environmental bins). (C) In contrast, lower-level novelties, such as the post–Paleozoic genera of crinoids, do not show preferential onshore origination. Vertical axis in A is millions of years before present (Ma). From Jablonski (2005a), who gives a complete listing of orders included in A, along with first occurrences of poorly preserved orders that serve as a sampling control.

population dynamics onshore; and the many hypothesized but poorly understood links between environmental stress and developmental perturbations. Potential extrinsic factors include the diversity and variability of selection pressures onshore, and higher nutrient inputs in onshore settings, which might buffer populations bearing evolutionary novelties against stochastic extinction. The discordance across levels does not mean that orders must arise by major saltations (or, for that matter, that ordinal rank per se has an objective reality), but it does suggest that the phylogenetic nodes accorded the ordinal rank are initiated in a fashion that differs significantly in environmental context, and perhaps mechanism, from nodes phenotypically less distinct and higher in each tree that represent genus and species origination (and for an encouraging concordance between molluscan and mammalian genera defined on morphological and molecular grounds, see Jablonski and Finarelli 2009). The ability of the purely speculative mechanisms just mentioned to account for that discordant pattern remains unclear.

A more familiar spatial dynamic can be seen with respect to latitude. The same marine invertebrate orders discussed above tend to originate preferentially in the tropics, and to spread to higher latitudes while retaining their tropical presence, when the strong sampling biases of the fossil record are taken into account (Jablonski 2005a; Martin et al. 2007). In contrast to the bathymetric pattern, this latitudinal dynamic is also seen at the lower taxonomic levels, most fully documented in marine bivalve genera (Jablonski et al. 2006; Roy and Goldberg 2007; Mittelbach et al. 2007, Krug et al. 2009); terrestrial groups may follow a somewhat different path dictated by high-latitude glaciation, but this contrast is uncertain. Thus, the first-order global diversity pattern in the modern (and, evidently, most ancient) marine biotas is underlain by an evolutionary and spatial dynamic that involves a spatial bias in the creation and spread of novelty at several taxonomic levels. Here, too, evolutionary mechanisms are poorly known. In contrast to the temporal and bathymetric dynamics, a simple probabilistic model—more species yield more orders—can be fitted to the "out of the tropics" pattern. Is this a fundamental contradiction that requires a different mechanism? That seems unlikely (given the discordance between species richness and other measures of diversity in so many other contexts) or at least highly inelegant, but until a broader and more robust theory of biological form is in place, it may be difficult to decide.

Hierarchy and Clade Dynamics

The fates of novelties and clades are also shaped by processes operating at multiple levels. Organismic-level processes are still important, of course, but effects of scale and hierarchy clearly influence long-term, large-scale evolutionary dynamics. All possible combinations of evolutionary tempo and mode have been documented for species in the fossil record, but Hunt (2007a) confirmed earlier views that sustained directional evolution of species is exceedingly rare (approximately 5% of fossil sequences in Hunt's compilation of 251 traits in 53 lineages) (see also Gould 1982, 2002; Jackson and Cheetham 1999; Jablonski 2000; Eldredge et al. 2005). The remaining 95% were split nearly equally between random walks and stasis in the strict sense, with variables related to body size less likely than shape traits to remain static. Many of the "random walk" examples had little spatial control, so that at least some of these sequences almost certainly involved shifting geographic variation within static species (see Stanley and Yang 1987; Jablonski 2000). Potential explanations for the pervasiveness of stasis are plentiful, and of course many fall squarely within the Modern Synthesis. However, the stark contrast between the great responsiveness of present-day populations to directional selection over the short run, and the paleontological data showing a strong tendency toward nondirectional or bounded phenotypic change over million-year scales, suggests that we do not fully understand the processes at work here.

The scarcity of directional evolutionary change within species over their geologic history is not a prerequisite for differential speciation and extinction to drive large-scale evolutionary changes (e.g., Slatkin 1981; contra Gould 2002; see also Jablonski 2008b). However, that scarcity drives home our need for a deeper understanding of multilevel evolutionary processes. Extrinsic physical and biotic factors must play a role in differential diversifications and the proliferation or decline of phenotypes: the uplift of the Andes or the Himalayas promoted speciation in many local lineages (and doubtless hastened the demise of others), and the formation of coral reefs boosted speciation of many resident clades (see references in Jablonski 2008a). However, the most fertile ground for the expansion of the synthesis probably lies in developing theory for the role of intrinsic biotic factors in setting differential origination and extinction rates, and how those fundamental macroevolutionary variables sum to net diversification.

Species selection in the broad sense—clade-level patterns shaped by differential speciation and extinction within clades owing to intrinsic properties of the bodies, populations, or species within them—has been abundantly documented. Factors ranging from intensity of sexual selection to generation time to geographic range have been found to predict evolutionary dynamics in clades of living or fossil organisms; indeed, the massive comparative biology literature can be seen in this light (e.g., Freckleton et al. 2002; Coyne and Orr 2004; Paradis 2005; Jablonski 2000, 2008b), as can the somewhat more problematic literature on key innovations. The basic requirements are that (a) a trait exhibits little or no variation within species relative to the variation among species (see Gould 2002: 664–665), and (b) speciation and/or extinction covaries with that trait consistently across one or more clades, in what might be termed an emergent fitness surface.

The requirements for species selection in the strict sense are more stringent, involving the classical formula for evolution by selection outlined by Lewontin (1970): differential production or survival of units owing to interaction of heritable traits (at the focal level) with the environment. This definition requires both the identification of emergent species-level traits and their "heritability" among species. Identification of emergent traits is not straightforward, but most authors appear to agree on a few pervasive ones, such as geographic range and genetic population structure (Jablonski 2008b). All of these requirements are met by geographic range in marine mollusks: species within clades vary with respect to geographic ranges; geographic range has been abundantly demonstrated to influence extinction rates (and origination rates; see Jablonski and Roy 2003); and geographic range is heritable, in the sense that closely related species have more similar range sizes than expected by chance (a result also seen in several extant groups) (Hunt et al. 2005; Jablonski and Hunt 2006; Waldron 2007; Borregaard 2008; Jablonski 2008b; and see Roy et al. 2009 for phylogenetic signal in the *location* of range end points). Ranges of these taxa cannot simply be reduced to an organismic trait such as mode of larval development because (a) the crucial range–extinction risk relationship holds even within larval types; (b) marine invertebrates attaining broad ranges by other means (e.g., rafting as adults) show similar dynamics; and (c) general linear models and a model selection approach find that adding geographic range to models containing only larval mode significantly increased model fits, indicating that geographic range is not redundant with larval mode in predicting species survivorship, whereas adding larval mode to models containing geographic range provides only marginal and insignificant improvement in model fit (Jablonski and Hunt 2006; see also Bradbury et al. 2008 on the multivariate nature of dispersal in marine species). These results do not mean that earlier work finding an association between larval mode and geographic range size in snails was incorrect. Instead, they show that larval dispersal is just one vehicle for attaining broad ranges, and so does not adequately account for the pervasive relationship between range size and speciation or extinction.

The propensity to speciate or to resist extinction need not be a simple correlate of organismic fitness even if organismic traits lie at the root of the rate differentials: selection for large body size or high metabolic rates need not promote origination or damp extinction (see review by Jablonski 2008b). But even if selection on organismic traits and the effects of those traits on clade dynamics are in the same direction, the cross-level interaction is important because the combined forces will be more effective in driving large-scale change than either would be independently. Hierarchical expansion of the Price equation, path analysis and related methods, and contextual analysis have all been suggested as approaches to partitioning processes operating at multiple levels in this context (e.g., Arnold and Fristrup 1982; Rice 2004; Okasha 2006; Simpson 2008; see Jablonski 2008b for discussion). The potential for empirical and theoretical advances in this aspect of multilevel selection is high.

Multilevel processes, or at least a failure of simple extrapolation from short-term, local processes, are often evident in the dynamics of evolutionary trends, which can unfold via differential origination (i.e., with a shift in modal phenotype underlain by higher origination rates in one region of the tree), differential extinction (i.e., lower extinction rates in one region of the tree), or directional speciation (i.e., preferential branching in one direction in a phenotype space) (e.g., Gould 1982, 2002; Jablonski 2008b). Although clades may evolve in directions consistent with patterns of intraspecific variation (e.g., Hunt 2007b; Hansen and Houle 2008), simple extrapolation from short-term processes often breaks down—as might be expected, given the scarcity of directional species-level evolution over geologic timescales. For example, data suggesting that short-term selection consistently favors larger body sizes (Kingsolver and Pfennig 2004) are not matched by long-term patterns of size change at the species and clade levels in many groups (e.g., Jablonski 1996, 1997; Moen 2006), or can be seen in terms of cross-level conflict, as when short-term organism selection for large size produces species that are more extinction-prone (Jablonski 1996; Van Valkenburgh et al. 2004; Clauset and Erwin 2008; Liow et al. 2008). Even when overall size increases do occur, they are context-specific, for example, tracking directional climate changes (Hunt and Roy 2006; Millien et al. 2006), or are manifest at some levels but not others (as in brachiopod orders but not their constituent families; see Novack-Gottshall and Lanier 2008; see also Clauset and Erwin 2008).

The inadequacy of extrapolation from short-term, local effects to macroevolutionary outcomes can also be seen in the failure of positive and negative biotic interactions to translate simply into clade dynamics (see Jablonski 2008a). Thus, mutualisms that increase the fitness of participating organisms sometimes elevate extinction probabilities of one or both mutualistic species, as with corals that benefit in the short run from their partnership with Zooxanthellae but show higher extinction rates than azooxanthellate relatives (e.g., Kiessling and Baron-Szabo 2004). Conversely, predation unambiguously reduces organismic fitness of prey but may elevate speciation rates (Stanley 1986, 2008; Dieckmann and Doebeli 2004).

Such apparent conflicts do not mean that localized interactions inevitably yield contrary results at larger scales, of course. For example, the evolutionary bursts seen after mass extinctions (e.g., Jablonski 2001; Foote 2003) strongly suggest that negative biotic interactions (competition, predation, parasitism) play a significant role in damping diversification during "normal" times: ecological incumbency in the broad sense is evidently a potent macroevolutionary factor. However, we still cannot predict when the effects of biotic interactions will change sign at larger temporal and spatial scales, and when they will scale up more predictably (see Jablonski 2008a). The answers may simply depend on, for example, the magnitude of environmental or geographic breadth of species in a given mutualism compared with related taxa uninvolved in that partnership; or they may be idiosyncratically contingent on the biology of each individual clade. Here is an intriguing set of multilevel challenges for an enlarged synthesis.

The significant evolutionary role now recognized for differential origination and extinction among clades raises a fundamental point about the long-term fates of phenotypic traits: hitchhiking on other features may be rampant. A given trait may be borne by many species, or dwindle into extinction, not because it directly influences fitness to the point of promoting diversification or evolutionary decline, but because it happens to be associated with other organismic or clade-level traits that

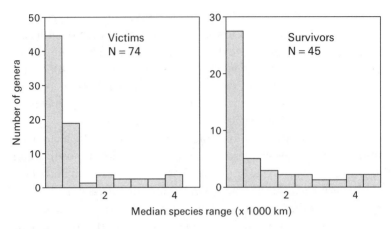

Figure 13.3
Marine gastropod genera show no significant difference in the geographic ranges of species between genera that are victims (left) and survivors (right) of the end–Cretaceous mass extinction. (From Jablonski (2005b). See Jablonski (1986) for similar results in marine bivalves.

govern clade dynamics. Such hitchhiking effects may underlie conflicting results of some comparative analyses, where different factors appear to drive diversification of extant species. For example, does sexual dichromatism actually drive bird diversification, or is it hitchhiking on such factors as dispersal or brain size (see Phillimore et al. 2006; Sol and Price 2008)?

Nowhere is the potential for hitchhiking greater than during mass extinctions, which appear to add another level to the process. Species-level geographic range (and a number of organismic traits) had no effect on molluscan clade survival of the end–Cretaceous (KT) extinction, the only available analysis at this level of detail (figure 13.3), whereas clade-level geographic range was one of the chief predictors of survivorship for this and other extinction events (e.g., Jablonski 2005b, 2008c; Powell 2007) (figure 13.4). Thus adaptations can be lost not because they are directly disfavored by selection, but because they happen to occur in one or more clades that lack the broad spatial deployment promoting survival. This decoupling effect, adding another level of sorting to the hierarchical equation, may help to explain a number of perplexing extinction patterns, such as the preferential extinction of complex colony forms in bryozoans during the late Ordovician extinction: complex colonies tended to occur in clades having narrow geographic ranges (Anstey et al. 2003). Similarly, the demise of the ammonites and the persistence

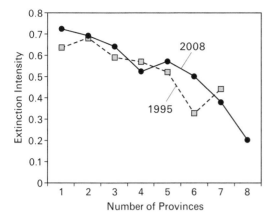

Figure 13.4
Significant inverse relation between extinction intensity and bivalve genus range (measured as the number of biogeographic provinces occupied) during the end–Cretaceous extinction (Spearman rank test, P < 0.01). Solid line indicates analysis of newly revised data set (n = 289 genera); broken line shows results of previous analyses. (From Jablonski 2008c)

of the nautiloids was probably not directly related to the complexity of the internal chamber walls of the ammonites and the simplicity of those walls in nautiloids (a point made by Seilacher 1988), but the KT extinction permanently changed the morphological range of the Cephalopoda nonetheless—presumably because the exquisite ammonite septa hitchhiked to oblivion on some other organismic or clade-level feature.

Hitchhiking via hierarchical effects comes into clearer focus when we compare the geographic ranges of clades with the ranges of their constituent species. One might expect widespread clades (in this case, molluscan genera) to consist mainly of environmentally tolerant, ecologically successful, and/or widely dispersing widespread species, and the converse to be true for restricted-range clades. But this expectation is not met for present-day marine mollusks, for which geographic ranges of genera and species are not significantly related (except in the trivial sense that a genus cannot be less widespread than its broadest-ranging species) (figure 13.5). Evidently broad ranges at the clade level can be achieved in too many ways—some clades with endemic species scattered across many provinces, others with widespread species encompassing broad regions, and every intermediate combination—to support simple extrapolation from species to clade level. Thus, wherever clade geographic range influences the persistence of evolutionary lineages (as

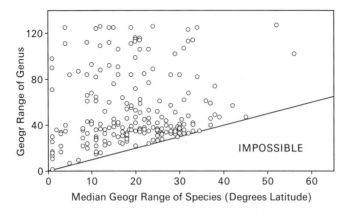

Figure 13.5
The geographic ranges of eastern Pacific bivalve genera (in degrees latitude from Point Barrow, Alaska, to Cape Horn, Chile) cannot be predicted by the median geographic ranges of their constituent species (n = 213 genera, Spearman's r = 0.17; not significant). The lower right corner of the graph represents a field of impossible combinations: species cannot be more widespread than their genera; statistical significance assessed by resampling the data, using 10,000 repetitions where impossible combinations were discarded with replacement. (From Jablonski 2005b)

figure 13.4 and much other work strongly suggest), the noncorrelation between species and genus range is arresting. It undermines any simple expectation drawn from the tolerances, competitive abilities, or other features of individual organisms in predicting clade persistence, and demands a hierarchical approach. Organismic traits associated with species-level ranges are subject to hitchhiking on clade-level geographic range, which is evidently a vehicle for clade sorting during background times as well as mass extinctions (see discussion in Jablonski 2008c). Now the question of heritability of clade-level geographic range becomes important; my impression is that phylogenetic conservation of clade ranges is weak, undermining operation of a protracted process analogous to organismic or species selection, but formal analyses are lacking.

This hierarchical view of evolution requires a basic change in how we interpret the occurrence of phenotypic traits within and among clades. Again, it does not mean that selection at the organismic level is ineffective or unimportant, but selectivity at a higher level can yield a predictable, coherent pattern that may override or even mimic processes at lower levels as it drags along organismic traits. Approaches that explicitly tackle the problem of cause and effect in taxon survivorship as a set of alternative models across hierarchical levels are needed to tease apart these processes (Jablonski and Hunt 2006; Jablonski 2008c).

Conclusion

Calls for expanding the Evolutionary Synthesis have been plentiful and at least partially successful (e.g., in a sampling of old and new statements; Stanley 1979; Gould 1980, 1982, 2002; Raff and Kauffman 1983; Eldredge 1985; Jablonski 1986, 2000, 2007, 2008b; Arthur 2002; Müller 2007; Pigliucci 2007; Carroll 2008). Much hard work lies ahead if we are to evaluate and integrate the novel or neglected elements emphasized by these and other authors, and it is crucial to move beyond consistency arguments to confront theory with data in a critical fashion.

Documentation and analysis of macroevolutionary patterns in living and fossil organisms amply demonstrate the need to more fully incorporate scale and hierarchy into the Evolutionary Synthesis. The predictive power of conventional short-term observations is not strong over larger temporal and spatial scales, which demand novel and expanded approaches to the origin of evolutionary novelty, and to the dynamics of novelties and the clades they define. One of the key contributions of evolutionary developmental biology has been to undermine an operating assumption of the Evolutionary Synthesis in scaling population genetics up to macroevolution: the presumed one-to-one mapping between magnitude of genotypic change and magnitude of phenotypic change. New work on the structure and evolution of gene regulatory networks, epigenetics, modularity, and plasticity converges on a more complex view of the potential for nonlinear and coordinated phenotypic change that makes the nonrandom origins of evolutionary novelties in time and space especially intriguing. Simple probabilistic models, in which major groups originate stochastically according to random mutation, number of individuals, or species richness across times and environments, all fail when tested against the regularities that emerge from the fossil record. Such large-scale patterns in the generation of form, which were barely an issue in the founding documents of the Evolutionary Synthesis, represent a little-explored macroevolutionary intersection between development and ecology. If we wish to understand the origins and dynamics of major groups, this intersection needs to be explored in far greater detail, and in an explicitly hierarchical framework, from genomes through organisms to species and clades.

Spatial and temporal regularities in the origin of phenotypic novelties are just one aspect of macroevolution that appears to require an expansion of evolutionary concepts and methods. The recent corroboration of evolutionary stasis as a pervasive species-level dynamic, and strong support for species selection in the broad sense (myriad examples) and in the strict sense (fewer well-documented cases, but potentially abundant), call for multilevel approaches to large-scale evolutionary dynamics. Indeed, regardless of whether traits, even complex ones such as evolvability or genomic modularity, arose as a by-product of processes at lower levels (e.g., Lynch 2007) or by selection at higher levels (e.g., Pigliucci 2007), if they affect speciation rates, extinction rates, or a clade's behavior in morphospace, then a hierarchical approach to evolution is required. This need is seen even more clearly in the long-term phenotypic effects of differential survival stemming from species-level and clade-level geographic ranges, which are in turn demonstrably uncoupled, not only from one another but also from the many organismic features that can hitchhike along with differential origination and extinction at the species and clade levels. Further, the failure of positive and negative biotic interactions to translate simply and consistently into clade dynamics attests to the need to get beyond extrapolation from local, short-term observations. Contrary to the past focus by the Evolutionary Synthesis on a single hierarchical level (bodies within populations), evidence is now strong that multilevel approaches are essential to understanding long-term evolutionary processes.

Acknowledgments

I thank Massimo Pigliucci and Gerd Müller for inviting me to participate in this stimulating workshop, and for their thoughtful reviews. In addition to interactions in the workshop, discussions with Susan M. Kidwell, Kaustuv Roy, and James W. Valentine were especially valuable. My research and attempts at synthesis have been supported by NSF, NASA, and the John Simon Guggenheim Foundation.

References

Adamowicz SJ, Purvis A, Wills MA (2008) Increasing morphological complexity in multiple parallel lineages of the Crustacea. Proceedings of the National Academy of Sciences of the USA 105: 4786–4791.

Anstey RL, Pachut JF, Tuckey ME (2003) Patterns of bryozoan endemism through the Ordovician–Silurian transition. Paleobiology 29: 305–328.

Arnold AJ, Fristrup K (1982) The theory of evolution by natural selection: A hierarchical expansion. Paleobiology 8: 113–129.

Arthur W (2002) The emerging conceptual framework of evolutionary developmental biology. Nature 415: 757–764.

Bambach RK, Bush AM, Erwin DH (2007) Autecology and the filling of ecospace: Key metazoan radiations. Palaeontology 50: 1–22.

Borregaard MK (2008) Range size heritability may explain current range size distributions. In Joint Meeting of the Center for Macroecology (University of Copenhagen) & the Biodiversity and Global Change Lab (Museo Nacional de Ciencias Naturales, CSIC), Monasterio San Juan de la Peña, Huesca, Spain. Abstract.

Bradbury IR, Laurel B, Snelgrove PVR, Bentzen P, Campana SE (2008) Global patterns in marine dispersal estimates: The influence of geography, taxonomic category and life history. Proceedings of the Royal Society of London B275: 1803–1809.

Carroll SB (2008) Evo-devo and an expanding evolutionary synthesis: A genetic theory of morphological evolution. Cell 134: 25–36.

Clauset A, Erwin DH (2008) The evolution and distribution of species body size. Science 321: 399–401.

Coyne JA, Orr HA (2004) Speciation. Sunderland, MA: Sinauer.

Dieckmann U, Doebeli M (2004) Adaptive dynamics of speciation: Sexual populations. In Adaptive Speciation. U. Dieckmann, M. Doebeli, J.A.J. Metz, D Tautz, eds.: 76–111. Cambridge: Cambridge University Press.

Eble GJ (1998) The role of development in evolutionary radiation. In Biodiversity Dynamics: Turnover of Populations, Taxa, and Communities. ML McKinney, JA Drake, eds.: 132–161. New York: Columbia University Press.

Eble GJ (1999) Originations: Land and sea compared. Geobios 32: 223–234.

Eble GJ (2000) Contrasting evolutionary flexibility in sister groups: Disparity and diversity in Mesozoic atelostomate echinoids. Paleobiology 26: 56–79.

Eldredge N (1985) Unfinished Synthesis: Biological Hierarchies and Modern Evolutionary Thought. Oxford: Oxford University Press.

Eldredge N, Thompson, JN, Brakefield PM, Gavrilets S, Jablonski D, Jackson JBC, Lenski RE, Lieberman BS, Mcpeek MA, Miller W (2005) Dynamics of evolutionary stasis. Paleobiology 31 (suppl. to no. 2): 133–145.

Erwin DH (1992) A preliminary classification of evolutionary radiations. Historical Biology 6: 133–147.

Erwin DH (2007) Disparity: Morphologic pattern and developmental context. Palaeontology 50: 57–73.

Foote M (2003) Origination and extinction through the Phanerozoic: A new approach. Journal of Geology 111: 125–148.

Freckleton RP, Harvey PH, Pagel M (2002) Phylogenetic analysis and comparative data: A test and review of evidence. American Naturalist 160: 712–726.

Gavrilets S, Losos JB (2009) Adaptive radiation: Contrasting theory with data. Science 323: 732–737.

Gould SJ (1980) Is a new and general theory of evolution emerging? Paleobiology 6: 119–130.

Gould SJ (1982) The meaning of punctuated equilibria and its role in validating a hierarchical approach to macroevolution. In Perspectives on Evolution. R. Milkman, ed.: 83–104. Sunderland, MA: Sinauer.

Gould SJ (2002) The structure of evolutionary theory. Cambridge, MA: Belknap Press of Harvard University Press.

Hansen TF, Houle D (2008). Measuring and comparing evolvability and constraint in multivariate characters. Journal of Evolutionary Biology 21: 1201–1219.

Hollander J (2008) Testing the grain-size model for the evolution of phenotypic plasticity. Evolution 62: 1381–1389.

Hunt G (2007a) The relative importance of directional change, random walks, and stasis in the evolution of fossil lineages. Proceedings of the National Academy of Sciences of the USA 104: 18404–18408.

Hunt G (2007b) Evolutionary divergence in directions of high phenotypic variance in the ostracode genus *Poseidonamicus*. Evolution 61: 1560–1576.

Hunt G, Roy K (2006) Climate change, body size evolution, and Cope's Rule in deep-sea ostracodes. Proceedings of the National Academy of Sciences of the USA 103: 1347–1352.

Hunt G, Roy K, Jablonski D (2005) Heritability of geographic range sizes revisited. American Naturalist 166: 129–135.

Jablonski D (1980) Adaptive radiations: Fossil evidence for two modes. In Evolution Today: Proceedings of the 2nd International Congress on Systemic and Evolutionary-Biology. Abstracts: 243.

Jablonski D (1986) Background and mass extinctions: The alternation of macroevolutionary regimes. Science 231: 129–133.

Jablonski D (1996) Body size and macroevolution. In Evolutionary Paleobiology. D Jablonski, DH Erwin, JH Lipps, eds.: 256–289. Chicago: University of Chicago Press.

Jablonski D (1997) Body-size evolution in Cretaceous molluscs and the status of Cope's rule. Nature 385: 250–252.

Jablonski D (2000) Micro- and macroevolution: Scale and hierarchy in evolutionary biology and paleobiology. Paleobiology 26 (suppl. to no. 4): 15–52.

Jablonski D (2001) Lessons from the past: Evolutionary impacts of mass extinctions. Proceedings of the National Academy of Sciences of the USA 98: 5393–5398.

Jablonski D (2002) Survival without recovery after mass extinctions. Proceedings of the National Academy of Sciences of the USA 99: 8139–8144.

Jablonski D (2005a) Evolutionary innovations in the fossil record: The intersection of ecology, development and macroevolution. Journal of Experimental Zoology B304: 504–519.

Jablonski D (2005b) Mass extinctions and macroevolution. Paleobiology 31 (suppl. to no. 2): 192–210.

Jablonski D (2007) Scale and hierarchy in macroevolution. Palaeontology 50: 87–109.

Jablonski D (2008a) Biotic interactions and macroevolution: Extensions and mismatches across scales and levels. Evolution 62: 715–739.

Jablonski D (2008b) Species selection: Theory and data. Annual Review of Ecology, Evolution and Systematics 39: 501–524.

Jablonski D (2008c) Extinction and the spatial dynamics of biodiversity. Proceedings of the National Academy of Sciences of the USA 105 (suppl. 1): 11528–11535.

Jablonski D, Bottjer DJ (1990) The origin and diversification of major groups: Environmental patterns and macroevolutionary lags. In Major Evolutionary Radiations. PD Taylor, GP Larwood, eds.: 17–57. Oxford: Clarendon Press.

Jablonski D, Finarelli JA (2009) Congruence of morphologically-defined genera with molecular phylogenies. Proceedings of the National Academy of Sciences of the USA 106: 8262–8266.

Jablonski D, Hunt G (2006) Larval ecology, geographic range, and species survivorship in Cretaceous mollusks: Organismic vs. species-level explanations. American Naturalist 168: 556–564.

Jablonski D, Lidgard S, Taylor PD (1997) Comparative ecology of bryozoan radiations: Origin of novelties in cyclostomes and cheilostomes. Palaios 12: 505–523.

Jablonski D, Roy K (2003) Geographic range and speciation in fossil and living molluscs. Proceedings of the Royal Society of London B270: 401–406.

Jablonski D, Roy K, Valentine JW (2006) Out of the Tropics: Evolutionary dynamics of the latitudinal diversity gradient. Science 314: 102–106.

Jackson JBC, Cheetham AH (1999) Tempo and mode of speciation in the sea. Trends in Ecology and Evolution 14: 72–77.

Kiessling W, Baron-Szabo RC (2004) Extinction and recovery patterns of scleractinian corals at the Cretaceous–Tertiary boundary. Palaeogeography, Palaeoclimatology and Palaeoecology 214: 195–223.

Kingsolver JG, Pfennig DW (2004) Individual-level selection as a cause of Cope's rule of phyletic size increase. Evolution 58: 1608–1612.

Krug AZ, Jablonski D, Valentine JW, Roy K (2009) Generation of Earth's first-order biodiversity pattern. Astrobiology 9: 113–124.

Lewontin RC (1970) The units of selection. Annual Review of Ecology and Systematics 1: 1–18.

Liow LH, Fortelius M, Bingham E, Lintulaakso K, Mannila H, Flynn L, Stenseth NC (2008) Higher origination and extinction rates in larger mammals. Proceedings of the National Academy of Sciences of the USA 105: 6097–6102.

Lynch, M (2007) The frailty of adaptive hypotheses for the origins of organismal complexity. Proceedings of the National Academy of Sciences of the USA 104 (suppl. 1): 8507–8604.

Marks CO, Lechowicz MJ (2006) Alternative designs and the evolution of functional diversity. American Naturalist 167: 55–66.

Marshall CR (2006) Explaining the Cambrian "explosion" of animals. Annual Review of Earth and Planetary Sciences 34: 355–384.

Martin PR, Bonier F, Tewksbury JJ (2007) Revisiting Jablonski (1993): Cladogenesis and range expansion explain latitudinal variation in taxonomic richness. Journal of Evolutionary Biology 20: 930–936.

Millien V, Lyons SK, Olson L, Smith FA, Wilson T, Yom-Tov Y (2006) Ecotypic variation in the context of global climate change: Revisiting the rules. Ecology Letters 9: 853–869.

Mittelbach, GG, Schemske DW, Cornell HV, Allen AP, Brown JM, Bush MB, Harrison SP, Hurlbert AH, Knowlton N, Lessios HA, McCain CM, McCune AR, McDade LA, McPeek MA, Near TJ, Price TD, Ricklefs RE, Roy K, Sax DF, Schluter D, Sobel JM, Turelli M (2007) Evolution and the latitudinal diversity gradient: Speciation, extinction and biogeography. Ecology Letters 10: 315–331.

Moen DS (2006) Cope's rule in cryptodiran turtles: Do the body sizes of extant species reflect a trend of phyletic size increase? Journal of Evolutionary Biology 19: 1210–1221.

Müller GB (2007) Evo-devo: Extending the evolutionary synthesis. Nature Reviews Genetics 8: 943–949.

Niklas KJ (1994) Morphological evolution through complex domains of fitness. Proceedings of the National Academy of Sciences of the USA 91: 6772–6779.

Niklas KJ (2004) Computer models of early land plant evolution. Annual Review of Earth and Planetary Sciences 32: 47–66.

Novack-Gottshall PM, Lanier MA (2008) Scale-dependence of Cope's rule in body size evolution of Paleozoic brachiopods. Proceedings of the National Academy of Sciences of the USA 105: 5430–5434.

Okasha S (2006) Evolution and the Levels of Selection. New York: Oxford University Press.

Paradis E (2005) Statistical analysis of diversification with species traits. Evolution 59: 1–12.

Peterson KJ, Cotton JA, Gehling JG, Pisani D (2008) The Ediacaran emergence of bilaterians: Congruence between the genetic and the geological fossil records. Philosophical Transactions of the Royal Society of London B363: 1435–1443.

Phillimore AB, Freckleton RP, Orme CDL, Owens IPF (2006) Ecology predicts large-scale patterns of phylogenetic diversification in birds. American Naturalist 168: 220–229.

Pigliucci M (2008) Is evolvability evolvable? Nature Reviews Genetics 9: 75–82.

Powell MG (2007) Geographic range and genus longevity of late Paleozoic brachiopods. Paleobiology 33: 530–546.

Raff RA, Kaufman TC (1983) Embryos, Genes, and Evolution. New York: Macmillan.

Reznick D, Ricklefs RE (2009) Darwin's bridge between microevolution and macroevolution. Nature 457: 837–842.

Rice SH (2004) Evolutionary Theory: Mathematical and Conceptual Foundations. Sunderland, MA: Sinauer.

Roy K, Goldberg EE (2007) Origination, extinction and dispersal: Integrative models for understanding present-day diversity gradients. American Naturalist 170 (suppl.): S71–S85.

Roy K, Hunt G, Jablonski D, Krug AZ, Valentine JW (2009) A macroevolutionary perspective on species range limits. Philosophical Transactions of the Royal Society of London B276: 1485–1493.

Seilacher A (1988) Why are nautiloid and ammonoid sutures so different? N. Jb. Geol. Paläont. 177: 41–69.

Shen B, Dong L, Xiao S, Kowalewski M (2008) The Avalon Explosion: Expansion and saturation of Ediacara morphospace. Science 319: 81–84.

Simpson, C. (2008) Species selection and driven mechanisms jointly generate a large-scale morphological trend in monobathrid crinoids. Paleobiology 34 (in press).

Slatkin M (1981) A diffusion model of species selection. Paleobiology 7: 421–425.

Sol D, Price TD (2008) Brain size and the diversification of body size in birds. American Naturalist 172: 170–177.

Stanley SM (1979) Macroevolution. San Francisco: W. H. Freeman.

Stanley SM (1986) Population size, extinction, and speciation: The fission effect in Neogene Bivalvia. Paleobiology 12: 89–110.

Stanley SM (2008) Predation defeats competition on the sea floor. Paleobiology 34: 1–22.

Stanley SM, Yang X (1987) Approximate evolutionary stasis for bivalve morphology over millions of years: A multivariate, multilineage study. Paleobiology 13: 113–139.

Valentine JW (1969) Taxonomic and ecological structure of the shelf benthos during Phanerozoic time. Palaeontology 12: 684–709.

Valentine JW (1980) Determinants of diversity in higher taxonomic categories. Paleobiology 6: 444–450.

Valentine JW (1995) Why no new phyla after the Cambrian? Genome and ecospace hypotheses revisited. Palaios 10: 190–194.

Valentine JW (2004) On the Origin of Phyla. Chicago: University of Chicago Press.

Van Valen, LM (1982) Why misunderstand the evolutionary half of biology? In Conceptual Issues in Ecology. E Saarinen, ed.: 323–343. Dordecht: Reidel.

Van Valkenburgh B, Wang X, Damuth J (2004) Cope's Rule, hypercarnivory, and extinction in North American canids. Science 306: 101–104.

Wagner A (2008) Robustness and evolvability: A paradox resolved. Proceedings of the Royal Society of London B275: 91–100.

Waldron A (2007) Null models of geographic range size reaffirm its heritability. American Naturalist 170: 223–231.

Webster M (2007) A Cambrian peak in morphological variation within trilobite species. Science 317: 499–502.

Witmer LM, Rose KD (1991) Biomechanics of the jaw apparatus of the gigantic Eocene bird *Diatryma*: Implications for diet and mode of life. Paleobiology 17: 95–120.

Wright S (1949) Adaptation and selection. In Genetics, Paleontology and Evolution. GL Jepson, GG Simpson, E Mayr, eds.: 365–389. Princeton, NJ: Princeton University Press.

14 Phenotypic Plasticity

Massimo Pigliucci

The concept of phenotypic plasticity has just turned 100 (Woltereck 1909), and yet it is common both in the published literature and especially in the halls of scientific meetings to hear professional biologists make pronouncements that betray their lack of understanding of what plasticity actually is. By far the most common misunderstanding is that plasticity is simply a fancy word to indicate the old "environmental component" of a phenotype (Falconer 1952), and that therefore it still makes sense to think in terms of genetics versus plasticity. Next in line among the misconceptions about plasticity is the idea that one can tell whether an organism is "more" or "less" plastic than another one, across the board. In fact, ever since Woltereck coined the term "reaction norm," it should have been clear that plasticity is a property of a genotype, and that it is specific to particular traits within a given range of environments. Before we come to a discussion of the possible roles of phenotypic plasticity as a causal factor in evolution, therefore, it may be appropriate to review the basics.

A norm of reaction is a genotype-specific function that relates the range of environments experienced during ontogeny to the range of phenotypes that the particular genotype produces in that range of environments (figure 14.1). A population of reaction norms, therefore, is characterized by three general properties (Pigliucci 2001): there can be genetic variation for the focal trait across the environments being considered, which means that some norms of reaction are positioned higher than others on the diagram; there can be overall plasticity (i.e., across genotypes) if on average the reaction norms show a nonzero slope; and there can be genotype-by-environment interaction if the slopes of distinct reaction norms are significantly different from each other.

From a reaction norm perspective, therefore, it simply makes no sense to attempt to separate genetic from environmental effects, because the

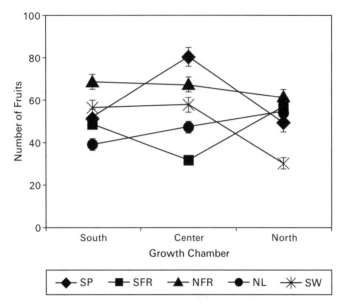

Figure 14.1
A simple set of norms of reaction from a plasticity experiment conducted on the model plant system *Arabidopsis thaliana*. The horizontal axis represents the environment (in this case the simulated photoperiods of northern, central, and southern latitudes in Europe), and the vertical axis is a measure of phenotype. Each line represents a distinct genotype's reaction norm. The diagram illustrates the properties of genetic variation (different heights of the various norms), environmental variation (when individual norms have a slope different from zero) and genotype-by-environment interaction (when different norms are not parallel to each other). Symbols to the right indicate the provenances of individual populations: SP = Spain; SFR = Southern France; NFR = Northern France; NL = Netherlands; SW = Sweden (Data from Josh Banta, Pigliucci Lab)

"genetic," "environmental," and interaction variances are all properties of the specific genotype–environment combinations that are characteristic of a given population of organisms. To use a metaphor first introduced by Richard Lewontin, if you were building a house with bricks and lime, it would make little sense to estimate the different contributions of the two components by weighing them: the house is made of the specific pattern of brick-and-lime layering, though of course if you really insisted, you could count the bricks and weigh the lime; it just would not tell you much of interest. This has strong implications for the concept of heritability, which Lewontin (1974)—on the basis of a reaction norm perspective—showed to be entirely dependent on the specific combination of genotypes and environments one is considering: change the environment, and the "genetic" variance in the population might, counterintuitively, change; alter the frequency of different genotypes and, perhaps even more counterintuitively, the "environmental" variance is altered as well. Since heritability is by definition the ratio of genetic to total (i.e., including environmental) variance, one can easily see how heritability is anything but a fixed feature of a population, and even less so of a species.

While norms of reaction are usually depicted as continuous functions of the environment, the phenomenon of phenotypic plasticity is more general, and some of its most spectacular manifestations are the well-known "polyphenisms" that characterize plant and animal species. For instance, different larval diets produce male horned beetles with or without the horn (Moczek 2006), a trait that dramatically influences their ability to compete for mates. In plants, one of the earliest examples of adaptive phenotypic plasticity ever studied is a type of polyphenism: heterophylly in semiaquatic plants is an instance of developmental plasticity in which the organism produces different types of leaves depending on whether it finds itself below or above water (Cook and Johnson 1968; Wells and Pigliucci 2000; see figure 14.2), the advantage being that each type of leaf is best suited—both morphologically and physiologically—to its own environment.

Phenotypic plasticity is now the paradigmatic way of thinking about gene–environment interactions (the so-called nature–nurture problem), and one of the best studied biological phenomena in the evolutionary literature, with knowledge steadily advancing about its genetic–molecular underpinnings (Schlichting and Smith 2002; Suzuki and Nijhout 2008), ecological role (Callahan and Pigliucci 2002; Nussey et al. 2007), and evolution (Pigliucci et al. 2003; Paenke et al. 2007). In the rest of this chapter I will explore how evolutionary biologists are now using the concept of plasticity to expand the horizons of the Modern Synthesis of the 1930s and 1940s (Mayr and Provine 1980), and what role plasticity may play in the shaping of an Extended Evolutionary Synthesis.

Plasticity, Buffering, and Capacitance

For a long time in the literature the idea of phenotypic plasticity has been linked to those of homeostasis, canalization, and buffering (Flatt 2005). Although plasticity is often portrayed as the opposite of canalization, it is easy to see why this cannot strictly be the case: a canalized phenotype is one that is reliably produced by the developmental system;

Figure 14.2
An example of adaptive plasticity in plants, the polyphenism known as heterophylly (the ability to produce distinctly shaped leaves depending on whether the stem of the plant is below or above water). The species photographed here is *Proserpinaca palustris*. (Photo by Carolyn Wells, Pigliucci Lab)

but a reaction norm can also be a reproducible (set of) phenotype(s), given a particular genotype and range of environments. In other words, there is no contradiction in speaking of canalized plastic norms of reaction; hence, canalization is not the opposite of plasticity. On the other hand, there does seem to be an inverse relationship between plasticity and (environmental) homeostasis, if one understands homeostasis as the maintenance of a given phenotype regardless of (external) conditions. Finally, "buffering" is a generic term for whatever mechanism leads to genetic or environmental homeostasis.

To summarize the differences and relationships among these biological phenomena, then:

• Phenotypic plasticity is a trait- and environment-specific property of the genotype which may or may not be advantageous and may or may not be the result of adaptive evolution (i.e., natural selection).

• Canalization is a property of the developmental system that allows it to reliably reproduce the same phenotype under the same set of conditions. Plasticity can be canalized, and canalization is usually thought of as a derived condition, resulting from natural selection.

• Homeostasis comes in two flavors: environmental and genetic. Either way, it means that the developmental system is resilient to changes in the external or internal environment. A plastic reaction norm is not (environmentally) homeostatic, by definition. Homeostasis is also usually thought of as a derived evolutionary outcome, brought about by natural selection.

• Buffering refers to the range of mechanisms that cause homeostasis, which means that a breakdown in buffering may lead to (probably nonadaptive) phenotypic plasticity.

Several authors have argued that nonadaptive plasticity must be considered a default attribute of biological systems, because a variety of environmental factors, such as temperature, pH, and others, alter the functionality of biomolecules by default, simply as a result of standard biochemistry. As Newman et al. (2006: 290) put it: "Much morphological plasticity reflects the influence of external physico-chemical parameters on any material system and is therefore an inherent, inevitable property of organisms." According to Nijhout (2003: 9): "Temperature, nutrition, photoperiod, and so on … affect the underlying chemical and metabolic processes of development directly, without the intervention of a specially evolved mechanism … phenotypic plasticity is obtained gratis, as a by-product of the physics and chemistry of development." A typical example is represented by the well-known temperature–activity curves that are characteristic of enzymes: while the specific shapes of these curves vary with the protein being studied, enzymes simply cannot avoid having an optimal (and usually narrow) range of temperatures, flanked by temperatures at which they can still function, but suboptimally, and finally by temperatures at which they have no detectable biological activity. This kind of phenomenon helps us make sense of the above-mentioned idea that environmental homeostasis, where it exists, must be an evolutionarily derived state of things.

By the same token, continuous reaction norms like those depicted in figure 14.1 may not necessarily be adaptive, but simply the inevitable result of a developmental system exposed to different environmental conditions. On the other hand, more structured plastic responses, such as the polyphenisms mentioned earlier, are usually thought of as the result of natural selection because their sharply distinct phenotypes are often associated with clearly advantageous functions. The obvious inference is that polyphenisms evolve from preexisting, continuous, reaction norms.

It is the interplay between inherent plasticity and the necessity to maintain functionality over a range of environments that links the evolution of phenotypic plasticity to that of buffering mechanisms, as was first glimpsed by Waddington (1942, 1961) with his discussion of what at the time were still termed "acquired characters." (Waddington did not use the reaction norm terminology or framework, although today we can see that his famous mechanism of genetic assimilation is really an example of selection on the shape of an organism's reaction norm. I will return to this point later in the chapter.) Waddington was interested in the generation of novel phenotypes, using an environmental stimulus simply as a trigger. The end product of his process of genetic assimilation was a "canalized" trait that would no longer require the environmental trigger. Though Waddington was successful in demonstrating by experiments on Drosophila that his hypothetical mechanism could in fact work, the standing criticism of that body of work has always been that genetic assimilation had not been demonstrated to occur under natural conditions (of course the fact that few people bothered to look for it should have significantly weakened that objection, but somehow failed to do so).

The situation changed recently with the onset of research on so-called "capacitors" of phenotypic evolution, an approach that in a sense turned Waddington's interest on its head, focusing on how the disruption of a buffering system may yield an explosion of phenotypic forms and a surge in phenotypic plasticity, thereby providing new raw material for natural selection to work on (Rutherford and Lindquist 1998; Queitsch et al. 2002). This body of work—ironically, still conducted by means of experimental protocols on model systems—has prompted a new appreciation of Waddington's ideas, now cast in the modern language of reaction norms and capacitance. It is an entirely open question whether and to what extent these phenomena are relevant to organismic evolution, but there are at least compelling arguments based on computer models that natural selection may in fact favor capacitance (Masel 2005), and therefore plasticity, as an intermittently major player in the generation of phenotypic novelties.

Whether one looks at the problem of the generation of novel phenotypic variation from the point of view of genetic assimilation or from that of capacitance—which, I am arguing, are in fact two sides of the same coin—phenotypic plasticity emerges as a key player, either in allowing the initial steps leading to assimilation or in providing the raw material for renewed evolution after the disruption of a buffering system. In either case, the idea is that phenotypic evolution may occur surprisingly fast, within the span of a few generations. This has led Murren and myself (Pigliucci and Murren 2003) to suggest one counterintuitive reason why it may be difficult to find natural examples of genetic assimilation (other than the already mentioned fact that few people have been looking for them): the telltale signs may be gone from a natural population so quickly as to induce the investigator to think that all that is going on is standard selection on genetic variants, just as prescribed by the theoretical framework of the Modern Synthesis. There are ways around this problem, as discussed in Pigliucci and Murren (2003), once one knows what to look for, but this issue will come back a fortiori when we consider West-Eberhard's ideas about phenotypic and genetic accommodation below.

The possibility that plasticity may facilitate the evolution of fast phenotypic changes has consequences in at least two areas of evolutionary biology: the study of invasive species and the question of speciation. Richards et al. (2006) have pointed out that phenotypic plasticity may allow a potential invader to establish populations at low demographic levels, essentially buying time until the standard genetic variation–selection mechanism kicks in and allows the invader to fine-tune its adaptation to the novel environment. This would neatly account for an oft-observed pattern characterizing biological invasions in which the invader colonizes multiple locations at low population densities and then survives at those densities for a "lag time" that may last decades. Suddenly one or more of the established populations then experiences an aggressive bout of growth, and the devastating part of the invasion begins. This is precisely the pattern that a "plasticity first" hypothesis would predict, and there is an increasing interest in the invasive biology community in empirically testing this role of phenotypic plasticity in ongoing demographic changes of alien species.

As for the link between plasticity and speciation, here is what Levin (2004: 808) has to say: "An ecological shift most often involves the occupation of novel habitats in the physical and genetic vicinity of the source population ... plasticity buys populations time to adapt, in that they may persist across generations without genetic alteration. ... Long-term population survival of the newly founded populations is conditional on genetic refinement." This is precisely the idea I have articulated a few lines above in the context of invasions, which Levin independently applies to the possibility of ecological speciation. This mode of specia-tion has received increasing attention recently (Fournier and Giraud 2008; Rasanen and Hendry 2008), but to my knowledge little has yet been done to integrate a reaction norm perspective into the empirical study of speciation driven by natural selection.

The Mechanics of Plasticity: Development, Genetics, and Epigenetics

In order to understand how deep a role phenotypic plasticity plays in the restructuring of evolutionary theory, one has to consider issues related to the mechanics of plasticity in terms of molecular basis (genes, proteins, and hormones involved), as well as of epigenetic effects and development more broadly construed. Let me start with the latter.

If plasticity, especially inherent (i.e., not necessarily adaptive) plastic-ity, is about anything, it is about the direct influence of the environment on the developing phenotype. As Moore (2003) pointed out, the mechani-cal environment in particular (meaning whatever mechanical forces may be applied to the organism during development) can play four distinct functions: it can (a) act as a selective environment, essentially discrimi-nating between developmental processes that do and do not work for a particular type of organism; (b) provide cues for the developmental processes themselves; (c) be itself modified by the organism; or (d) alter the morphogenetic process by means of the inherent plasticity of the developing organism. Examples of the latter possibility in particular are easy to find, and are beginning to play an increasingly important role in our understanding of the evolution of adaptive organismal forms. Just think of the shapes of colonies of scleractinian corals, which are directly molded by water currents, or of the instantaneous transition to bipedal-ism in some mammals effected by developmental defects in the fore-limbs—the so-called "bipedal goat effect." Figure 14.3 shows a less well known, but not for that less spectacular, case involving the model system Arabidopsis thaliana. Genotypes of this plant react dramatically to expo-sure to mechanical stimulation, which in nature can be generated by both abiotic and biotic factors (wind, rain, snow, trampling, or herbivory, to

Figure 14.3
A spectacular example of mechanically induced plasticity in the model system *Arabidopsis thaliana*. The two plants have identical genotype, but the one on the left has been exposed to gentle mechanical stimulation throughout its growth (in nature this may be caused by abiotic factors such as wind and rain, or by biotic ones such as herbivory). Notice not just the very different sizes of the two organisms, but also the distinct branching architectures. (Photo by Janet Braam, used with permission)

mention some). The genetic basis of this response has been investigated (Braam 2005), but—as is often the case—we are far behind in terms of understanding its ecology and evolution. Nonetheless, it is tempting to see simple plasticity to mechanical stimuli as the vehicle through which specialized alpine ecotypes of a variety of plant species may have evolved all over the world (Alokam et al. 2002).

Newman et al. (2006) go much further and actually propose an elaborate hypothesis that places plasticity to mechanical (and chemical) stimulation at the center of the origin of nothing less than the variety of animal body plans. These authors identify two key cell properties, differential adhesion and cell polarity, as well as four types of patterning mechanisms: diffusion gradients, sedimentation gradients, chemical oscillation, and reaction–diffusion. They then show how different combinations of these elements—all mediated by inherent physical–chemical plasticity—can generate the fundamental body plans found in the animal kingdom. The point, of course, is not that genes had nothing to do with the evolution of developmental processes, but rather that the initial facilitation may have come through nongenetic mechanisms which would lead the way, so to speak, in a manner analogous to what I have discussed above in the cases of invasions and ecological speciation.

Moving from development to genetics, one simply cannot write about the mechanistic bases of plasticity without asking whether there are genes "for" it. Famously, Via (1993) answered in the negative, suggesting that plasticity evolves as a by-product of selection within environments. I have elsewhere (Schlichting and Pigliucci 1995) made the argument that Via is fundamentally mistaken here, apparently not realizing that the two possibilities are certainly not mutually exclusive, and that there are plenty of examples in the literature of genes that cannot possibly be conceived as being for anything other than adaptive phenotypic plasticity, for instance, light-sensitive phytochromes in plants (Ballaré 1999). Whether a gene can reasonably be considered "for" a trait is actually a complex question that philosophers of science have debated for some time, but it seems clear that if anything satisfies the rather stringent set of requirements laid out by Kaplan and Pigliucci (2001), then genes underlying known adaptive plasticity syndromes such as the shade avoidance response in plants certainly qualify. The important point is that in order to make a case for a gene being "for" something, one has to know quite a bit about both the molecular–developmental biology of the system and its ecology–evolution. Well-studied instances of adaptive plasticity provide precisely this intellectually satisfying conjunction of molecular and organismal biological research.

Of course the mechanistic bases of phenotypic plasticity are not limited to genes, and in fact here, too, studying plasticity is providing increasingly convincing examples of how to do so-called "integrative biology" while at the same time highlighting the limitations of the Modern Synthesis's simplistic view of genes-to-phenotypes mapping. The functionally flexible hormonal systems of both plants and animals (though the two do work very differently) have offered a starting point from which to understand how environmental signals are translated, interpreted, and reacted to by the organism (Friml and Sauer 2008). Nijhout (2003: 9) concluded that "the development of alternative phenotypes in reaction norms and polyphenisms can be caused by especially evolved mechanisms that are regulated by variation in the patterns of hormone secretion," while according to Badyaev (2005: 880), "phenotypic assimilation of the appropriate stress response is . . . facilitated by a common involvement of neural and endocrine pathways of the stress response in other organismal functions." Finally, Crespi and Denver (2005: 50) commented that "the neuroendocrine stress axis represents a phylogenetically ancient signaling system that allows the fetus or larva to match its rate of development to the prevailing environmental conditions."

More recently, yet another layer of investigation of the mechanisms of phenotypic plasticity has been brought into the discussion, with the possibility that epigenetic processes such as methylation patterns and interference RNAs may be involved in mediating plastic responses (Bossdorf et al. 2008; Jablonka and Raz 2008). Epigenetic inheritance has long been suspected to play some role in the connection between environmental stimuli and heritable phenotypic responses, as in the classic case of floral characteristics in flax (Cullis 1986). But it has been only recently, with the conceptual articulation of multiple "dimensions" of the evolutionarily relevant inheritance systems (Jablonka and Lamb 2005), that we begin to see the coalescence of a coherent theoretical picture encompassing the entire spectrum of phenotypically plastic responses, from their mechanistic generation to their inheritance, to their evolutionary ecology. Much more work needs to be done, of course, but we now have in place a sufficient number of pieces of the puzzle to turn to an examination of the most complete and current model of how plasticity plays a potentially major role in the evolution of phenotypic novelties (see chapter 12 in this volume), a role that simply could not have figured in the conceptual arsenal of the Modern Synthesis.

Phenotypic and Genetic Accommodation

The current buzzword in discussions of the macroevolutionary implications of phenotypic plasticity is "accommodation," a term introduced by West-Eberhard (2003) and much discussed since (Crispo 2007). West-Eberhard's ideas clearly are historically related to those of a series of scholars, chief among them James Marc Baldwin, Conrad Hal Waddington, and Ivan Ivanovich Schmalhausen. In this section I will briefly summarize these historical antecedents and then present a discussion of the modern sense of phenotypic and genotypic accommodation. I suggest that contemporary writers both be aware of the historical precedents (and accordingly recognize intellectual priorities) and adopt West-Eberhard's more modern and compact terminology.

Baldwin (1861–1934) wrote a historical paper on what is now known as the "Baldwin effect" (a term introduced by Simpson in 1953 to criticize it), titled "A New Factor in Evolution" (Baldwin 1896). In it, he presented ideas similar to those developed independently by Morgan (1896) and Osborn (1896), aimed at explaining the role of behavior—and of what today we would call phenotypic plasticity—in evolution. The discussion was framed in the then still relevant context of the possibility of Lamarckian processes, with which Darwin himself had flirted. Baldwin's idea was that behavior can affect the action of natural selection, in some instances facilitating it. The result would be something that would look like acquired inheritance, but that in fact was due to this additional "factor" that simply interacted with, but did not invalidate, the role of selection (Baldwin was no Lamarckian). It is actually difficult to read Baldwin unambiguously, because he was writing in a pre–Mendelian world (Mendel's work had been published, but was yet to be broadly acknowledged). Nonetheless, the Baldwin effect has been explored more recently in works dealing with the interaction between learning and evolution (e.g., Hinton and Nowlan 1987; Mayley 1996). The most sensible modern interpretation of Baldwin is that phenotypic plasticity can facilitate evolution by natural selection, depending on the particular combination of shape of the reaction norms and of the selection pressures in a given population of organisms: in particular, if some of the reaction norms happen to produce a viable (if suboptimal) phenotype in a novel environment, then those genotypes will have a chance to survive, and the population, to establish itself. After that, as discussed above, natural selection will fine-tune the reaction norm by its standard filtering of existing and novel genetic variation.

Waddington's (1905–1975) conceptual and experimental work on related subject matters spanned several decades (Waddington 1942, 1961), and it is a bit easier to frame in modern terms because it was published during the genetic–molecular revolution, as well as largely after the Modern Synthesis itself—although, interestingly, Waddington, too, framed his ideas in terms of a Darwinian explanation of alleged "acquired" characteristics, against Lamarckism. We have already briefly examined Waddington's idea of genetic assimilation, which is closely related (though not identical) to the Baldwin effect. As in the Baldwin effect, it is preexisting variation for plasticity that makes it possible for a fraction of the population to produce a novel phenotype in response to an environmental stimulus. In contrast to the cases of interest to Baldwin, however, Waddington focused on the evolution of a newly canalized trait, which would eventually be stabilized regardless of the continued presence of certain environmental circumstances (evolution of environmental homeostasis). In this sense, although Waddington actually showed experimentally that his mechanism could work, genetic assimilation is probably of less broad interest than the original Baldwin effect.

Schmalhausen (1884–1963), despite having done his work contemporaneously with Waddington's early production, is more obscure and difficult to read—partly because we have to rely on Dobzhansky's translation from the original Russian, and partly because he was more isolated from mainstream science and developed his own terminology. Regardless, his process of "stabilizing selection" (Schmalhausen 1949) should not be confused with what we mean today by that term (i.e., with selection for maintenance of the current population mean, having the result of lowering the population's variance for the trait under selection). Rather, stabilizing selection sensu Schmalhausen is closely related (but, again, not identical) to both genetic assimilation and the Baldwin effect, especially the latter. Schmalhausen envisioned a process by which a shift in environmental conditions would trigger selection for a new "norm" (of reaction), essentially describing a mode of evolution of phenotypic plasticity. Eventually, the new norm is "stabilized" if the environmental shift persists (differing in this from genetic assimilation), producing a population adapted to the novel conditions. In a sense, Schmalhausen's ideas can be interpreted as a generalization of the Baldwin effect that applies to any plastic trait, not just to behavioral ones.

West-Eberhard (2003) has updated the concepts and terminology accumulated by Baldwin, Waddington, Schmalhausen, and a number of others throughout the twentieth century, and essentially distilled it into two key phenomena: phenotypic accommodation and genetic accommodation. In her words:

Phenotypic accommodation is the adaptive mutual adjustment, without genetic change, among variable aspects of the phenotype, following a novel or unusual [external or internal] input during development. (West-Eberhard 2005a: 610)

Genetic accommodation is simply quantitative genetic change in the frequency of genes that affect the regulation or form of a new trait. (emphasis added; West-Eberhard 2005b: 6547)

Phenotypic accommodation, therefore, is a direct consequence of the inherent plasticity of developmental systems, and—in West-Eberhard's definition—is related to the concepts of environmental and genetic homeostasis. Genetic accommodation, as should be clear from the above quote, simply refers to the standard mechanism of genetically enabled evolutionary change envisioned by the Modern Synthesis, applied to the specific case of phenotypic novelties; the reason for the introduction of a new term here is simply to provide a unified model of evolutionary change under the broad rubric of "accommodation."

Indeed, West-Eberhard goes on to present what she sees as a four-step recipe for evolutionary change in general:

1. A novel input affects one (in the case of mutation) or several (in the case of an environmental change) individuals in a population.

2. Because of inherent developmental plasticity, we observe phenotypic accommodation of the novel input; consequently, a novel phenotype emerges.

3. The initial spread of the novel phenotype may be rapid (if it is due to an environmental effect) or slow (if it is the result of genetic input).

4. If the novel phenotype is advantageous, natural selection "fixes" it by stabilizing its appearance through an alteration of the genetic architecture; genetic accommodation has occurred.

There are two key conceptual points in this list that need to be appreciated. First, the fact that phenotypic change always begins with the plastic reaction of the developmental system to a genetic or environmental perturbation. Of course, there is no claim that such plastic response will be adaptive (in the sense of advantageous to the organism under current conditions), but there is in fact a good chance that it will be adaptive because—the idea is—developmental systems have been selected in the past to maintain functionality in the face of a broad range of perturba-tions. Moreover, the "novel" environment (if the change is environmen-tal in nature) will often not be entirely novel at all, but will be some variant of the sort of environment that has been common through the history of the species in question. This makes it even more likely that the existing developmental norm of reaction will produce at least a sub-optimal phenotype.

Second, notice that—if the novel stimulus is environmental—there will likely be several developmental systems that will respond in similar fashions, because the corresponding reaction norms will be similar. This is crucial, because it may lead not only to the appearance, but also to the prevalence (or at least the non-rarity), of the new phenotype in the population. This will in turn facilitate the process of genetic accommoda-tion that follows, because several individuals simultaneously will provide the raw material for selection to act on.

By far the most famous example of phenotypic accommodation is the so-called bipedal goat effect mentioned earlier. It refers to the fact that mammals' (and perhaps other organisms') developmental plasticity allows animals that are born with nonfunctional forelimbs to adjust their muscular and skeletal system and adopt a bipedal posture. The first recorded case is that of a goat with such condition, studied in detail by Slijper (1942), but similar cases are common in dogs and even in pri-mates. Figure 14.4 shows the case of a bipedal macaque found in an Israeli zoo; the animal had suffered from a severe and life-threatening infection that had paralyzed its upper limbs. It is hard to stare at pictures like this one and not think that perhaps this sort of phenotypic plasticity is what first opened the way to the evolution of bipedalism in hominids. This is, of course, speculation, but surely a very tempting one.

More generally, and on more solid developmental and evolutionary ground, some authors have called attention to the role of bone morpho-genetic proteins in structuring the inherent plasticity of vertebrate devel-opmental systems. These proteins are involved in the shaping of spectacular phenotypes such as the turtle's carapace, bat wings, the spotted hyena's sagittal crest (from chewing food), the jaws of cichlid fish (in response to diet), and the shape of bird bills (e.g., in Darwin's finches, again in response to diet) (Young and Badyaev 2007). Most of these examples, presumably, are of genetic accommodation, because the original plastic response has been stabilized (in the sense of Schmalhausen) in the currently existing populations, and it has a clearly adaptive meaning. The difficult task, of course, is to uncover convincing examples

Figure 14.4
(Left) Phenotypic accommodation in a macaque suffering from a crippling disease that made its upper limbs nonfunctional from early on in development, and led—through the inherent plasticity of the developmental system—to a bipedal posture. (right) Reconstruction of two austrolopithecines, likely members of the line of ancestry that led to humans. Could they have started their evolution toward bipedalism by a similar process of phenotypic accommodation? (Original attribution of left photo unknown, various copies on Internet; right drawing, Wikipedia Commons)

of transition from phenotypic to genetic accommodation and, moreover, to show that the phenomenon is sufficiently common to be evolutionarily relevant. Studies of this type are beginning to appear in the literature (Gomez-Mestre and Buchholz 2006; Suzuki and Nijhout 2008), though it is too early to draw broad conclusions on their generality.

Consequences for an Extended Evolutionary Synthesis

What are the consequences of contemporary views on phenotypic plasticity for the possibility of expanding the Modern Synthesis into an Extended Evolutionary Synthesis? There are at least six that should help provide a blueprint for future research on both the theoretical and the empirical fronts:

1. As West-Eberhard (2003) pointed out, genes could come to be seen as "followers" rather than leaders in the evolutionary process, a change that may have little impact on, say, research in molecular genetics, but that would represent a major conceptual shift in evolutionary theory. As

West-Eberhard (2005b: 6547) puts it: "I consider genes followers, not leaders, in adaptive evolution....We forget that...environmental factors constitute powerful inducers and essential raw materials whose geographically variable states can induce developmental novelties as populations colonize new areas." This, of course, is true if the Baldwin effect, genetic assimilation, and related phenomena are in fact frequent enough in nature.

2. Phenotypic–genetic accommodation could come to be considered a major explanation behind the well-known phenomenon of mosaic evolution. The latter's textbook definition is along the following lines: "Evolution of different characters at different rates within a lineage is called mosaic evolution....It says that an organism evolves not as a whole, but piecemeal" (Futuyma 1998). Except that if the "two-legged goat" effect and similar phenomena are frequent, we would have the appearance of mosaic evolution in the fossil record, even though most of the changes would have occurred simultaneously, made possible by the inherent phenotypic plasticity of developmental systems.

3. Phenotypic plasticity should also be seriously considered as a candidate mechanism for another well-known evolutionary phenomenon, pre-adaptation. Futuyma's (1998) definition of the term is "Possession of the necessary properties to permit a shift into a new niche or habitat. A structure is pre-adapted for a new function if it can assume that function without evolutionary modification." But the concept of pre-adaptation of a structure to a new function may appear rather spooky if there is no further elaboration about exactly how such pre-adaptation comes about. As I mentioned earlier, however, this is clarified once we realize that most new environments are in fact correlated to historical environments, and that therefore the variation for phenotypic plasticity existing in a given population is not altogether unlikely to include reaction norms that will work in at least a suboptimal fashion in the new environment (or for the new "function"). This is what Baldwin called "organic selection."

4. Another consequence stressed by West-Eberhard (2005a: 611) is the pre-eminent role that behavior plays in directing evolutionary change: "Behavior is, of course, a common mediator of normal skeletal and muscle development because it is especially flexible in response to environmental contingencies. It follows that behavior must often be an important mechanism in the origins of novel morphological traits. So we have to list behavior and its neuro-endocrinological underpinnings, alongside genomic changes, as among the primary developmental causes

of morphological novelty." This is, of course, what Baldwin was interested in to begin with. The point can be further broadened to all life forms if we consider phenotypic plasticity as a generalized equivalent of behavior (as is in fact done by several authors: Mayley 1996; Novoplansky 2002; Paenke et al. 2007).

5. Phenotypic plasticity should also be considered as a major player in the process of niche construction (Odling-Smee et al. 2003; Okasha 2005; Laland and Sterelny 2006), which, of course, is itself still a somewhat controversial concept (but see chapter 8 in this volume). According to Stamps (2003): "animals frequently select their own environments, or modify their environments through their own actions," which is an obvious reason why plasticity (in this case in the form of behavior) is important in this context. However, as Jablonka (2007) put it: "because it is difficult to recognize the role that persistent environmental or developmental inputs play in ontogeny, their effects are usually attributed to genetic inputs," which means that there is a built-in tendency by biologists who work within the framework of the Modern Synthesis to simply attribute phenotypic change to genes without further consideration of the developmental and epigenetic alternatives (but see chapter 7 in this volume).

6. Finally, phenotypic plasticity probably plays a hitherto largely ignored part in one of the fundamental phenomena of the biological world: speciation. Again Jablonka (2007): "Heritable nongenetic variations may initiate population divergence and lead to speciation. Gottlieb interpreted the well-known case of sympatric speciation in apple maggots (Rhagoletis pomonella) in these terms." And, along similar lines, West-Eberhard (2005b): "Geneticists may end up describing the results of speciation rather than its causes." Then again, it is still hard enough to get some biologists to take seriously even the possibility of sympatric speciation (Coyne and Orr 2004), despite documented examples in a variety of organisms (e.g., Doebeli and Dieckmann 2000; Fournier and Giraud 2008).

All of the above directly implies, it seems to me, specific steps within a research program on phenotypic plasticity and its macroevolutionary impact. This work would cover three major areas: empirical research in organismal biology, empirical research in mechanistic biology, and theoretical/conceptual investigations.

With respect to empirical research in organismal biology, we need further—and better characterized—examples of genetic accommodation. This may not be easy, because, as I mentioned above, the process may be too quick for evolutionary biologists to be able to detect, if they do not know what they are looking for. Awareness of the stages of the process, of course, is crucial. As Darwin put it long ago while discussing the relationship between theory and "data": "How odd it is that anyone should not see that all observation must be for or against some view if it is to be of any service!" (from a letter to Henry Fawcett). This should be remembered by anyone who dismisses genetic accommodation (or sympatric speciation, or epigenetic inheritance, or niche construction) on the simple basis that "we do not have compelling data in its favor." Compelling data, as Darwin understood, do not simply emerge from a collection of facts about the world. No new theory has ever declared itself from beneath a heap of facts.

Similarly, we need more examples of phenotypic accommodation in order to better assess its short-term ecological impact and long-term evolutionary relevance. This, however, is actually simpler to achieve, as the literature on phenotypic plasticity is very rich in potential examples, and the empirical and analytical methods to study the phenomenon are well established (Pigliucci 2001). Also, there needs to be more integration of research programs on phenotypic plasticity, on the one hand, and on behavioral ecology and genetics, on the other. While students of behavior are certainly aware of phenotypic plasticity, and vice versa for researchers in the plasticity field, good examples of how the two relate to each other are still surprisingly uncommon in the literature (but see Charmantier et al. 2008).

Empirical research on the mechanistic aspects of phenotypic plasticity has, of course, been carried out for several years, and it has reached a good level of sophistication, especially when conducted on so-called model systems (e.g., Feng et al. 2008). Still, two areas of particular attention for future studies concern the role of hormones and of epigenetic inheritance systems as intermediaries between the genetic level and the actual deployment of developmentally plastic responses by the organism. Research on the role of hormones in evolutionarily relevant plasticity (e.g., Emlen et al. 2007) is comparatively much more advanced than research on heritable epigenetics, a phenomenon still largely, and I think erroneously, regarded as irrelevant to evolutionary questions (but see Bossdorf et al. 2008; Jablonka and Raz in press). The difference between the two is that we know beyond reasonable doubt that hormones (both in plants and in animals) do play a crucial role in the deployment of plastic responses, whereas the role and frequency of heritable epigenetic effects are still much more debatable, since so comparatively few examples are well characterized. Nonetheless, this may change quite rapidly as molecular techniques become readily available for population-level screenings of epigenetic markers and, again, as conceptual awareness pushes more researchers in that direction.

As far as the theoretical evolutionary biology of macroevolutionary plasticity is concerned, several lines of inquiry can be pursued. Perhaps one of the most neglected so far has been the bridge to some interesting literature in computational science, where the concepts of learning and plasticity (broadly defined) have been explored in terms of their consequences for the evolution of artificial life systems, neural networks, and genetic algorithms (e.g., Mills and Watson 2005; Wiles et al. 2005). There is much to be gained for evolutionary biologists from a less occasional interaction with theoretical scientists in computational and cognitive sciences, as this will make it possible both to generalize the concepts arising from within each discipline, and to adopt theoretical and computational methodologies that may help biologists think outside of the box imposed by standard population–quantitative genetic models.

In turn, population and quantitative genetics, which are the theoretical–mathematical backbone of the Modern Synthesis, need to be reevaluated in light of a broader concept of what modeling in biology means (Laubichler and Müller 2007) and how it is to be pursued. Population genetics is notoriously limited in its analytic treatments to a small number of loci/alleles and to overly simplifying assumptions, without which its problems rapidly become mathematically intractable. Quantitative genetics itself, which was originally developed to address (statistically, as opposed to analytically) precisely the sort of complex problems that are outside the scope of population genetics, is beginning to show signs of reaching its own limits in terms of generality and applicability to biologically relevant questions (Pigliucci and Kaplan 2006). Indeed, referring to the limited usefulness of quantitative genetic models of adaptive landscapes (see chapter 3 of this volume), Gavrilets (1997) concluded that a model's predictive ability—the gold standard in disciplines such as physics—is not necessarily its most important contribution to science, as models in biology are more often useful as metaphors and tools to sharpen one's thinking about a given problem. Modeling in computational science and complexity theory (Toquenaga and Wade 1996) is inherently of a different kind from the approaches that have been standard in evolutionary biology for the past century or so, which may explain why today many biologists still consider results from complexity theory as "too vague and metaphorical" to be biologically informative. That may need to change in the near future.

As I mentioned above, taking a macroevolutionary role of phenotypic plasticity seriously also broadens or reopens the discussion on rather controversial topics such as niche construction and sympatric speciation. The two cases are, of course, distinct in a variety of ways. Niche construction is something that most biologists think probably does happen, and the question is how it should be explicitly incorporated into theoretical treatments of evolution, and what such incorporation might accomplish. The case of sympatric speciation is different, since a number of biologists still think it is either impossible or, in any case, a very rare event with no broad consequences for evolutionary theory. Coyne and Orr's (2004) book, for instance, treats allopatric speciation as the null hypothesis (and reinterprets the available evidence accordingly) on the ground that it is much more likely a priori. This is not the place for a discussion of the role and limitations of null models in biology (but see Pigliucci and Kaplan 2006; Sober 2008), but at the very least the possibility of additional mechanisms leading to speciation, such as the phenotypic–genotypic accommodation sequence proposed by West-Eberhard, should sound a note of caution about premature dismissal of alternative speciation models.

Phenotypic plasticity as a concept and an area of study has had a convoluted history since its inception with Woltereck (1909). Although the idea is contemporaneous with the formal recognition of the distinction between genotype and phenotype (indeed, it arguably precedes it, in the work of Baldwin, Morgan, and Osborn), it was neglected and considered a nuisance by evolutionary biologists until the latter part of the 20th century. Nowadays it is a concept that most practicing biologists recognize as important, and yet with which many still wrestle in terms of what it means for evolutionary studies. If at least part of what I have outlined in this chapter comes to fruition, in the way of development of ideas and empirical research programs, phenotypic plasticity will significantly expand the way we think organic evolution takes place.

References

Alokam S, Chinnappa CC, Reid DM (2002) Red/far-red light- mediated stem elongation and anthocyanin accumulation in Stellaria longipes: Differential response of alpine and prairie ecotypes. Canadian Journal of Botany 80: 72–81.

Badyaev AV (2005) Stress-induced variation in evolution: From behavioural plasticity to genetic assimilation. Proceedings of the Royal Society B 272: 877–886.

Baldwin JM (1896) A new factor in evolution. American Naturalist 30: 441–451.

Ballaré CL (1999) Keeping up with the neighbours: Phytochrome sensing and other signalling mechanisms. Trends in Plant Science 4: 97–102.

Bossdorf O, Richards CL, Pigliucci M (2008) Epigenetics for ecologists. Ecology Letters 11: 106–115.

Braam J (2005) In touch: Plant responses to mechanical stimuli. New Phytologist 165: 373–389.

Callahan H, Pigliucci M (2002) Shade-induced plasticity and its ecological significance in wild populations of Arabidopsis thaliana. Ecology 83: 1965–1980.

Charmantier A, McCleery RH, Cole LR, Perrins C, Kruuk LEB, Sheldon BC (2008) Adaptive phenotypic plasticity in response to climate change in a wild bird population. Science 320: 800–803.

Cook SA, Johnson MP (1968) Adaptation to heterogeneous environments. I. Variation in heterophylly in Ranunculus flammula L. Evolution 22: 496–516.

Coyne JA, Orr HA (2004) Speciation. Sunderland, MA: Sinauer.

Crespi EJ, Denver RJ (2005) Ancient origins of human developmental plasticity. American Journal of Human Biology 17: 44–54.

Crispo E (2007) The Baldwin effect and genetic assimilation: Revisiting two mechanisms of evolutionary change mediated by phenotypic plasticity. Evolution 61: 2469–2479.

Cullis CA (1986) Unstable genes in plants. In Plasticity in Plants. D.H. Jennings and A. J.Trewavas, eds.: 77–84. Cambridge, UK: Pindar.

Doebeli M, Dieckmann U (2000) Evolutionary branching and sympatric speciation caused by different types of ecological interactions. American Naturalist 156(suppl.): S77–S101.

Emlen DJ, Lavine LC, Ewen-Campen B (2007) On the origin and evolutionary diversification of beetle horns. Proceedings of the National Academy of Sciences of the USA 104: 8661–8668.

Falconer DS (1952) The problem of environment and selection. American Naturalist 86: 293–298.

Feng S, Martinez C, Gusmaroli G, Wang Y, Zhou J, et al. (2008) Coordinated regulation of Arabidopsis thaliana development by light and gibberellins. Nature 451: 475–480.

Flatt T (2005) The evolutionary genetics of canalization. Quarterly Review of Biology 80: 287–316.

Fournier E, Giraud T (2008) Sympatric genetic differentiation of a generalist pathogenic fungus, Botrytis cinerea, on two different host plants, grapevine and bramble. Journal of Evolutionary Biology 21: 122–132.

Friml J, Sauer M (2008) In their neighbour's shadow. Nature 453: 298–299.

Futuyma DJ (1998) Evolutionary Biology. Sunderland, MA: Sinauer.

Gavrilets S (1997) Evolution and speciation on holey adaptive landscapes. Trends in Ecology and Evolution 12: 307–312.

Gomez-Mestre I, Buchholz DR (2006) Developmental plasticity mirrors differences among taxa in spadefoot toads linking plasticity and diversity. Proceedings of the National Academy of Sciences of the USA 103: 19021–19026.

Hinton GE, Nowlan SJ (1987) How learning can guide evolution. Complex Systems 1: 495–502.

Jablonka E (2007) The developmental construction of heredity. Developmental Psychobiology 49: 808–817.

Jablonka E, Lamb MJ (2005) Evolution in Four Dimensions: Genetic, Epigenetic, Behavioral, and Symbolic Variation in the History of Life. Cambridge, MA: MIT Press.

Jablonka E, Raz G (2008) Transgenerational epigenetic inheritance: Prevalence, mechanisms, and implications for the study of heredity and evolution. Quarterly Review of Biology 84: 131–176.

Kaplan JM, Pigliucci M (2001) Genes "for" phenotypes: A modern history view. Biology and Philosophy 16: 189–213.

Laland KN, Sterelny K (2006) Seven reasons (not) to neglect niche construction. Evolution 60: 1751–1762.

Laubichler MD, Müller GB, eds. (2007) Modeling Biology: Structures, Behaviors, Evolution. Cambridge, MA: MIT Press.

Lerner IM (1954) Genetic Homeostasis. New York: Dover.

Levin DA (2004) Ecological speciation: Crossing the divide. Systematic Botany 29: 807–816.

Lewontin RC (1974) The analysis of variance and the analysis of causes. American Journal of Human Genetics 26: 400–411.

Masel J (2005) Evolutionary capacitance may be favored by natural selection. Genetics 170: 1359–1371.

Mayley G (1996) Landscapes, learning costs, and genetic assimilation. Evolutionary Computation 4: 213–234.

Mayr E, Provine WB (1980) The Evolutionary Synthesis: Perspectives on the Unification of Biology. Cambridge, MA: Harvard University Press.

Mills R, Watson RA (2005) Genetic assimilation and canalisation in the Baldwin Effect. In Proceedings of the 8th Conference on Artificial Life. LNCS 3630: 353–362. Heidelberg: Springer.

Moczek AP (2006) Integrating micro- and macroevolution of development through the study of horned beetles. Heredity 97: 168–178.

Moore SW (2003) Scrambled eggs: Mechanical forces as ecological factors in early development. Evolution and Development 5: 61–66.

Morgan CL (1896) On modification and variation. Science 4: 733–740.

Newman SA, Forgacs G, Müller GB (2006) Before programs: The physical origination of multicellular forms. International Journal of Developmental Biology 50: 289–299.

Nijhout HF (2003) Development and evolution of adaptive polyphenisms. Evolution and Development 5: 9–18.

Novoplansky A (2002) Developmental plasticity in plants: Implications of non-cognitive behavior. Evolutionary Ecology 16: 177–188.

Nussey DH, Wilson AJ, Brommer E (2007) The evolutionary ecology of individual phenotypic plasticity in wild populations. Journal of Evolutionary Biology 20: 831–844.

Odling-Smee FJ, Laland KN, Feldman MW (2003) Comments on Niche Construction: The Neglected Process in Evolution. Princeton, NJ: Princeton University Press.

Okasha S (2005) On niche construction and extended evolutionary theory. Biology and Philosophy 20: 1–10.

Osborn HF (1896) A mode of evolution requiring neither natural selection nor the inheritance of acquired characters. Transactions of the New York Academy of Sciences 15: 141–142.

Paenke I, Sendhoff B, Kawecki TJ (2007) Influence of plasticity and learning on evolution under directional selection. American Naturalist 170: E47–E58.

Pigliucci M (2001) Phenotypic Plasticity: Beyond Nature and Nurture. Baltimore: Johns Hopkins University Press.

Pigliucci M, Kaplan J (2006) Making Sense of Evolution: The Conceptual Foundations of Evolutionary Biology. Chicago: University of Chicago Press.

Pigliucci M, Murren CJ (2003) Genetic assimilation and a possible evolutionary paradox: Can macroevolution sometimes be so fast as to pass us by? Evolution 57: 1455–1464.

Pigliucci M, Pollard H, Cruzan MB (2003) Comparative studies of evolutionary responses to light environments in Arabidopsis. American Naturalist 161: 68–82.

Queitsch C, Sangster TA, Lindquist S (2002) Hsp90 as a capacitor of phenotypic variation. Nature 417: 618–624.

Rasanen K, Hendry AP (2008) Disentangling interactions between adaptive divergence and gene flow when ecology drives diversification. Ecology Letters 11: 624–636.

Richards CL, Bossdorf O, Muth NZ, Gurevitch J, Pigliucci M (2006) Jack of all trades, master of some? On the role of phenotypic plasticity in plant invasions. Ecology Letters 9: 981–993.

Rutherford SL, Lindquist S (1998) Hsp90 as a capacitor for morphological evolution. Nature 396: 336–342.

Schlichting CD, Pigliucci M (1995) Gene regulation, quantitative genetics and the evolution of reaction norms. Evolutionary Ecology 9: 154–168.

Schlichting CD, Smith H (2002) Phenotypic plasticity: Linking molecular mechanisms with evolutionary outcomes. Evolutionary Ecology 16: 189–211.

Schmalhausen II (1949) Factors of Evolution: The Theory of Stabilizing Selection. Chicago: University of Chicago Press.

Simpson GG (1953) The Baldwin effect. Evolution 7: 110–117.

Slijper EJ (1942) Biologic–anatomical investigations on the bipedal gait and upright posture in mammals, with special reference to a little goat, born without forelegs. Proceedings of the Koninklijke Nederlandse Akademie van Wetenschappen 45: 288–295, 407–415.

Sober E (2008) Evidence and Evolution: The Logic Behind the Science. Cambridge: Cambridge University Press.

Stamps J (2003) Behavioural processes affecting development: Tinbergen's fourth question comes of age. Animal Behaviour 66: 1–13.

Suzuki Y, Nijhout HF (2008) Genetic basis of adaptive evolution of a polyphenism by genetic accommodation. Journal of Evolutionary Biology 21: 57–66.

Toquenaga Y, Wade MJ (1996) Sewall Wright meets artificial life: The origin and maintenance of evolutionary novelty. Trends in Ecology and Evolution 11: 478–482.

Via S (1993) Adaptive phenotypic plasticity: Target or by-product of selection in a variable environment? American Naturalist 142: 352–365.

Waddington CH (1942) Canalization of development and the inheritance of acquired characters. Nature 150: 563–565.

Waddington CH (1961) Genetic assimilation. Advances in Genetics 10: 257–290.

Wells C, Pigliucci M (2000) Heterophylly in aquatic plants: Considering the evidence for adaptive plasticity. Perspectives in Plant Ecology, Evolution and Systematics 3: 1–18.

West-Eberhard MJ (2003) Developmental Plasticity and Evolution. New York: Oxford University Press.

West-Eberhard MJ (2005a) Phenotypic accommodation: Adaptive innovation due to developmental plasticity. Journal of Experimental Zoology B304: 610–618.

West-Eberhard MJ (2005b) Developmental plasticity and the origin of species differences. Proceedings of the National Academy of Sciences of the USA 102(suppl. 1): 6543–6549.

Wiles J, Watson J, Tonkes B, Deacon T (2005) Transient phenomena in learning and evolution: Genetic assimilation and genetic redistribution. Artificial Life 11: 177–188.

Woltereck R (1909). Weiterer experimentelle Untersuchungen über Artveränderung, speziell über das Wesen quantitativer Artunterschiede bei Daphniden. Verhandlungen der Deutsche Zoologische Gesellschaft 19: 110–172.

Young RL, Badayev AV (2007) Evolution of ontogeny: Linking epigenetic remodeling and genetic adaptation in skeletal structures. Integrative and Comparative Biology 47: 234–244.

15 Evolution of Evolvability

Günter P. Wagner and Jeremy Draghi

Over the last half century evolutionary biology has been a highly active and successful field of biological research with an increasing number of journals and online outlets supporting a rapidly growing scientific literature. Many of these publication organs have citation rates equal to or higher than many established journals in molecular biology, and thus it should not come as a surprise that our knowledge of evolution is rapidly expanding. The current understanding of evolution is much different, of greater reach and depth than either that of Darwin or that of the architects of the "Evolutionary Synthesis" from the 1950s. The fact that evolutionary biology has extended beyond that of the original Evolutionary Synthesis is therefore obvious. Of course there are different ways by which a science can expand and change. Rapid expansion of the factual knowledge base is one way that science makes progress, and another is change in the conceptual makeup of a discipline. When Massimo Pigliucci recently asked in the journal *Evolution* "Do we need an extended evolutionary synthesis?" he certainly had the second mode of change in mind (i.e., the question of whether the way we explain and understand evolution has changed or should change, given what we know [Pigliucci 2007]). Here we want to discuss one specific aspect of evolutionary biology that represents a break from the research tradition of the synthesis: namely, research on evolvability and its evolution. This subject is still considered by some prominent biologists as suspect (Sniegowski and Murphy 2006; Lynch 2007). Nevertheless, a PubMed search with the keyword "evolvability" yielded 236 papers on September 4, 2008; some of them refer to evolvability in a generic sense, but a large fraction also talk about evolution of evolvability. We think that the idea of evolution of evolvability is not as radical a break from the tradition of population genetics theory as some population geneticists may think. The neglect of evolution of evolvability by the research program of the synthesis is more a self-inflicted blind spot rather than dictated by a real limitation of the conceptual framework of population genetic theory itself. Incorporating research on the evolution of evolvability into mainstream evolutionary biology, however, can greatly enhance the conceptual reach of evolutionary biology and resolve long-standing theoretical disputes.

Understanding evolvability and the evolution of evolvability can have both far-reaching conceptual and practical benefits. The most fundamental reason why research into evolvability is important for evolutionary biology is that it addresses one of the most basic assumptions of contemporary evolutionary theory: that complex organisms can arise from selection on random genetic variation. This is the part of evolutionary biology that is most consistently challenged from outside of biology, most notably by creationists. It is easy to sympathize with those whose intuition does not accommodate the idea of random variation leading to increased complexity in evolution. The reasons why most trained biologists do not have these problems is that there are good experimental and theoretical arguments why our intuition is misleading in that respect. But this body of evidence does not yet amount to a comprehensive understanding of what and how genomic and developmental features of organisms affect (positively or negatively) their ability to evolve. In particular we do not have a good understanding of how and why these features evolve (Pigliucci 2008). Intellectual honesty requires that evolutionary biologists develop a deeper understanding of evolvability and how it arose and changed during evolution.

On a somewhat more specialized level, but still of fairly general importance, is the possibility that at least some features of genomic organization may have a deep connection to evolvability and the evolution of evolvability: for instance, the way cells are put together and how development is orchestrated (Rutherford and Lindquist 1998; Rutherford 2000; Gerhart and Kirschner 2007; Hendrikse et al. 2007; see also chapter 10 in this volume). These features include conserved core components used in different contexts, weak linkage among components, modularity, robustness, and many more. Many of these features are global organizational features of organisms that have not received serious consideration within evolutionary biology, which often is narrowly focused on individual traits or genes. A different aspect of the same way of thinking is Rupert Riedl's idea that differences in body plan organization might be understood as different ways of answering the challenge of evolving complex organisms from randomly generated genetic variation (Riedl 1978; for a brief summary of these ideas see GP Wagner and Laubichler

2004). There are vast areas of biological reality that plausibly can relate to issues of evolvability and the evolution of evolvability; and a deeper understanding of these issues may benefit biology in general.

But evolvability is also involved in questions of practical importance. We want to mention only two to make our point. One is the problem of invasive species. A species is called invasive if it can succeed outside its native context. It is not yet clear exactly what makes a species invasive, but there is a plausible link between invasion success and evolvability (Gilchrist and Lee 2007). Emerging disease agents are just one form of invasive species, and our ability to understand and manage evolvability of those agents may turn out to be important. On the other hand, biotechnology is based on manipulating biological macromolecules, which in themselves are complex systems with many interacting components. For instance, in protein engineering the evolvability of a protein is called designability, which determines the ease with which proteins can be manipulated for a given purpose. Understanding, in general, what evolvability is and how it can be managed could have important consequences for biotechnology (Bloom et al. 2006).

Here we want to focus on the most controversial aspect of evolvability, namely, the factors that lead to evolutionary changes in evolvability. Specifically, we want to examine the most frequently cited arguments against the possibility that evolution of evolvability might be the result of selection favoring more evolvable genotypes. We will argue that all these arguments have not been rigorously examined by their proponents, and are thus a self inflicted blind spot in evolutionary biology.

Evolvability: Definition and Measurement

On its most basic level evolvability refers to the ability of a species to evolve (Altenberg 1994; GP Wagner and Altenberg 1996; Gerhart and Kirschner 1997; Love 2003; Gerhart and Kirschner 2007; Pigliucci 2008); no surprise there. Usually this notion is specified further as the ability to evolve in response to natural selection. On the face of it, this seems to be a non-issue, since all extant species are the result of evolution, and thus at least their ancestors were certainly evolvable. But the issue becomes less trivial if we ask whether the collection of species we have with us today is not only the product of the survival of the fittest, but also that of the survival of the most evolvable. Posed that way, the question becomes one of quantitative differences in evolvability, and more precise definitions and measurements are required.

There are three timescales at which evolvability has been considered; for a review see Pigliucci (2008). At the shortest timescale the ability of a population to respond to natural selection is determined by the amount of segregating additive genetic variation. This is uncontroversial, and the study of the factors that determine the amount of additive genetic variation in a population is an old but still active area of research within evolutionary biology (Barton and Turelli 1989; Bürger 2000). In short, the amount of genetic variation in a population is influenced by the mutation rate, the effect of mutations, the intensity of selection, population size, and migration rates, and only marginally by recombination rate; for a review see Bürger (2000). For quantitative characters the amount of additive genetic variation is conventionally summarized in the additive genetic variance–covariance matrix, **G**, which predicts the short-term response to selection (Lande 1979).

At the next level, evolvability is determined by the genetic variability of the phenotype (*sensu* GP Wagner and Altenberg 1996; i.e., the ability to vary, as opposed to the amount of variation realized in a population). Variability is measured by the mutation rate and the mutational variance V_m, that is, the amount of additive genetic variation for a trait created by mutation in a generation (Lynch and Walsh 1998). For multivariate phenotypes the mutational variance can be generalized to the mutational covariance matrix, **M**, which is a table of the additive genetic variances and covariances generated per generation (Lande 1980). This measure is more relevant for exploring the medium-term evolutionary responses after the standing genetic variation has been exhausted. Mutational variances and covariances are much harder to measure than segregating variation, but with appropriate model organisms it is at least feasible (Camara and Piglucci 1999; Brakefield et al. 2003). One noteworthy point is that **G** is a population measure, summarizing the amount of genetic variation in a given population, whereas **M** is strictly a property of a genotype.

The third notion of evolvability is the least developed, and refers to the ability of a genotype (or a lineage) to generate truly novel phenotypes (Maynard Smith and Szathmáry 1995; Brookfield 2001). To our knowledge there is no operational measure for this concept, and we will not further discuss this idea here; but see Pigliucci (2008).

Although this latter notion of evolvability is hard to quantify, the former, more restricted concepts are central objects in quantitative genetics; measuring evolvability is already part of mainstream evolutionary biology. One important point about measuring evolvability, and the concept of evolvability in general, directly follows from standard quantitative genetic theory. For instance, consider a population with a genetic covariance matrix $\mathbf{G}$; the response to selection will then be

$$\Delta\bar{z} = \mathbf{Gb}$$

where $\Delta\bar{z}$ is the vector of differences in mean values of all the traits and $\mathbf{b}$ is the selection gradient (Lande and Arnold 1983). If we assume that not all genetic covariances are zero and the genetic variances are also not all the same, then the selection response will depend on the relationship between $\mathbf{G}$ and $\mathbf{b}$. Intuitively, $\mathbf{b}$ is a vector that describes the direction in which selection pushes the population. If $\mathbf{b}$ points in a direction in which there is more additive genetic variation, then the selection response will be larger than if $\mathbf{b}$ pushes the population in a direction where there is little genetic variation. Hence, *evolvability is a relational concept and reflects how genetic variation, variability, and selection interact to determine evolution.* For instance, mammals are highly evolvable with respect to body size, but much less so in terms of the number of heads or number of limbs. There is much less genetic variation for limb number than there is for body size.

How to summarize evolvability from these measures of variation and variability is still an active area of research, but significant progress has been made recently (Cheverud and Marroig 2007; Hansen and Houle 2008; Kirkpatrick 2008). For instance, Kirkpatrick (2008) calculated the average selection response for a given $\mathbf{G}$ matrix averaged over all possible directions of $\mathbf{b}$, thereby measuring a type of mean evolvability. Hansen and Houle (2008), on the other hand, considered the conditional selection response (i.e., the response in the direction of the selection gradient $\mathbf{b}$) if all other variables are held constant. Hansen and Houle show that this measure estimates the short-term evolvabilities of multivariate phenotypes. Although these measures are useful summary statistics of overall evolvability and constraints, the evolvability in any specific situation is the product of $\mathbf{G}$ and $\mathbf{b}$ (i.e., an interaction between genetic variability and the environment that generates the selection pressure).

Extending the approaches by Kirkpatrick (2008) and Hansen and Houle (2008) to medium-term evolvability seems to be straightforward, because the asymptotic selection response of a quantitative character is proportional to the mutational variance times the effective population size (Hill 1982). In other words, as selection exhausts standing variation, $\mathbf{G}$ will increasingly resemble $\mathbf{M}$. Hence, replacing $\mathbf{G}$ with $\mathbf{M}$ in the equations derived for short-term evolvability seems to be a viable way to estimate overall medium-term evolvability.

The methods mentioned above all assume that the **G** matrix has full rank (i.e., that each trait has some potential to vary independently). Whether this is always the case is a difficult statistical question that does not seem to be fully resolved (Pavlicev et al. 2008). Published results of such estimates differ in their conclusions. For instance, Mezey and Houle (2005) estimated the genetic covariance among 20 wing traits and concluded that the **G** matrix has full rank (i.e., that there is genetic variation in all possible directions of trait space). In contrast, a study by Blows and collaborators (McGuigan and Blows 2007) has presented evidence that the rank of **G** is much less than the number of traits, suggesting strong constraints on evolvability in certain directions of the trait space. Some of these differences could be due to differences in the statistical approaches and more work on this problem is certainly necessary (Pavlicev et al. 2009).

In computational simulation studies, evolvability is more directly measurable because it is easy to set up replicate populations exposed to the same or a set of selection regimes, and measure the rate of change for different starting genotypes (Lenski et al. 1999). Straightforward as it seems, however, these measurements require further attention because it turns out that the rank order of genotype responses depends on the time-scale over which the response is measured (Draghi and Wagner 2008). That means that a genotype could have lower evolvability if measured over, say, 100 generations, but higher evolvability if measured over 1000 generations. What this implies for the biology and the measurement of evolvability needs to be explored. In principle, the same approach as in computational studies can be used in experimental evolution studies, although with greater investment of resources. To our knowledge, the first such study to measure evolvability of different genotypes in the lab is that of McBride et al. (2008). These authors compared the evolved response of different strains of the RNA virus Φ6. These genotypes differed in their genetic robustness (mutational variance for fitness) but were initially equivalent in their fitness in a high-temperature environment. It was shown that more robust genotypes evolved thermal tolerance quicker than less robust genotypes. Hence, evolvability can be measured in the lab with real organisms (if one considers viruses as "real" organisms, but similar experiments with bacteria seem to be feasible), and factors affecting evolvability can be experimentally identified.

Another approach to experimentally assessing evolvability is artificial selection. For instance, Beldade and Brakefield have shown that serially homologous eyespots on the butterfly *Bicyclus anyana* can be selected for different sizes (Beldade et al. 2002). Hence, size relationships are evolvable. However, in a more recent paper, Allen and collaborators (2008) show that color composition, although highly evolvable in itself, cannot be changed in different directions among serially homologous eyespots. This is interesting because the estimated genetic correlation in color composition of the two eyespots is less than 1, suggesting somewhat independent evolvability of color composition. This experimentally detected constraint (i.e., lack of evolvability) is consistent with the lack of interspecific variation in this trait (i.e., color composition of different eyespots on the same individual seems to be always the same).

Overall, there are multiple ways to assess short- and medium-term evolvability with data. All of these approaches are based on population genetic principles, and research in this area does not require a radical departure from the tradition of population genetics. Hence there is a chance that more interesting claims about evolvability-regulating traits are accessible to experimental investigation.

Why Is Evolution of Evolvability by Natural Selection Perceived as Impossible?

Here we want to examine the arguments that have been put forward to support the notion that natural selection cannot act on evolvability to enhance it (i.e., the argument that evolvability cannot be an adaptation). We will use two recent papers that summarized these arguments as guides (Sniegowski and Murphy 2006; Lynch 2007).

The Problem of Teleology

This is perhaps the least technical and the most philosophical argument against the idea that evolvability can be favored by natural selection. It posits that evolvability has its benefits in future environmental changes, and since natural selection cannot foresee the future, it also cannot select for evolvability. While it is true that nothing, including natural selection, prepares one for the totally unforeseeable, it is not true that all or even most environmental challenges are totally unpredictable. The physical environment is usually changing along stereotypic lines, with fluctuations along a limited number of dimensions, such as ambient temperature or the availability of water. Adaptation to recurrent environmental changes is thus no logical paradox, since the genotypes that adapted quicker to the last environmental change of the same kind can also be evolvable in response to a similar change in the future. The opportunity for evolvability to evolve therefore depends on the correlation between past and future environmental changes, and is an empirical question, not a logical one. Similarly, biotic interactions are also predictable in many cases, for instance, a parasite can "know" that the host immune system will most likely attack antigenic residues on its surface rather than, say, the GC content of its genomic DNA. Hence evolvability-enhancing genomic changes can be expected to affect the evolvability of surface proteins (Plotkin and Dushoff 2003).

Computational studies that show the evolution of evolvability in fact have always simulated an environment with varying selection forces. Kashtan and Alon (2005) studied the evolution of artificial networks adapting to two alternative "environments." In all cases the model "learned" to genetically adapt more quickly after a number of environmental changes. They also showed that the derived genotypes started to mimic the nature of the environmental change in their connectivity pattern, such that the patterns of the different environments were reflected in the structure of the network itself. This is identical to the idea of the "imitatory epigenotype" proposed by Rupert Riedl (1978): that genomes should evolve toward a structure of interactions that mimic the functional interdependencies imposed on the phenotype by the environment. In a recent paper (Draghi and Wagner 2008) we showed that a simple developmental model is able to adapt to random fluctuations in the environment by eliminating constraints on phenotypic variability. Hence this system "learned" by random mutation and selection what the scope of the random environmental fluctuations is.

The Unit of Selection Problem

Another, rather conceptual argument begins with the premise that since evolution is a population process, evolvability must be a population property. Therefore selection for evolvability seems to imply selection among populations, species, or lineages rather than between individuals. There is, however, a widespread belief that group selection is a weak evolutionary force, and thus, so the argument goes, evolution of evolvability is not possible on that account. The problem with this argument is that it overlooks the fact that conventional Darwinian selection requires only that the descendants of a genotype have on average higher fitness, than that of another genotype, to cause natural selection (GP

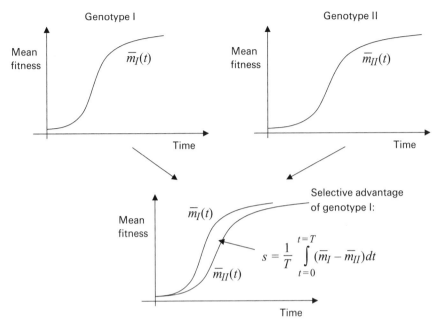

Figure 15.1
Selection of a genotype (genotype I) that has a higher evolvability than another genotype (genotype II). The difference in evolvability is reflected in a difference in the rate of increase in mean fitness. The population of genotype I increases in mean fitness faster than that of genotype II. If we superimpose these two curves of mean fitness, we can see that there is a time-dependent difference in the mean fitness of these two genotypes. This difference translates into the selection coefficient for genotype I. This simple model applies, for instance, to bacterial genotypes with different mutation rates. It shows that selection for evolvability does not require group selection, but can be understood as a case of straightforward Darwinian selection (after Wagner 1981).

Wagner 1981). For instance, consider two genotypes of the same bacterial species that differ in their evolvability for one reason or another (say, genetic robustness or mutation rate), but not in their initial fitness. Then, over time, the descendants of each genotype will increase in fitness, but at different rates, because of their difference in evolvability (figure 15.1). These different rates of increase in mean fitness will lead to a differential in mean fitness, and the genotype with the higher evolvability will outcompete the other. Hence, whatever the genetic property is that conveys higher evolvability to one of the strains, it will be selected because of the difference in mean fitness it causes over time (GP Wagner 1981):

$$\frac{dp}{dt} = pq(\bar{m}_I - \bar{m}_{II})$$

where p is the frequency of genotye I and $q = 1 - p$ is the frequency of genotype II, and $\bar{m}_I$ and $\bar{m}_{II}$ are the time-dependent mean fitnesses of the two genotypes. The only slightly unusual aspect of the selection equation that describes this process is that the selection coefficient is time-dependent rather than constant. The selection coefficient is the (time-dependent) difference in mean fitness of these two clones, ($\bar{m}_I - \bar{m}_{II}$). This shows that simple Darwinian selection, acting on phenotypic differences among individuals in a population, is sufficient to explain selection for evolvability.

Another issue in this context arises from the study of highly simplified artificial models of "development" which show evolution of evolvability under fluctuating selection pressures (Draghi and Wagner 2008, 2009). A detailed analysis of the population dynamics leading to the evolution of evolvability in the former study reveals that types with higher evolvabilities are more likely to invade evolving populations, and less likely to be displaced by other types. (For details see Draghi and Wagner 2008; however, the salient result is that natural selection can predictably shape evolvability through both direct and indirect population genetic mechanisms.) Still, these mechanisms are a consequence of first principles of population genetics, and do not require a radical departure from the conceptual basis of mainstream evolutionary biology; rather, they require a departure from a highly simplified notion of how selection and population genetics work.

Recombination and Selection

Another argument against the possibility of selection for evolvability is rooted in the results of selection experiments for mutation rate modifiers. In variable environments higher mutation rates are advantageous, as was shown in chemostat experiments with *E. coli* (Cox and Gibson 1974). The population genetic explanation is a straightforward extension of the previous equation. The chance of advantageous mutations increases with mutation rate, and a locus affecting mutation rate is linked to the mutations it facilitates, thus hitchhiking with the beneficial mutations to fixation. However—and this is the argument against evolution of evolvability by natural selection—this mechanism requires that the mutator gene be strongly linked to the loci at which advantageous alleles arise. In the case of chromosomal genes in bacteria, such tight linkage may sometimes occur, but relatively few loci are as tightly linked in the great majority of species with sexual reproduction. As recombination separates the mutator from beneficial alleles, the argument goes, positive selection on the mutator locus disappears. Extrapolation from this mutator model yields the conclusion that recombination prevents evolution of evolvability by natural selection. Though this argument accurately describes mutator genes in sexual species (Sniegowski and Murphy 2006), its applicability to other factors influencing evolvability is limited. There are two reasons why this verbal argument may not strongly restrict our expectations for the evolution of evolvability. The first is that recombination, even "free recombination," does not immediately lead to linkage equilibrium (a random association of alleles at different loci). This has been shown in the simplest example of an "evolvability" modifier, a model for dominance modification (GP Wagner and Bürger 1985). In a diploid population, the rate of selection of a gene depends on whether it is dominant or recessive relative to its alternative allele. Rare, dominant alleles are selected more quickly than recessive ones, because the heterozygote genotypes of dominant advantageous alleles have higher fitness. Consequently, dominance also leads to a faster increase in mean fitness. Hence, a dominance modifier can be seen as an evolvability modifier as well (if the population is far from mutation–selection equilibrium). Analysis of selection equations shows that even without linkage between the modifier and the primary locus, the modifier can be selected.

The example above establishes that selection can maintain the association between an evolvability allele and its beneficial by-products if epistasis contributes to the beneficial effect. Another case is the following example. Consider a locus A, with alleles a and A. Assume that the substitution of A for a increases the stability of a protein (following Bloom et al. 2006). Without this increase in stability, allele B at locus B causes deleterious misfolding; in combinations with A, however, this same allele produces beneficial changes in protein function. We might therefore regard allele A as conferring greater evolvability than allele a. Selection will therefore favor the combination AB, and the epistatic connection between the alleles requires that the change in evolvability fix with the beneficial mutation it facilitated. Thus, selection for evolvability is not limited to an allele at a single locus, but is an emergent property of certain allele combinations. What is selected, then, is not an isolated allele but a haplotype. The advantageous haplotype may get lost if recombination is above a certain rate, but if the loss rate by recombination is compensated by the selective advantage of the haplotype, the latter will still increase in frequency.

Several empirical examples also indicate that the diverse mechanisms of evolvability are not equally vulnerable to recombination. Sniegowski

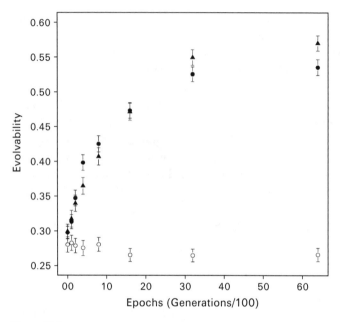

Figure 15.2
Evolvability increases during evolution in changing environments in a simulation model
for the evolution of a gene regulatory network (see Draghi and Wagner 2008b). Each
epoch represents an interval of 100 generations, where generations are discrete and population sizes constant. Filled symbols are populations in which the optimal phenotype
changes every 100 generations; the filled circles represent clonal reproduction; and the
filled triangles represent populations with recombination. The open circles represent
control populations that experience stabilizing selection for the entire 64 epochs. Each
point is the mean of 160 simulations, consisting of sets of 10 populations starting from each
of the 16 possible phenotypes. Bars are standard errors. Populations have 1000 individuals,
networks consist of 4 genes, and the genomic mutation rate is approximately 0.05. In this
model recombination does not impair the evolution of evolvability.

and Murphy (2006) suggest that contingency loci of microbial pathogens
are an example where an evolvability-enhancing allele, by virtue of its
low cost and tight linkage to its effects, is not lost by recombination.
Computational models of gene regulatory networks also illustrate the
potential robustness of evolvability mechanisms to recombination
(Draghi and Wagner 2008, 2009). If a model gene regulatory network is
allowed to evolve in a fluctuating environment, it evolves higher evolvability (Crombach and Hogeweg 2008; Draghi and Wagner 2009). This
is also the case when recombination is allowed (figure 15.2): there is
virtually no difference in the evolvability gains with and without recombination (Draghi and Wagner 2009). The reason is that at least one factor
determining the evolvability of these networks is the overall "network

excitation" (i.e., the sum of the positive and negative inputs per gene, averaged over all genes). It turns out that the network excitability determines (statistically) the accessibility of gene expression phenotypes. Crucially, in this case the evolvability trait is a distributed property of the genome, and evolvability cannot generally be genetically separated from beneficial mutations. This epistatic linkage should also describe the network models of Kashtan and Alon (2005) and Crombach and Hogeweg (2008), but in these studies the effect of recombination was not explicitly tested. These examples show that epistatic interactions among loci may not only provide opportunities for evolvability to evolve, but also can also shelter evolvability alleles from dissociation through recombination.

The Transient Benefit of Evolvability

Another argument against the evolution of evolvability by natural selection from first principles of population genetics is that features that enhance evolvability are likely to be selected against when the population reaches mutation–selection equilibrium. Again this intuition is best illustrated by the example of mutation rate modifiers. While it has been shown that increased mutation rate can be selected when the population encounters a new environment (e.g., Cox and Gibson 1974; GP Wagner 1981), it is also selected against when the population has adapted to the environment. The reason is that the mean fitness in mutation–selection equilibrium decreases with the genomic mutation rate; in fact, in mutation–selection equilibrium the mean relative fitness is $1-u$, where u is the genomic mutation rate. Selection is expected to favor lower mutation rates, leading to higher mean fitness at equilibrium. Again, this argument is true for mutation rate modifiers but cannot be generalized to other biological features that affect evolvability. For instance, Masel and Bergman (2003) investigated the claim that yeast prions could be selected for their effect on evolvability. Their analysis confirmed that such selection was plausible, and also showed that, at evolutionary equilibrium, the evolvability-affecting trait did not influence mean fitness. On a more general level, one must note that the mean fitness in mutation–selection equilibrium is affected only by the mutation rate, and not by the average effect of mutations. If evolution of evolvability selects for properties that influence the size and distribution of mutation effects (such as pleiotropy in Draghi and Wagner 2008, or network excitability in Draghi and Wagner 2009), then selection in mutation–selection equilibrium will not erase the changes caused by selection for evolvability. Furthermore, it has been shown that selection to decreased mutation effects at mutation–selection equilibrium (i.e., genetic canalization) is very weak (GP Wagner et al. 1997).

In addition to the theoretical arguments discussed above, which can be answered based on first principles and simulation studies, Michael Lynch (2007) also raises a number of empirical questions, pointing to a lack of direct empirical evidence. While it is true that there is very little empirical work on this subject, the lack of evidence is not the same as evidence against evolution of evolvability. The small amount of empirical work in this area is a direct consequence of the fact that evolution of evolvability has been a blind spot in the evolutionary synthesis, and thus has not been an integral part of the research program in mainstream evolutionary biology. We think that the theoretical studies cited in this section pave the way to targeted experimental investigations like the one by McBride and collaborators (2008) and Allen et al. (2008).

Evolution of Evolvability as a By-product

As with the evolution of all organismal traits, evolution of evolvability can also be the result of other evolutionary trends that selection is causing (Lynch 2007). For instance, there is a well-supported model that suggests that robustness could also lead to an enhancement of evolvability (A. Wagner 2005; Bloom et al. 2006; A. Wagner 2007). Hence selection for environmental and/or genetic robustness (Gavrilets and Hastings 1994; GP Wagner et al. 1997; Meiklejohn and Hartl 2002) can lead to evolution of evolvability as a by-product. This idea has been supported by an experimental study cited above (McBride et al. 2008) in which strains of the RNA virus $\Phi6$ that differ in their genetic robustness also differ in their evolvability. Many other scenarios are possible, and it will be a matter of empirical work to sort out directly selected evolution of evolvability and by-product scenarios. This is not a question that can be settled by theoretical arguments alone.

Conceptual Implications of Evolvability

We have argued that the concept of evolvability is entirely compatible with the mechanisms of evolution that form the basis of the Modern Synthesis, and that it remains controversial because it conflicts with some intuitions about the limits of those mechanisms. By examining a few major evolutionary questions where evolvability has been implicitly discussed, we can understand the roots of this conflict and demonstrate the benefits of a coherent evolutionary theory of evolvability. While many aspects of evolution have deep connections to the idea of evolvability, we will focus on two that have inspired debate since the field began: the costs and benefits of sex, and the concept of evolutionary constraint.

The costs of sex and recombination are many, and would seem to preclude the long-term success of sexual reproduction; yet, sex and recombination are rampant throughout nature. The magnitude and ubiquity of these costs have inspired a variety of attempts to explain the evolutionary success of sexual reproduction, and the search for a sufficient explanation remains a major preoccupation of evolutionary biology. The dominant theme of this search dates to Weismann, who proposed that reassortment and recombination, by producing a variety of allele combinations, serve to fuel adaptation through natural selection (Weismann 1904; reviewed in Burt 2000). This idea that sex is maintained by evolution, in the face of substantial costs, because it produces beneficial variation has been debated and extensively modeled since the 1930s (Bell 1982; Feldman et al. 1997; Burt 2000; Agrawal 2006; Hadany and Comeron 2008), and the development of this debate perfectly illustrates the blind spot of evolvability. Most attempts to translate Weismann's verbal argument into a quantitative model focus on the rate of evolution in a population and invoke group-level selection, often implicitly, to explain the success of recombination. These "optimality" models (Feldman et al. 1997) include the influential work of Fisher (1930), Muller (1932), Crow and Kimura (1965), Maynard Smith (1968), and Felsenstein (1974), and continue to directly motivate further theory (Peck 1993; Cohen et al. 2005) and experimentation (Zeyl and Bell 1997; Cooper 2007). The continued influence of the Fisher–Muller optimality argument is puzzling: the efficacy of group-level selection is generally considered to be totally inadequate to counter the costs of sex, and models applying individual-level selection on modifier loci present a superior alternative. Modifier models have coexisted with the Fisher–Muller argument for forty years (Nei 1967), and have been repeatedly invoked to demonstrate a plausible advantage for recombination (Felsenstein and Yokoyama 1976; Barton 1995; Barton and Otto 2005); yet the Fisher–Muller model, and its mechanism of group selection, still forms the implicit basis for many discussions of the advantages of sex (Roughgarden 1991; West et al. 1999; Agrawal 2006; Otto and Gerstein 2006).

One easy explanation for the persistence of optimality models is that optimality arguments are typically intuitive, while the arguments for modifier models are much more technical. Yet recent developments in the theory of evolvability reveal the simple intuition underlying modifier arguments. The key insight is that individual-level selection can act on heritable traits that are fully realized over the multigenerational course of a lineage. Modifier alleles are such traits, as are mutation rates (Tenaillon et al. 2001), patterns of pleiotropy and epistasis (GP Wagner et al. 1997; Draghi and Wagner 2008, 2009), sex ratios (Fisher 1930; Edwards 1998), and life histories (Stearns 1992). While the evolutionary significance of such traits is apparent in mathematical models and simulations, the concept of a trait with lineage-level phenotypes is still muddled in the literature with vague ideas of indirect, long-term, or group selection. We suggest that an immediate benefit from a coherent theory of evolvability would be the resolution of this conceptual tangle, and a clarifying simplification of the hypotheses for the evolution of sex.

While the evolutionary maintenance of sex is a vexing problem, the question itself fits within the boundaries of mainstream evolutionary biology. In contrast, the issue of constraint in evolution challenges fundamental notions of how adaptations are studied, and how evolution interacts with other biological disciplines. Evolutionary constraints are typically identified by the absence of expected adaptation (McKitrick 1993), or as a bias or limitation on the spectrum of phenotypic variants (Maynard Smith et al. 1985). Though some limits to adaptation are inevitable consequences of physics (animal locomotion cannot exceed the speed of light) or trade-offs (offspring number cannot vary completely independent of offspring size), many others are best described as developmental, and it is the role of these developmental constraints in evolutionary thought that demonstrates how biology can profit from a theory of evolvability.

Much has been written about how the Modern Synthesis entrenched a separation between evolutionary and developmental biologists, and discussion of constraint seems to be a promising bridge between these disciplines. But what is interesting is how the constraint concept biases this connection to be negative and one-directional. If, following McKitrick (1993), we define constraint as the absence of expected adaptive change, then we tacitly suggest that variation is expected to exist in all phenotypic directions (Salazar-Ciudad 2007). Several authors have criticized this deeply embedded idea that variability should be isotropic, and that apparent constraints are exceptions with special explanations (Gould

2002; Salazar-Ciudad 2005; Salazar-Ciudad 2007). The alternative is that variability, like any other trait, has both mechanistic and evolutionary causes, and should be studied for what it is, not as what is absent from an idealized distribution. This alternative viewpoint permits development to a make a positive, rather than a restrictive, contribution to evolution, but also establishes a causal connection between evolutionary processes and developmental systems.

The importance of this reciprocal link between evolution and development is not evident in discussions of constraint (Maynard Smith et al. 1985; Amundson 1994; Schwenk 1995), but is emphatically demonstrated by the evolution of evolvability. For example, Kashtan and Alon (2005) used a simple network model to explore how selection shapes the structure of the network, and hence the variability of the phenotype. A specific form of fluctuating selection can preserve modular networks with unconstrained patterns of variability, while other conditions allow the more likely, constrained network structures to dominate. In this example, the absence of constraint has a specific cause, while its presence merely reflects the preponderance of structures with biased variability. This explanation would be obscured from the perspective that constraints, as exceptions, have specific explanations. The development of the theory of evolvability, by illustrating the limitations of the concept of constraint, can therefore play a major role in motivating the reconciliation of evolution and development (Pigliucci 2007, 2008).

Conclusions

The factors affecting evolvability and the mechanism for the evolution of evolvability are poorly understood (Pigliucci 2008). The reason why we know so little about it is neither that the question is unimportant, nor that it is intrinsically alien to population genetic theory. We argue here that the question is important and that research into this question does not require a radical break from the traditions of population genetics. The hostility of some population geneticists toward these questions and concepts is due to a blind spot in the research program of evolutionary biology rather than the result of a deep conceptual divide.

In what way does the study of evolvability extend the evolutionary synthesis? In classical evolutionary biology, evolvability is an assumption rather than the subject of study. Availability of heritable phenotypic variation is considered like a boundary condition, rather than as in need of explanation. In that respect the study of evolvability is a link between the agenda of developmental evolution and population genetics (Hendrikse et al., 2007), as is research on canalization and robustness (A Wagner 2005). In addition, the focus of research is shifted to systemic properties of the organism, away from the single gene or single trait perspective. Hence, while we have argued that there are no deep conceptual obstacles for population genetic theory to explain the evolution of evolvability, the very focus on this subject will help shift the perspective of evolutionary biology toward a more holistic, systemic view of evolution.

References

Agrawal AF (2006) Evolution of sex: Why do organisms shuffle their genotypes? Current Biology 16: R696–R704.

Allen CE, Beldade P, Zwaan BJ, Brakefield PM (2008) Differences in the selection response of serially repeated color pattern characters: Standing variation, development, and evolution. BMC Evolutionary Biology 8: 94.

Altenberg L (1994) The evolution of evolvability in genetic programming. In Advances in Genetic Programming. K.E. Kinnear, ed.: 47–74. Cambridge, MA: MIT Press.

Amundson R (1994) Two concepts of constraint: Adaptationism and the challenge from developmental biology. Philosophy of Science 61: 556–578.

Barton NH (1995) A general model for the evolution of recombination. Genetic Research 65: 123–145.

Barton NH, Otto SP (2005) Evolution of recombination due to random drift. Genetics 169: 2353–2370.

Barton NH, Turelli M (1989) Evolutionary quantitative genetics: How little do we know? Annual Review of Genetics 23: 337–370.

Beldade P, Koops K, Brakefield PM (2002) Developmental constraints versus flexibility in morphological evolution. Nature 416: 844–847.

Bell G (1982) The Masterpiece of Nature: The Evolution and Genetics of Sexuality. Berkeley: University of California Press.

Bloom JD, Labthavikul ST, Sy T, Otey CR, Arnold FH (2006) Protein stability promotes evolvability. Proceedings of the National Academy of Sciences of the USA 103: 5869–5874.

Brakefield PM, Monteiro A, et al. (2003) The evolution of butterfly eyespot patterns. Butterflies: Ecology and Evolution Taking Flight: 243–258. Chicago: University of Chicago Press.

Brookfield JFY (2001) The evolvability enigma. Current Biology 11: R106–R108.

Bürger R (2000) The Mathematical Theory of Selection, Recombination, and Mutation. New York: Wiley.

Burt A (2000) Sex, recombination, and the efficacy of selection—was Weismann right? Evolution 54: 337–351.

Camara MD, Pigliucci M (1999) Mutational contributions to genetic variance-covariance matrices: An experimental approach using induced mutations in *Arabidopsis thaliana*. Evolution 53: 1692–1703.

Cheverud JM, Marroig G (2007) Comparing covariance matrices: Random skewers method compared to the common principal components model. Genetics and Molecular Biology 30: 461–469.

Cohen E, Kessler DA, Levine H (2005) Recombination dramatically speeds up evolution of finite populations. Physical Review Letters 94: 098102.

Cooper TF (2007) Recombination speeds adaptation by reducing competition between beneficial mutations in populations of Escherichia coli. PLoS Biology 5: 1899–1905.

Cox EC, Gibson TC (1974) Selection for high mutation rates in chemostats. Genetics 77: 169–184.

Crombach A, Hogeweg P (2008) Evolution of evolvability in gene regulatory networks. PLoS Computational Biology 4: e1000112.

Crow JF, Kimura M (1965) Evolution in sexual and asexual populations. American Naturalist 99: 439–450.

Draghi J, Wagner GP (2008) Evolution of evolvability in a developmental model. Evolution 62: 301–315.

Draghi J, Wagner GP (2009) The evolutionary dynamics of evolvability in a gene network model. Journal of Evolutionary Biology 22: 599–611.

Edwards AWF (1998) Natural selection and the sex ratio: Fisher's sources. American Naturalist 151: 564–569.

Feldman MW, Otto SP, Christiansen FB (1996) Population genetic perspectives on the evolution of recombination. Annual Review of Genetics 30: 261–295.

Felsenstein J (1974) The evolutionary advantage of recombination. Genetics 78: 737–756.

Felsenstein J, Yokoyama S (1976) Evolutionary advantage of recombination. 2. Individual selection for recombination. Genetics 83: 845–859.

Fisher RA (1930) The Genetical Theory of Natural Selection. Oxford: Clarendon Press.

Gavrilets S, Hastings A (1994) A quantitative genetic model for selection on developmental noise. Evolution 48: 1478–1486.

Gerhart J, Kirschner M (1997) Cells, Embryos, and Evolution. Malden, MA: Blackwell Science.

Gerhart J, Kirschner M (2007) The theory of facilitated variation. Proceedings of the National Academy of Sciences of the USA 104: 8582–8589.

Gilchrist GW, Lee CE (2007) All stressed out and nowhere to go: Does evolvability limit adaptation in invasive species? Genetica 129: 127–132.

Gould SJ (2002) The Structure of Evolutionary Theory. Cambridge, MA: Belknap Press of Harvard University Press.

Hadany L, Comeron JM (2008) Why are sex and recombination so common? Annals of the New York Academy of Sciences 1133: 26–43.

Hansen TF, Houle D (2008) Measuring and comparing evolvability and constraint in multivariate characters. Journal of Evolutionary Biology 21: 1201–1219.

Hendrikse JL, Parsons TE, Hallgrimsson B (2007) Evolvability as the proper focus of evolutionary developmental biology. Evolution & Development 9: 393–401.

Hill WG (1982) Rates of change in quantitative traits from fixation of new mutations. Proceedings of the National Academy of Sciences of the USA 79: 142–145.

Kashtan N, Alon U (2005) Spontaneous evolution of modularity and network motifs. Proceedings of the National Academy of Sciences of the USA 102: 13773–13778.

Kirkpatrick M (2008) Patterns of quantitative genetic variation in multiple dimensions. Genetica 136: 271–284.

Lande R (1979) Quantitative genetic analysis of multivariate evolution, applied to brain:body size allometry. Evolution 33: 402–416.

Lande R (1980) The genetic covariance between characters maintained by pleiotropic mutations. Genetics 94: 203–215.

Lande R, Arnold SJ (1983) The measurement of selection on correlated characters. Evolution 37: 1210–1226.

Lenski RE, Ofria C, Collier TC, Adami C (1999) Genome complexity, robustness, and genetic interactions in digital organisms. Nature 400: 661–664.

Love AC (2003) Evolvability, dispositions, and intrinsicality. Philosophy of Science 70: 1015–1027.

Lynch M (2007) The frailty of adaptive hypotheses for the origins of organismal complexity. Proceedings of the National Academy of Sciences of the USA 104(suppl. 1): 8597–8604.

Lynch M, Walsh B (1998) Genetics and Analysis of Quantitative Traits. Sunderland, MA: Sinauer.

Masel J, Bergman A (2003) The evolution of the evolvability properties of the yeast prion [PSI+]. Evolution 57: 1498–1512.

Maynard Smith J (1968) Evolution in sexual and asexual populations. American Naturalist 102: 469–473.

Maynard Smith J, Burian R, et al. (1985) Developmental constraints and evolution. Quarterly Review of Biology 60: 265–287.

Maynard Smith J, Szathmáry E (1995) The Major Transitions in Evolution. Oxford: Oxford University Press.

McBride RC, Ogbunugafor CB, Turner P (2008) Robustness promotes evolvability of thermotolerance in an RNA virus. BMC Evolutionary Biology 8: 231.

McGuigan K, Blows MW (2007) The phenotypic and genetic covariance structure of drosophilid wings. Evolution 61: 902–911.

McKitrick MC (1993) Phylogenetic constraint in evolutionary theory: Has it any explanatory power? Annual Review of Ecology and Systematics 24: 307–330.

Meiklejohn CD, Hartl DL (2002) A single model of canalization. Trends in Ecology and Evolution 17: 468–473.

Mezey JG, Houle D (2005) The dimensionality of genetic variation for wing shape in *Drosophila melanogaster*. Evolution 59: 1027–1038.

Muller HJ (1932) Some genetic aspects of sex. American Naturalist 66: 118–138.

Nei M (1967) Modification of linkage intensity by natural selection. Genetics 57: 625–641.

Otto SP, Gerstein AC (2006) Why have sex? The population genetics of sex and recombination. Biochemical Society Transactions 34: 519–522.

Pavlicev M, Wagner GP, et al. (2009) Measuring evolutionary constraints through the dimensionality of the phenotype: adjusted bootstrap method to estimate rank of phenotypic covariance matrices. Journal of Evolutionary Biology 59: 1027–1038.

Peck JR (1993) Frequency-dependent selection, beneficial mutations, and the evolution of sex. Proceeding of the Royal Soceity of London B 254: 87–92.

Pigliucci M (2007) Do we need an extended evolutionary synthesis? Evolution 61: 2743–2749.

Pigliucci M (2008) Is evolvability evolvable? Nature Reviews Genetics 9: 75–82.

Plotkin JB, Dushoff J (2003) Codon bias and frequency-dependent selection on the hemagglutinin epitopes of influenza A virus. Proceedings of the National Academy of Sciences of the USA 100: 7152–7157.

Riedl R (1978) Order in Living Organisms: A Systems Analysis of Evolution. New York: Wiley.

Roughgarden J (1991) The evolution of sex. American Naturalist 138: 934–953.

Rutherford SL (2000) From genotype to phenotype: Buffering mechanisms and the storage of information. BioEssays 22: 1095–1105.

Rutherford SL, Lindquist S (1998) Hsp90 as a capacitor for morphological evolution. Nature 396: 336–342.

Salazar-Ciudad I (2005) Developmental constraints vs. variational properties: How pattern formation can help to understand evolution and development. Journal of Experimental Zoology B306: 107–125.

Salazar-Ciudad I (2007) On the origins of morphological variation, canalization, robustness, and evolvability. Integrative and Comparative Biology 47: 390–400.

Schwenk K (1995) A utilitarian approach to evolutionary constraint. Zoology 98: 251–262.

Sniegowski PD, Murphy HA (2006) Evolvability. Current Biology 16: R831–R834.

Stearns SC (1992) The Evolution of Life Histories. Oxford: Oxford University Press.

Tenaillon O, Taddei F, Radman M, Matic I (2001) Second-order selection in bacterial evolution: Selection acting on mutation and recombination rates in the course of adaptation. Research in Microbiology 152: 11–16.

Wagner A (2005) Robustness and Evolvability in Living Systems. Princeton, NJ: Princeton University Press.

Wagner A (2007) Robustness and evolvability: A paradox solved. Proceedings of the Royal Society of London B 275: 91–100.

Wagner GP (1981) Feedback selection and the evolution of modifiers. Acta Biotheoretica 30: 79–102.

Wagner GP, Altenberg L (1996) Complex adaptations and the evolution of evolvability. Evolution 50: 967–976.

Wagner GP, Booth G, Bagheri-Chaichian N (1997) A population genetic theory of canalization. Evolution 51: 329–347.

Wagner GP, Bürger R (1985) On the evolution of dominance modifiers II: A non-equilibrium approach to the evolution of genetic systems. Journal of Theoretical Biology 113: 475–500.

Wagner GP, Laubichler MD (2004) Rupert Riedl and the re-synthesis of evolutionary and developmental biology: Body plans and evolvability. Journal of Experimental Zoology B302: 92–102.

West SA, Lively CM, Read EF (1999) A pluralist approach to sex and recombination. Journal of Evolutionary Biology 12: 1003–1012.

Zeyl C, Bell G (1997) The advantage of sex in evolving yeast populations. Nature 388: 465–468.

VII PHILOSOPHICAL DIMENSIONS

16 Rethinking the Structure of Evolutionary Theory for an Extended Synthesis

Alan C. Love

Is contemporary evolutionary theory adequate? Does it contain gaps or inconsistencies? Do we need an expanded or extended evolutionary synthesis (Müller 2007; Pigliucci 2007)? There is something presumptuous involved in asking these questions about the scope and status of evolutionary theory, especially from the perspective of philosophy of science. Jacques Monod expressed a natural suspicion that accompanies this presumption:

[A] curious aspect of the theory of evolution is that everybody thinks he understands it. I mean philosophers, social scientists, and so on. While in fact very few people understand it, actually, as it stands, even as it stood when Darwin expressed it, and even less as we now may be able to understand it in biology. (Monod 1975: 12)

Keeping this in mind, the epistemological perspective offered here in service of answering these questions is explicitly *pluralist*. By this I mean that my aim is not to provide the one and only correct way of thinking about evolutionary theory, but rather to offer one fruitful possibility in direct comparison with others (cf. Kellert et al. 2006). There is more than one way to represent evolutionary theory, and among these alternatives there are different advantages and disadvantages.[1] A particular construal or "presentation" (Griesemer 1984) of evolutionary theory is recommended when it produces advantages to some particular end. How we use evolutionary theory guides our preferences about its representation; choices are made based on the goals of our contexts of inquiry (scientific, philosophical, and otherwise). Different representations of evolutionary theory can be considered analogous to the tools in a toolbox. There is no question of whether a hammer is better than a wrench *in general*; only whether one is more appropriate to the task at hand. Therefore, an assessment of the fruitfulness of a representation of evolutionary theory will be keyed to the task that motivates the inquiry.

I begin by characterizing this task in terms of a distinction between the content and the structure of a scientific theory. The controversy over how the inclusion of developmental *content* leads to a revised or expanded evolutionary synthesis, as seen in the visible agitation around the significance of EvoDevo (on display in other chapters of this volume), can be used as a template for isolating a new perspective on representational *structure* for evolutionary theory. Next, I turn to the methodological options available for identifying aspects of theoretical structure. I adopt a "bottom-up" approach focused on evolutionary theory in particular, as opposed to a "top-down" strategy that attempts to characterize the structure of all scientific theories. Pursuing a bottom-up strategy differs from many previous philosophical approaches and leads to the identification of alternative structures for evolutionary theory in terms of different circumscriptions of content, which I label "narrow" and "broad" representations.

Narrow representations of content that are primarily focused on evolutionary genetics have been the main province of philosophers, but broad ones are more favorable to isolating structure for an evolutionary synthesis, modern or extended, because they include more of the biological knowledge available for explaining evolution. Scrutinizing broad representations in textbooks reveals consistent and stable domains of problems ("problem agendas") that are being investigated by combinations of diverse life science disciplines. This suggests an *erotetic* (pertaining to questioning) structure for evolutionary theory, which characterizes a synthesis in terms of multiple problem agendas exhibiting complex but coordinating relationships. Understanding structure in this new way broadens the philosophical discussion about theory structure while clarifying how developmental biology and other disciplinary approaches provide necessary contributions to our understanding of evolution. As a consequence, it yields a useful perspective on theoretical structure for an Extended Synthesis.

Evolutionary Theory: Content versus Structure

When analyzing epistemological aspects of science we can make a distinction between the structure and content of a scientific theory. Much of the literature discussing the adequacy of evolutionary theory has been about whether its *content* is sufficient. Does the Modern Synthesis, as developed in the post–World War II context, adequately capture the different facets of evolutionary processes? Did it exclude developmental biology (née embryology) or misrepresent and overlook particular domains of life, such as bacteria and archaea (née prokaryotes)? For example, Ulrich Kutschera and Karl Niklas treat the possibility of an expanded evolutionary synthesis through specific topics such as rates of evolution, mass extinction, species selection, endosymbiosis, EvoDevo, phenotypic plasticity, epigenetic inheritance, and experimental microbial evolution (Kutschera and Niklas 2004; cf. RL Carroll 2000; Müller 2007; Pigliucci, 2007). In these cases, a revision to evolutionary theory comes in the guise of augmented or modified content; the endeavor is one of either "enlarging" the previous synthesis or somehow revising particular claims within it.[2]

A distinct issue is the *structure* of evolutionary theory, which involves questions about how the content is organized. Instead of concentrating on whether explanations generalize to all species or whether specific biological disciplines have been sidelined, a focus on theory structure suggests different questions: How is knowledge referred to as "evolutionary theory" organized (Tuomi 1981)? Does it have a structure that is similar to or different from other scientific theories, especially those in physics (cf. Rosenberg 1985: ch. 5; Ruse 1973: ch. 4)? These descriptive questions can be supplemented with prescriptive ones: Should we organize "evolutionary theory" in a particular way? Should it be similar in structure to other theories? All of these questions introduce philosophical topics pertaining to the nature of scientific theories (Suppe 1977). None of them are disconnected from theoretical content, but instead demand viewing evolutionary theory from an angle that is less frequently adopted by working biologists. The potential gain of paying attention to structure is a novel perspective on the contours of an extended evolutionary synthesis and a more precise characterization of what its formulation requires.

Although evolutionary theory has been treated in philosophical debates about theory structure, the epistemic organization of the Modern Synthesis has generated different concerns. Dudley Shapere's commentary on historical dimensions of the Modern Synthesis highlighted its incongruence with other scientific theories. This raised the issue of whether existing philosophical analyses of structure were adequate (Shapere 1980; cf. Beckner 1959; Caplan 1978). John Beatty posed the question of whether evolutionary theory should be explicitly distinguished from the Modern Synthesis (Beatty 1986). Given that the Modern Synthesis includes numerous commitments beyond genetic models of evolution by natural selection, it might be more appropriate to reserve the label "evolutionary theory" for this more circumscribed domain. If this distinction is necessary, then it raises special questions about the relationship between the "Modern Synthesis" and its frequent synonym, "the synthetic theory of evolution." Jean Gayon has expressed this succinctly: "it is very difficult to define the Modern Synthesis as a 'theory' ... it is indeed questionable whether the Modern Synthesis should be considered as one single theory" (Gayon 1990: 3).

When Wolf-Ernst Reif and colleagues set out to delineate German contributions to the Modern Synthesis, they devoted attention to discerning the right "structure" of concepts, results, and research programs for their analysis (Reif et al. 2000). They distinguished the Modern Synthesis as a process from the synthetic theory of evolution as a product. Other attempts have been made to describe the structure of evolutionary theory qua synthesis, such as a "hypertheory" that subsumes subordinate theories of evolutionary mechanisms (Wasserman 1981) or as a "multifield" theory (Darden 1986). Together, these diverse descriptions of Modern Synthesis "structure," and by implication evolutionary theory, constitute nagging questions that are related to, but distinct from, issues of content.

Although the title of Stephen Jay Gould's final book on evolutionary theory seems to indicate that "structure" is the primary topic (Gould 2002), this is only partially true. His account of theoretical structure derives from three principles pertaining to the central concept of natural selection (agency, efficacy, and scope), and asks how the content of these three structural "branches" needs to be revised with varying severity.[3] Gould's argument is not focused on whether the number of major branches should be increased or decreased, or whether the idea of tripartite branches is the right structural conception in the first place, but rather what content should be included within this structure and how that content is interpreted. A large portion of the book (Part II: "Towards a Revised and Expanded Evolutionary Theory") is devoted to a critical review of empirical findings and conceptual revisions surrounding species selection and developmental constraints, which Gould argues should be incorporated into the content of evolutionary theory with attendant consequences for our views of evolutionary processes.

The reaction to Gould's argument for revising evolutionary theory has been mixed and sometimes decidedly negative. But even in the latter cases, the worries about the adequacy of evolutionary theory return to issues of content, not structure: "Given how hard it is to understand why evolution has taken the paths it has, and how little we understand about how structure and form develop, discussion of how developmental systems have constrained evolution from taking the paths that it has not seems at best premature" (Zimmerman 2003: 458). A similar dialectic on the topic of including developmental findings in evolutionary biology is widely observable, especially with respect to how EvoDevo is supposed to transform evolutionary theory. Sean Carroll has labeled EvoDevo the "third revolution," which follows after the idea of evolution introduced in the nineteenth century and the establishment of genetics in the twentieth century:

Over the past two decades, a new revolution has unfolded in biology. Advances in developmental biology and . . . "Evo Devo" have revealed a great deal about the invisible genes and some simple rules that shape animal form and evolution. Much of what we have learned has been so stunning and unexpected that it has profoundly reshaped our picture of how evolution works. (SB Carroll 2005: x).

This "profound reshaping" implies modifications to evolutionary theory in order to accommodate this revolution: "The Modern Synthesis established much of the foundation for how evolutionary biology has been discussed and taught for the past sixty years. However, despite the monikers of 'Modern' and 'Synthesis,' it was incomplete" (SB Carroll 2005: 7).

Unsurprisingly, these claims have been met with resistance. In his review of the book for *Nature*, the evolutionary geneticist Jerry Coyne expressed skepticism: "Carroll presents his vision of the field without admitting that large parts of that vision remain controversial. . . . it proclaim[s] a clever but still unproved hypothesis as central to the evolutionary process" (Coyne 2005: 1029–1030). Coyne has followed up this suspicion with purported empirical refutation (Hoekstra and Coyne 2007). The disagreement revolves around the *content* of evolutionary theory, not its *structure* (see Pennisi 2008 for an overview). A multitude of analogous examples could be selected to make a similar point.[4] The vigor of these kinds of controversies has led one researcher to ponder whether an extended evolutionary synthesis is just around the corner or simply an impossible chimera (Pigliucci 2003).

What these disputes provide for the present philosophical analysis is a template; they motivate and guide the isolation of kinds of structure relevant to conceptualizing an Extended Synthesis. Ongoing negotiations over the import of EvoDevo content for evolutionary theory offer a measure of fecundity for the particular account of evolutionary theory structure discussed herein. Can we isolate structure for evolutionary theory that better accommodates the features emphasized in these debates about the place of EvoDevo, revolutionary or not?[5] A representation of evolutionary theory is useful if it assists us in the task of understanding how developmental content "fits in," and thereby yields structural materials for resolving these disputes.[6]

This philosophical strategy is not guaranteed to adjudicate disputes about content. These may continue even if a structure meeting this desideratum is found; there is no logical relationship between theory content and structure, even though they have intimate relations with one another. Various recommendations to revise or augment the content of evolutionary theory can be consistent with different structures. An alternative representation of evolutionary theory's structure may or may not demand differential inclusion or exclusion of empirical content. Similarly, exploring the structural possibilities of evolutionary theory does not by itself demand a revision of its empirical content. But if an expanded synthesis is deemed necessary, then considerations of structure may be just as relevant as those of content.

The goal, then, is to take just one prominent element pertaining to content that spurs controversy about an extended evolutionary synthesis and provide a structural conception that is sensitive to it. My suggestion of an alternative representation of evolutionary theory is motivated by the controversy over how to fit developmental components into evolutionary theory, but it does not aspire to the immediate arbitration and resolution of these differences—consensus about these matters takes time. But explicitly treating structure also may have more immediate benefits, such as the facilitation of ongoing research, regardless of the trajectory for this controversy. These added benefits emerge from the particular methodology utilized to identify alternative structural conceptions.

Methodology: Isolating Structural Features of Evolutionary Theory

Two different methods are available for isolating theory structure. The first can be labeled a "top-down" approach, and proceeds by asking what is true of all scientific theories. This strategy has been popular in philosophy of science, in part because a general account of scientific theory structure would yield a unified account of how theories change over time and how they are confirmed by evidence. A second approach can be labeled "bottom-up," and is motivated by a different question: What is an adequate account of this particular scientific theory (e.g., special relativity)? An answer to this question may not be useful in understanding how every scientific theory is subject to change or confirmed by evidence, but it is fruitful when trying to produce a philosophical account that closely tracks the details of a specific theory. It also may (1) facilitate ongoing research, (2) systematically organize the knowledge (which can be useful both for research and for pedagogy), (3) illuminate the theory's historical development, and (4) serve as a means to respond to internal or external challenges.

As might be expected from a pluralist stance, these methods can be complementary, and neither is inherently superior. Their utility as methodologies is related to the questions they are meant to address. The bottom-up strategy is natural when focused on evolutionary theory alone, especially because a central aim of formulating an Extended Synthesis is to stimulate continued research on evolution. This strategy directs our attention to clues from the community of researchers for isolating epistemic materials relevant to a reformulation. These clues are not necessarily found in theory content:

Similarity in substantive content ... is not sufficient for individuating scientific theories [because it] changes too rapidly and sporadically for that. Instead the continuing commitment on the part of scientists to certain procedures, goals, problems, and metaphysical presuppositions supply [*sic*] most of the continuity in science. (Hull 1976: 656)

Thus, we can ask from the bottom-up perspective how ongoing commitments to particular goals or problems *structure* the theoretical content and potentially clarify its historical development (Hull 1988). One natural place to find these clues is in different textbooks on evolution. By linking philosophical analysis with the language biologists use to present the theory, we achieve a direct engagement with aims relevant to scientific investigation, such as the facilitation of ongoing research (Love 2008a). Despite the natural propensity to adopt a bottom-up strategy in this context, recent literature on theory structure reinforces its advantages and testifies in favor of a distinct approach to philosophically analyzing evolutionary theory.

Within debates about the nature of scientific theories using a top-down methodology (Suppe 1977), there have been two rival options. The first is the "syntactic view," which argues that a scientific theory is best reconstructed as an axiom system in first-order predicate logic along with an empirical interpretation of its terms. One of the distinctive features of the syntactic view is the emphasis on the presence of some (small) set of universal, exceptionless laws as the core principles of the theory.[7] The second option has been termed the "semantic view," and treats scientific theories as families of models, where these models are indispensable to or constitutive of the meaning of the theory. Variations of the semantic view have been preferred for interpreting evolutionary theory, in part because they do not require the universal laws central to the syntactic view and perceived to be absent from biology (e.g., Lloyd 1988).[8]

And yet the semantic view of theories has several outstanding issues. One is related to the individuation of scientific theories (Suppe 2000). The semantic view does not indicate the boundaries of a scientific theory, and can be used to represent similar theoretical content in different ways.[9] This is complementary to the pluralist stance, but implies that any details about individuation will be purchased via a bottom-up strategy. Another difficulty for the semantic view is the specification of theoretical content involved in applying models (Morrison 2007). Again, this can be mitigated using a bottom-up strategy because the actual specification in scientific practice is close at hand. Finally, no top-down solution adequately accounts for the historical development and change exhibited by scientific theories. This lends further credence to a bottom-up strategy that concentrates on isolating structure for an evolutionary synthesis from goals, procedures, or problems, rather than from content. Therefore, approaching evolutionary theory using a philosophical methodology distinct from that found in traditional debates over theory structure not only generates a critical component for ongoing discussions about the formulation of an extended evolutionary synthesis, but also may uncover neglected aspects of scientific knowledge.

Broad versus Narrow Representations of Evolutionary Theory

One general way to produce distinct representations of evolutionary theory is to draw different boundaries around its content. For example, if we focus on evolutionary mechanisms (process) and exclude phylogenetic relationships (pattern), then diverse types of natural selection (mortality, fecundity, fertility) will be included, but the distinction between monophyly and paraphyly will not. Alternatively, if we include only those components that yield formal mathematical models, then the evolutionary stable strategies of game theory will be in, but patterns of *Hox* gene expression in arthropods, echinoderms, and chordates will be out. Depending on the principles used to draw the boundaries, there can be greater or lesser degrees of content inclusion. The structure of evolutionary theory in these cases will be correlated with the principles used for demarcation.

From a bottom-up approach, we can identify demarcating principles for different representations in the material elements of science. Textbooks are one of these material elements, and quite germane for isolating presentations of evolutionary theory. Although they clearly simplify actual practices and distort historical development (Kuhn 1996), they also must capture substantial consensus in order to be effective and widely adopted.[10] This consensus often includes predominant biases or norms, in addition to an acceptable coverage of concepts and empirical content (cf. Winther 2006: 478–479). The role of textbooks is to codify knowledge in an area of science and provide a mechanism for transferring it to novices. Stasis and change in textbooks with respect to thematic coverage, methodological emphases, or explanatory preferences can indicate deep knowledge commitments and organization precisely because many researchers vet them. Areas undergoing rapid change are often substantially transformed between editions (e.g., EvoDevo–related content), whereas some material is nearly identical across all textbooks.

One difference that can be identified in textbook presentations of evolutionary biology is narrow versus broad representations of evolutionary theory. A narrow representation focuses on models thought central to elucidating mechanisms of change in populations, especially those from evolutionary genetics along with game theory. While these incorporate diverse explanatory factors, natural selection is usually the most salient. A good example of this representation of evolutionary theory is Sean Rice's *Evolutionary Theory: Mathematical and Conceptual Foundations* (2004). Chapters focus on population genetics, quantitative genetics, and game theory. Rice aims to show that "evolutionary theory is not just a collection of separately constructed models, but is a unified subject in which all of the major results are related to a few basic biological and mathematical principles" (xiii). This includes the relationship between development and evolution (cf. Rice 1998), but many issues often considered part of evolutionary biology are missing: the history of life, ecology, systematics, and biogeography. A narrow representation is often presumed when other scientists or those external to the scientific community call into question the adequacy of evolutionary theory.[11] Treating material in the narrow representation as "foundational" or "fundamental" encourages this, suggesting that other elements are peripheral or ancillary to evolutionary theory.

Broad representations of evolutionary theory are more inclusive with respect to content, and embrace the full explanatory power of evolutionary biology that is associated with the idea of a "synthetic" theory.[12] They incorporate evolutionary genetics and hierarchical selection theory as well as systematics, biogeography, speciation, paleobiology, ecology, and comparative development, as seen in introductory textbooks. One textbook discusses differential gene expression in limb buds and the evolutionary origin of the vertebrate foot as an affair of paleontology, comparative anatomy, developmental biology, and molecular genetics, labeled together as "evo-devo" (Freeman 2002: 475–476). Other textbooks follow a nearly identical pattern (cf. Campbell and Reece 2002: 476–480). Notably, the broad representation often includes purportedly excluded content, such as development.[13] Patterns congruent with a broad representation of evolutionary theory are ubiquitous in textbooks aimed at different levels.

Most philosophers interested in evolutionary theory have focused on a narrow representation: "[Philosophers] seem to agree on one thing: if you can characterize formal population genetics, then you have characterized the 'guts' or 'core' of evolutionary theory" (Lloyd 1988: 8).[14] While many analyses incorporate more than population genetics, they primarily range over topics central to a narrow representation (e.g., Brandon 1981). This holds for advocates of both top-down options for theory structure: syntactic (Rosenberg 1985) and semantic (Lloyd 1988; Thompson 1989; Beatty 1981). Some recent work incisively examines quantitative genetic models (Pigliucci and Kaplan 2006), which have been largely neglected, but this still falls within a narrow representation. Topics such as systematics are often treated separately from evolutionary theory (e.g., Sober 1988). The consensus in favor of the semantic view accommodating evolutionary theory better than the syntactic view is predicated on a narrow representation. Appropriate structure for a broad representation is relatively unexplored territory.

The narrow representation is unlikely to be appropriate when searching for structure relevant to an Extended Synthesis. The Modern Synthesis included far more content than the narrow representation circumscribes (e.g., speciation, systematics, and paleontology), even if that content was primarily molded around population genetics with a definition of evolution in terms of allele frequency change. An evolutionary theory akin to the broad representation was the stated goal of Modern Synthesis proponents when institutionalizing their research outlook: "the Society for the Study of Evolution, the aim of which is integration of the contributions to this study made by comparative anatomists, ecologists, geneticists, paleontologists, physiologists, systematists, and others" (Dobzhansky 1949: 376).[15] Therefore, the narrow representation is not well suited for capturing the heterogeneous disciplinary contributors and topics usually associated with the Modern Synthesis, regardless of whether it needs expansion. The contours of this multidisciplinary synthesis in a broad representation of evolutionary theory will emerge as we explore the composition of textbook presentations.

Scrutinizing Broad Representation in Textbook Structure

One way to ascertain structure for evolutionary theory using the bottom-up strategy is through a comparative evaluation of evolution textbooks (tables 16.1–16.3). Can we discern shared components and relations for a broad representation? It is apposite to begin with the third edition of John Maynard Smith's *Theory of Evolution* (Maynard Smith 1993), which came out prior to a resurgence of interest in the importance of development for evolution. It begins with chapters on adaptation, the theory of natural selection, and heredity, but the method used to order the chapters is unclear (table 16.1). Although "Weismann, Lamarck, and the Central Dogma" follows "Heredity," "The Origin and Early Evolution of Life" is sandwiched between "Molecular Evolution" and "The Structure of Chromosomes and the Control of Gene Action." Chapters on variation, artificial selection, and natural selection in the wild are not juxtaposed with earlier discussions of adaptation, natural selection, and heredity. An overarching structure is not evident, but the book is not amenable to a narrow representation of evolutionary theory. The origin and early evolution of life, speciation, the fossil record, evolution and development, and the specifics of human evolution fit a broader representation. This pattern is comparable with those in contemporary textbooks (table 16.1), although several items are left out, such as systematics and biogeography. An allegiance to the outlook of the Modern Synthesis remains from the first edition's 1958 origins (evidenced in the "Further Reading" suggestions), but there are also comments that resonate with contemporary content disputes: "an understanding of evolution requires an understanding of development" (326). Although the chapter on development is predictably focused on heterochrony, it considers problems that are salient in EvoDevo research today, such as the origins of evolutionary novelty (Love 2003).

Table 16.1
A comparison of chapter headings in four different evolution textbooks

#	Maynard Smith 1978	Ridley 2004	Futuyma 2005	Barton et al. 2007
1	Adaptation	The Rise of Evolutionary Biology	Evolutionary Biology	The History of Evolutionary Biology: Evolution and Genetics
2	The Theory of Natural Selection	Molecular and Mendelian Genetics	The Tree of Life: Classification and Phylogeny	The Origin of Molecular Biology
3	Heredity	The Evidence for Evolution	Patterns of Evolution	Evidence for Evolution
4	Weismann, Lamarck, and the Central Dogma	Natural Selection and Variation	Evolution in the Fossil Record	The Origin of Life
5	Molecular Evolution	The Theory of Natural Selection	A History of Life on Earth	The Last Universal Common Ancestor and the Tree of Life
6	The Origin and Early Evolution of Life	Random Events in Population Genetics	The Geography of Evolution	Diversification of Bacteria and Archaea I: Phylogeny and Biology
7	The Structure of Chromosomes and the Control of Gene Action	Natural Selection and Random Drift in Molecular Evolution	The Evolution of Biodiversity	Diversification of Bacteria and Archaea II: Genetics and Genomics
8	Variation	Two-Locus and Multi-Locus Population Genetics	The Origin of Genetic Variation	The Origin and Diversification of Eukaryotes
9	Artificial Selection: Some Experiments with Fruitflies	Quantitative Genetics	Variation	Multicellularity and Development
10	Natural Selection in Wild Populations	Adaptive Explanation	Genetic Drift: Evolution at Random	Diversification of Plants and Animals
11	Protein Polymorphism	The Units of Selection	Natural Selection and Adaptation	Evolution of Developmental Programs
12	Altruism, Social Behaviour, and Sex	Adaptations in Sexual Reproduction	The Genetical Theory of Natural Selection	Generation of Variation by Mutation and Recombination

Table 16.1
(Continued)

#	Maynard Smith 1978	Ridley 2004	Futuyma 2005	Barton et al. 2007
13	What are Species?	Species Concepts and Intraspecific Variation	Evolution of Phenotypic Traits	Variation in DNA and Proteins
14	The Origins of Species	Speciation	Conflict and Cooperation	Variation in Genetically Complex Traits
15	What Keeps Species Distinct?	The Reconstruction of Phylogeny	Species	Random Genetic Drift
16	The Genetics of Species Differences	Classification and Evolution	Speciation	Population Structure
17	The Fossil Evidence	Evolutionary Biogeography	How to Be Fit: Reproductive Success	Selection on Variation
18	Evolution and Development	The History of Life	Coevolution: Evolving Interactions among Species	The Interaction between Selection and Other Forces
19	Evolution and History	Evolutionary Genomics	Evolution of Genes and Genomes	Measuring Selection
20	—	Evolutionary Developmental Biology	Evolution and Development	Phenotypic Evolution
21	—	Rates of Evolution	Macroevolution: Evolution above the Species Level	Conflict and Cooperation
22	—	Coevolution	Evolutionary Science, Creationism, and Society	Species and Speciation
23	—	Extinction and Radiation	—	Evolution of Genetic Systems
24	—	—	—	Evolution of Novelty

One of the most widely used textbooks is Douglas Futuyma's *Evolutionary Biology*/*Evolution* (Futuyma 1979, 1986, 1998, 2005). Although the chapter components are relatively stable across the different editions, the chapter order and structuring of Futuyma's textbook has changed over time (table 16.2). In addition to the treatment of variation and natural selection in the context of evolutionary genetics, other topics such as biodiversity, rates of evolution, the origin of novelties, species, speciation, and co-evolution receive distinct attention. The place where the potential significance of development (including saltation, allometry, heterochrony, and recapitulation) is routinely treated is the chapter on the origins of evolutionary novelty.[16] Sometimes the significance of development for evolutionary theory is considered more generally:

Like most scientific theories, evolutionary theory is incomplete in several respects, most conspicuously in that ... it lacks a sufficient body of principles for translating between genes and phenotypes. In other words, it requires a theory of developmental biology. (Futuyma 1998: 649)[17]

Subsequent editions increase the number of developmental considerations taken as relevant for evolution (e.g., morphogenesis, modularity, and pattern formation), and partition discussions of saltation into a distinct chapter. Eventually this leads to a unique feature in the fourth edition—two of the chapters have been subcontracted to other researchers: "Evolution of Genes and Genomes" (chap. 19; Scott Edwards) and "Evolution and Development" (chap. 20; John True). Although neither of these is a new area of content, which is remarkably stable across all four editions, it does indicate the way in which it is increasingly difficult or unlikely for one researcher to synthesize all of the material considered relevant for evolutionary theory.

In Futuyma's second edition chapter on the origin of novelties, a brief defense of the Modern Synthesis perspective reveals a commitment to a broad rather than a narrow representation of evolutionary theory:

The power of neo-Darwinism lies in its generality of explanation. But like most general theories, it is highly abstract. It gains full explanatory power when concepts such as gene frequencies and selection are given empirical content by applying them to real features of real organisms: behavior, life histories, breeding systems, physiology and morphology. When this is done, however, new questions appropriate to those particular features emerge and context-specific factors must be added to the theory. (Futuyma 1986: 440).[18]

This summary statement has three emphases that reinforce the utilization of a broad representation of evolutionary theory. First, there is a

Table 16.2
A comparison of four editions of Douglas Futuyma's textbook

#	Futuyma 1979	Futuyma 1986	Futuyma 1998	Futuyma 2005
I	*Introduction*	The Origin and Impact of Evolutionary Thought	*I. Background to the Study of Evolution*	Evolutionary Biology
1	The Origin and Impact of Evolutionary Thought		Evolutionary Biology	
II	*Prelude to the Study of Evolution*	The Ecological Context of Evolutionary Change	A Short History of Evolutionary Biology	The Tree of Life: Classification and Phylogeny
2	A Synopsis of Evolutionary Theory			
3	Heredity and Development	Heredity: Fidelity and Mutability	Genetics and Development	Patterns of Evolution
4	The Ecological Context of Evolutionary Change	Variation	Ecology: The Environmental Context of Evolutionary Change	Evolution in the Fossil Record
III	*Historical Evolution*	Population Structure and Genetic Drift	*II. Patterns and History*	A History of Life on Earth
5	The History of Biological Diversity		The Tree of Life: Classification and Phylogeny	
6	Spatial Patterns of Diversity	Effects of Natural Selection on Gene Frequencies	Evolving Lineages in the Fossil Record	The Geography of Evolution
7	Rates and Directions of Evolution	Selection on Polygenic Characters	A History of Life on Earth	The Evolution of Biodiversity
8	The Origin of Evolutionary Novelties	Speciation	The Geography of Evolution	The Origin of Genetic Variation
IV	*The Mechanisms of Evolution*	Adaptation	*III. Evolutionary Processes in Populations and Species*	Variation
9	Variation		Variation	
10	Genetic Variation and its Origins	Determining the History of Evolution	The Origin of Genetic Variation	Genetic Drift: Evolution at Random
11	Population Structure	The Fossil Record	Population Structure and Genetic Drift	Natural Selection and Adaptation
12	The Nature of Natural Selection	The History of Biological Diversity	Natural Selection and Adaptation	The Genetical Theory of Natural Selection

Table 16.2
(Continued)

#	Futuyma 1979	Futuyma 1986	Futuyma 1998	Futuyma 2005
13	The Effects of Selection on Gene Frequencies	Biogeography	The Theory of Natural Selection	Evolution of Phenotypic Traits
14	Selection on Polygenic Characters	The Origin of Evolutionary Novelties	Multiple Genes and Quantitative Traits	Conflict and Cooperation
15	Responses to Selection	Evolution at the Molecular Level	Species	Species
16	The Origins of Species	The Evolution of Interactions Among Species	Speciation	Speciation
17	Microevolution and Macroevolution	Human Evolution and Social Issues	*IV. Character Evolution* Form and Function	How to Be Fit: Reproductive Success
V	*Postlude: Special Topics in Evolution*	—	The Evolution of Interactions among Species	Coevolution: Evolving Interactions among Species
18	Coevolution			
19	Social Issues in Human Evolution	—	The Evolution of Life Histories	Evolution of Genes and Genomes*
20	—	—	The Evolution of Behavior	Evolution and Development*
21	—	—	The Evolution of Genetic Systems	Macroevolution: Evolution above the Species Level
22	—	—	Molecular Evolution	Evolutionary Science, Creationism, and Society
23	—	—	*V. Macroevolution: Evolution Above the Species Level* Development and evolution	—
24	—	—	Pattern and Process in Macroevolution	—
25	—	—	The Evolution of Biological Diversity	—
26	—	—	Human Evolution and Variation	—

*Not authored by Futuyma.

loss of explanatory power when only the apparatus of evolutionary genetics is available. Second, there are distinct sets of questions not directly addressed by evolutionary genetics. This bolsters the importance of attending to the different domains of questions that evolutionary theory purports to address. Third, full explanatory power is achieved only when these other components are included in evolutionary theory. Futuyma's textbook components and structure not only support a broad representation of evolutionary theory but also highlight its erotetic elements.

Another prominent textbook is Mark Ridley's *Evolution*, which has gone through three editions over the past fifteen years (Ridley 1993, 1996, 2004). Ridley consistently groups the different sections of his textbook into four parts: (1) "Evolutionary genetics"; (2) "Adaptation and Natural Selection"; (3) "Evolution and Diversity"; and (4) "Paleobiology and Macroevolution" (table 16.3). The rationale is offered explicitly in his first edition (and reappears with stylistic modification in later editions): "The theory of evolution . . . has four main components. Population genetics provides the fundamental theory of the subject. . . . The second component is the theory of adaptation . . . [Third] Evolution is the key to understanding the diversity of life. . . . Finally, evolution on the grand scale" (Ridley 1993: vii). Even though other textbooks did not always provide meta-structure, Ridley tackles nearly identical subject matter within the vein of a broad representation of evolutionary theory. Adaptation, natural selection, and evolutionary genetics corresponding to a narrow construal of content are supplemented with systematics and phylogeny, speciation, biogeography, history of life, origin of novelties, co-evolution, and extinction. The most variable parts of the organization concern genomics and EvoDevo, which account for the major additions and reorganization, with the last section, "Macroevolution," being expanded in length and depth of coverage (e.g., "Coevolution" gets its own chapter).

The final textbook to consider is a new multi-authored text, *Evolution* (Barton et al. 2007). Co-authorship (termed an "interdisciplinary approach") is a significant feature for a broad representation because each author brings a distinct specialty: evolutionary genetics, paleontology, evolutionary genomics of microorganisms, human genetic diversity and disease, and developmental evolution. The rationale for this textbook is to foreground the powerful molecular tools that have been applied to different facets of evolutionary biology. "Our aim . . . is to integrate these and other modern breakthroughs into a complete

Table 16.3
A comparison of three editions of Mark Ridley's textbook

#	Ridley 1993	Ridley 1996	#	Ridley 2004
I	*Introduction*	*Introduction*	*I*	*Introduction*
1	The Rise of Evolutionary Biology	The Rise of Evolutionary Biology	1	The Rise of Evolutionary Biology
2	Molecular and Mendelian Genetics	Molecular and Mendelian Genetics	2	Molecular and Mendelian Genetics
3	The Evidence for Evolution	The Evidence for Evolution	3	The Evidence for Evolution
4	Natural Selection and Variation	Natural Selection and Variation	4	Natural Selection and Variation
II	*Evolutionary Genetics*	*Evolutionary Genetics*	*II*	*Evolutionary Genetics*
5	The Theory of Natural Selection	The Theory of Natural Selection	5	The Theory of Natural Selection
6	Random Events in Population Genetics	Random Events in Population Genetics	6	Random Events in Population Genetics
7	Molecular Evolution and the Neutral Theory	Molecular Evolution and the Neutral Theory	7	Natural Selection and Random Drift in Molecular Evolution
8	Two-Locus and Multi-Locus Population Genetics	Two-Locus and Multi-Locus Population Genetics	8	Two-Locus and Multi-Locus Population Genetics
9	Quantitative Genetics	Quantitative Genetics	9	Quantitative Genetics
10	Genome Evolution	Genome Evolution		
III	*Adaptation and Natural Selection*	*Adaptation and Natural Selection*	*III*	*Adaptation and Natural Selection*
11	The Analysis of Adaptation	The Analysis of Adaptation	10	Adaptive Explanation
12	The Units of Selection	The Units of Selection	11	The Units of Selection
13	Adaptive Explanation	Adaptive Explanation	12	Adaptations in Sexual Reproduction
IV	*Evolution and Diversity*	*Evolution and Diversity*	*IV*	*Evolution and Diversity*
14	Classification and Evolution	Classification and Evolution	13	Species Concepts and Intraspecific Variation
15	The Idea of a Species	The Idea of a Species	14	Speciation
16	Speciation	Speciation	15	The Reconstruction of Phylogeny

Table 16.3
(Continued)

#	Ridley 1993	Ridley 1996	#	Ridley 2004
17	The Reconstruction of Phylogeny	The Reconstruction of Phylogeny	16	Classification and Evolution
18	Evolutionary Biogeography	Evolutionary Biogeography	17	Evolutionary Biogeography
V	*Paleobiology and Macroevolution*	*Paleobiology and Macroevolution*	V	*Macroevolution*
19	Rates of Evolution	The Fossil Record	18	The History of Life
20	Macroevolutionary Change	Rates of Evolution	19	Evolutionary Genomics
21	Macroevolutionary Trends, Coevolution, Species Selection	Macroevolutionary Change	20	Evolutionary Developmental Biology
22	Extinction and Mass Extinction	Coevolution	21	Rates of Evolution
	—	Extinction	22	Coevolution
	—	—	23	Extinction and Radiation
	Appendix: The Fossil Record	—		—

evolutionary perspective that forms the central theme of the book" (xi). The book is structured into four unequal parts: (1) "An Overview of Evolutionary Biology"; (2) "The Origin and Diversification of Life"; (3) "Evolutionary Processes"; and (4) "Human Evolution." Although the textbook strives to distinguish itself from others, the same core topics appear here as well, including variation, speciation, phylogeny, the evolution of novelty, and adaptation (table 16.1).

One unmistakable feature emerging from these textbooks, which all consistently opt for a broad representation, is that evolutionary theory is the product of multiple disciplinary contributors. Evolutionary theory is structured as a synthesis of different concepts, methods, and disciplines. Maynard Smith presciently described this in his 1958 preface: "Work in many different fields of biology is relevant. . . . Unhappily no one biologist can hope to be familiar with all these fields, and I am no exception to this rule. Yet it seemed worth while to attempt a book covering the whole subject of modern evolution theory" (Maynard Smith 1993: xvii).[19] This multidisciplinarity is what makes the term "synthesis" so applicable. Evolutionary theory needs to be a synthesis of disciplinary approaches in order to produce an integrated or cohesive body of scien-

tific knowledge.[20] It was an explicit goal of the Modern Synthesis architects, and is embodied in broad representations of evolutionary theory
to this day.

Other features are also apparent when comparing these textbooks.
Natural selection and adaptation are central thematic elements. This
includes, but is not limited to, the various aspects of evolutionary genetics. The narrow representation of evolutionary theory is a necessary
component of any broad representation. But there is also remarkable
consistency across the textbooks in terms of other recurring thematic
elements, such as systematics, history of life and paleontology, biogeography, speciation, and co-evolution. As expected, there is stasis (e.g.,
population genetics coverage) and change (e.g., evolution and development coverage). There are also persistent associations, such as the
relevance of development for understanding the origin of evolutionary
novelties. In the midst of these shared ingredients of a broad representation of evolutionary theory, there is little in the way of an explicit, overall
organization or structuring of these topics. Even though Ridley's meta-
structure is stable, he does not offer a characterization of how all the
components fit together. The consensus concerns the thematic coverage,
but not necessarily how that material is organized. This suggests that a
philosophical contribution about structure in a broad representation
may be applicable regardless of whether an Extended Synthesis is ultimately formulated.

An Erotetic Structure for Evolutionary Theory

Our analysis thus far has left us with two main results. First, there are
multiple stable components contained within a broad representation
of evolutionary theory. These include variation, adaptation, diversity,
heredity, novelty, classification/phylogeny, biogeography, co-evolution,
and speciation. A natural question, given the absence of shared structure
across the textbook presentations, is how to provide organization for
these components in a broad representation.[21] Second, these components are the province of a large and diverse group of life science
subdisciplines that invoke a variety of concepts and methods. This is
why the terminology of synthesis continues to be suitable; a broad
representation of evolutionary theory is a *synthesis* of contributions from
multiple disciplinary approaches aimed at producing an integrated body
of scientific knowledge about evolution. Multidisciplinarity was over-
looked by philosophical analyses centered on narrow representations

because only one or a few cognate disciplines were involved. But without some account of organization we have not yet isolated how development fits in, which was our desideratum on structure for an extended evolutionary synthesis.

Wallace Arthur has offered an interdisciplinary "wheel" of evolutionary theory that aims to indicate how developmentally oriented fields complement more standard areas of evolutionary biology (figure 16.1). There are several difficulties with this picture (cf. Love 2007). First, the wheel structure has a restrictive juxtaposition of disciplinary approaches. It allows different disciplines to be related directly to only two neighbors, which is implausible for the actual multidisciplinary explanations required in evolutionary studies. Second, the disciplinary structuring mixes methodology, relations, scale, and phenomena. This lack of commensurability makes the structure problematic. A better strategy is to recover the multidisciplinary aspects of evolutionary theory based on what needs explaining, that is, the complex phenomena that recur as topics in the textbook chapters: adaptation, variation, biogeography, co-evolution, and so on.

It has long been recognized that complex phenomena require multiple disciplinary contributions in order to provide satisfactory explanatory accounts. This was true even before the complete disciplinary professionalization of the sciences had occurred: "Complex problems ... can be solved only by the ultimate combination of several sciences that are at present cultivated in a wholly independent manner" (Comte 1988: 27). We now see this recognition as a demand for "interdisciplinary research" because complex phenomena elude straightforward scientific explanations from existing disciplines (Hansson 1999). Biologists recognize it as a situation they routinely face: "many features of nature are directly influenced by multiple mechanisms that fall into disparate scientific disciplines" (McPeek 2006: S1). This style of argument is also observable in discussions of evolutionary research, where a synthesis provides explanatory power unavailable to biological subdisciplines in isolation.[22] The process of evolution is a complex set of phenomena posing a diverse array of questions that requires dissection from many different angles simultaneously.

This returns us to the relatively stable topics identified in the survey of textbooks. Because these correspond to arrays of research questions about complex phenomena that are concurrently tackled by multiple disciplinary approaches, it is apt to label them "problem agendas" because they are "lists" of things to be done (Love 2008c). Explaining

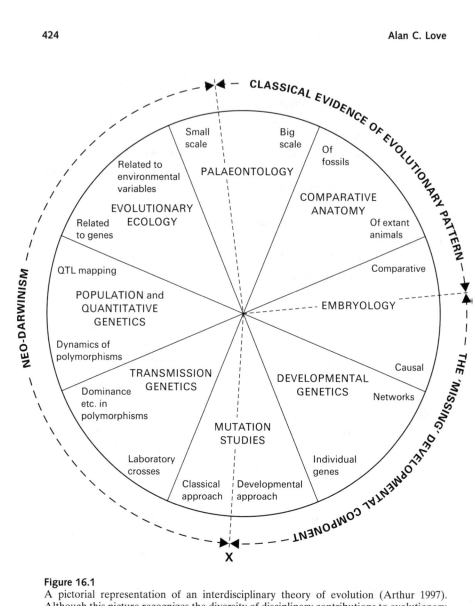

Figure 16.1
A pictorial representation of an interdisciplinary theory of evolution (Arthur 1997). Although this picture recognizes the diversity of disciplinary contributions to evolutionary theory, the wheel structure is restrictive in how it conceptualizes their relations. For example, each approach is directly related only to its two adjoining neighbors. It also mixes phenomena (individual genes vs. networks; fossils vs. extant animals), methodology (QTL mapping, laboratory crosses; classical approach; developmental approach; mutation studies), scale (small vs. big), relations (related to genes vs. related to environmental variables), and explanatory styles (comparative vs. causal). This restrictiveness and lack of commensurability between units makes the relations it represents problematic and unable to provide theoretical structure that explicates the required contributions of different disciplinary approaches in an extended evolutionary synthesis. (From Arthur 1997: 297, fig. 12–1. © 1997 by Cambridge University Press. Reprinted with the permission of Cambridge University Press)

them requires answering numerous empirical and conceptual questions. Problem agendas are exactly what disciplinary syntheses coalesce around. For example, a full comprehension of human cognition (minimally) requires contributions from anatomy, physiology, cognitive science, psychology, artificial intelligence, and neurobiology. As a unit of analysis, problem agendas allow for a common domain of inquiry with changing questions. Some questions associated with the problem agenda of adaptation or speciation have changed while there has remained a commitment to generating an overarching explanatory framework for "adaptation" or "speciation" phenomena. This feature of problem agendas provides for historical continuity and demonstrates how scientists can be tackling related issues despite having different theoretical commitments in disparate geographic places at distinct periods of history.

Additionally, problem agendas direct our attention to characterizing the criteria of explanatory adequacy, that is, what needs to be done. Implicit or explicit agreement about the acceptability of answers is what fuels the demand for distinct disciplinary contributions to explanations. Criteria of explanatory adequacy are based on the questions researchers think need to be addressed concerning the phenomena under scrutiny in a particular problem agenda. They can be identified by abstraction through an analysis of descriptions found in ongoing research (Love 2008a). Once made explicit, they can play the role of indicating how different disciplinary contributions are required (Love 2006a; 2008c). Identification and articulation of these criteria does not imply possession of an explanation or consensus about an explanation currently on offer. Two explanations may meet criteria of explanatory adequacy but not be empirically supported to the same degree.[23]

In previous work I have characterized the problem agenda of evolutionary innovation and novelty, following the emphasis placed on it by EvoDevo practitioners (Love 2003, 2006a, 2008c). It is common to see the following descriptions in textbook discussions of novelty origins: "Paleontologists, comparative anatomists, developmental biologists, and molecular geneticists are all contributing data aimed at clarifying the genetic basis for novel structures like heads, tails, and limbs" (Freeman 2002: 473). The difficulty is that there is no discussion of exactly how all these contributions are made or how they fit together. By examining the discourse about evolutionary innovation and novelty (e.g., chapters 10, 11, 12, and 15 in this volume), we can summarize the problem agenda as concerned with the origin of qualitatively distinct variation at particular phylogenetic junctures, which is separate from comprehending the relative fit between organism and environment and its transgenerational implications, that is, the problem agenda of adaptation (Love 2006a; 2008c). The criteria of explanatory adequacy can be summarized via three abstract concerns: form versus function features; the level of biological hierarchy in view; and degree of generalization (Love 2006a; 2008c). When they are combined with further details, we arrive at a better understanding of what kinds of disciplinary approaches are needed to sufficiently address these criteria and thereby offer an adequate explanation (figure 16.2).

Using erotetic units such as problem agendas has several advantages when contrasted with those commonly used in philosophy of science (paradigms, research programs, or models).[24] First, erotetic units highlight the goals of research, those things that scientists are aiming to explain. This is part of the reason why they are more stable than theo-

Explanation Component	PATTERN	PROCESS	Level of Organization
	Non-homology Phylogenetic Juncture	Mechanistic Account of Variation	
Disciplinary Approaches	Paleontology/Morphology Comparative Development Systematics (Morphological)	Behavioral Biology Quantitative Genetics Functional Morphology	HIGHER
			STRUCTURES PARTS ORGANS TISSUES
		Epigenetics Phenotypic Plasticity	
			LOWER
	Systematics (Molecular) Developmental Genetics	Quantitative Genetics Developmental Genetics Molecular Biology/Genetics	CELL ORGANELLE GENETIC BIOCHEMICAL

Figure 16.2
A schematic representation of different disciplinary contributions required to explain evolutionary novelties (Love 2008c). Once criteria of explanatory adequacy have been explicitly identified for a problem agenda (see Love 2006a; 2008c), we can specify how different biological disciplines make their explanatory contribution. This schematic pertains to the problem agenda of innovation and novelty, but does not capture all dimensions of the criteria of adequacy because it represents only form features (novelties). Another aspect of the problem agenda is the origin of novel functions (innovations). (For further discussion, see Love 2006a; 2008c.) (From Love 2008c: 882, fig. 1. © 2008 by the Philosophy of Science Association)

retical content meant to address them, which can help in comprehending the historical development of theories (cf. Kitcher 1993; Laudan 1977).[25] Second, erotetic units can clarify particular methodological choices and the standards adopted to judge the adequacy of explanations. They draw attention to criteria of explanatory adequacy, especially the need to make them explicit. Third, erotetic units are useful for illuminating disciplinary interactions. Paying attention to the research questions that are salient to different biological disciplines can be especially instructive, as opposed to asking about models or paradigms.

This leaves us with an interpretation of the components of a broad representation of evolutionary theory as problem agendas, as well as an explanation of why they would demand multidisciplinary explanation strategies. The final question concerns how these problem agendas are related to one another or organized. How do we structure the erotetic units? The problem agendas of evolutionary theory exhibit structural relationships with one another by way of *constraining principles*.[26] These principles "constrain" by limiting conceptual possibilities although they are often not controversial. For example, there is widespread agreement that three conditions are necessary for natural selection to operate: variation exists, it positively contributes to fitness, and is heritable (Lewontin 1970). Debate focuses on the degree to which they require supplementation in order to be sufficient. This constraint can operate organizationally because the problem agenda of adaptation (Why did natural selection preserve trait variation?) requires addressing three other problem agendas: (1) variation (Why did this variation exist?); (2) fitness (Why did it positively contribute to survival and/or reproduction?); (3) inheritance (Why was this variation heritable?). A similar procedure can be pursued for other relations. The problem agenda of co-evolution requires addressing adaptation, fitness, and biogeography, whereas explaining diversity demands attention to classification/phylogeny, the history of life, adaptation, and speciation. The problem agenda of innovation and novelty is related to variation, classification/phylogeny, and form/function (Love 2006a, 2008c; cf. Müller and Newman 2005).

These and other organizational relations can be represented in an idealized visualization of the erotetic structure of evolutionary theory (figure 16.3a). The structural relations can be overlain with major topical divisions often identified for biological science (genetics, cell, and developmental biology; ecology; systematics), which makes perspicuous why particular problem agendas might be more or less associated with disciplinary approaches falling within these divisions (figure 16.3b). It also can explain why mistaking one or more problem agendas for the whole of evolutionary theory, or collapsing one problem agenda into another, can lead to a devaluing of development, systematics, or ecology in formulating an account of evolutionary processes. Thus, we have achieved our goal because this provides us with structure that illuminates two places for EvoDevo in an extended synthesis: (1) with respect to a subset of problem agendas, such as evolvability and innovation/novelty, and (2) with respect to meeting particular criteria of adequacy for specific problem agendas (figure 16.2).

An erotetic structure for evolutionary theory shares one key feature with earlier philosophical discussions.[27] Beckner (1959), Caplan (1978), Ruse (1973), Tuomi (1981), and Wasserman (1981) all recognized that evolutionary theory was in some sense composed of "subunits"—a reticulate model. But none of them understood the subunits erotetically (e.g., Ruse treats them as disciplines; figure 16.4a), nor did they include all of the problem agendas recovered here from textbook presentations.[28] They also construed the reticulation hierarchically, with natural selection theory (e.g., population genetics) unifying the subunits (e.g., "metatheory" of Tuomi 1981 or "hypertheory" of Wasserman 1981; see figures 16.4a and 16.4b).[29] The erotetic structure offered here is not necessarily hierarchical, though it can be with the use of further constraining principles. This allows for the variability and complexity of relations among these diverse problem agendas. Additionally, it is clear why multiple disciplinary perspectives are needed within each of the problem agenda subunits of evolutionary theory. Once their criteria of explanatory adequacy are made explicit, we have a picture of how different disciplines synthesize to produce integrated explanatory frameworks. Finally, unlike other accounts, my pluralist stance makes room for the inclusion or exclusion of different subunits to occur as a function of different aims governing a chosen representation of structure in evolutionary theory.

The conception offered here is not an immutable conclusion, but rather a revisable vision of how an erotetic structure of evolutionary theory works in principle. Clearly, many missing links remain. We require an account of each of the problem agendas, an identification of their criteria of explanatory adequacy, a delineation of the required disciplinary contributions, and a more robust description of how the constraining principles generate organizational relations among the problem agendas. The relations among problem agendas are subject to debate among biologists. Some of this debate will be the controversy over content,

(a)

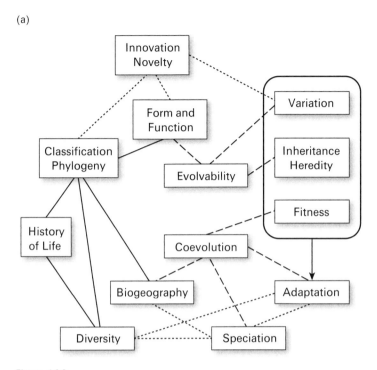

Figure 16.3
(a) Erotetic structure for evolutionary theory. This is an idealized picture of the problem agendas that compose evolutionary theory from a broad representation perspective, along with their relations to each other. The solid line around variation, inheritance–heredity, and fitness indicates their combined connection to the problem agenda of adaptation. The square-dotted lines at the top link the problem agenda of innovation and novelty with variation, form–function, and classification–systematics. The problem agenda of diversity is connected by solid lines to the history of life and classification–systematics, as well as to adaptation and speciation with circular dotted lines. Co-evolution is linked with fitness, adaptation, and biogeography by broken lines. Other relations are also indicated (see discussion in text). (b) Erotetic structure and the major areas of biological knowledge. The same idealized picture of the problem agendas that compose evolutionary theory from a broad representation perspective, along with their relations to one another, but overlaid with three major areas of biological knowledge: (1) systematics, (2) genetics, cell, and development biology, and (3) ecology. This illustrates the association of particular problem agendas with these major areas and reveals why an inordinate focus on some subset of problem agendas might lead to a devaluing of the importance of development, systematics, or ecology (see discussion in text).

(b)

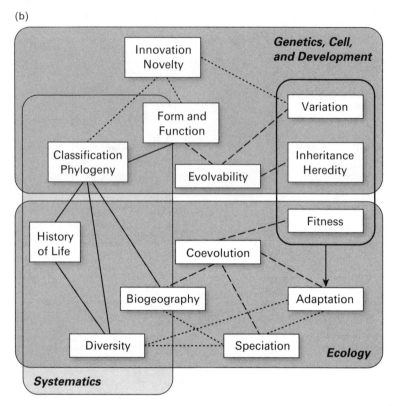

Figure 16.3
(Continued)

EvoDevo and otherwise, but now there is a new framework in which this debate can occur. First, evolutionary theory (on the broad interpretation relevant to "synthesis") is composed of multiple, distinct problem agendas with various disciplinary contributors. This highlights questions about whether a problem agenda is genuinely distinct or whether we have correctly characterized its criteria of explanatory adequacy, thereby delineating the required contributions of different disciplinary approaches. Second, these problem agendas are organized or exhibit relationships with one another (i.e., structure) by way of constraining principles, such as the necessary conditions for the operation of natural selection. Whether specific relations exist, how they are fleshed out with different principles, or how criteria of explanatory adequacy are articulated can be, and is, part of ongoing discussion. The need for further detailed elucidation of this perspective also indicates a philosophical task

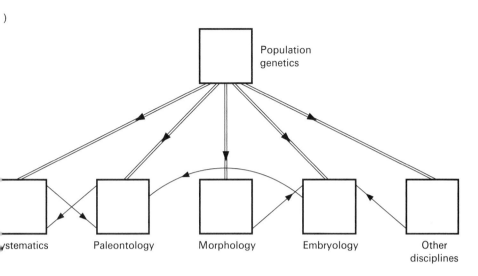

Figure 16.4
(Top) Ruse's representation of the structure of evolutionary theory (Ruse 1973). Rectangles represent biological disciplines. Double lines show the actual hierarchical relationship of population genetics as being fundamental to all other disciplines: "All the different disciplines are unified in that they presuppose a background knowledge of genetics, particularly population genetics" (Ruse 1973: 48). Single lines pick out possible links among the "subsidiary" disciplines ("many of the disciplines borrow from other disciplines"), but Ruse stresses that those represented in the diagram are purely illustrative (even though links such as these are presumed to exist). (Redrawn from Ruse 1973: 49, fig. 4.1. © 1973 by Hutchinson University Library. Reproduced with the permission of Taylor & Francis Books, UK) (Bottom) Tuomi's dynamic multilevel structure of evolutionary theory (Tuomi 1981). Panel (A) represents the hierarchical aspect of structure, and panel (B) represents the reticulate structure. Both in combination yield the overarching structure of evolutionary theory in Tuomi's dynamic multilevel account. Solid lines pick out deductive inferences, and dotted lines refer to inductive inferences. T_g = metatheory; T_s = specific theory; M = theoretical model; s = specific ancillary assumptions; ST = sub-theory; E = empirical level. Tuomi understands T_g to be the theory of natural selection for evolutionary theory. The nature of the subtheories is unspecified, and the relations among them are wholly abstract in this picture. They have no concrete instantiation with respect to the standard topics of evolutionary biology observed in textbooks. (Redrawn from Tuomi 1981: 27, fig. 1. © 1981 by the Society for Systematic Biology. Reproduced with the permission of Juha Tuomi, the Society for Systematic Biology, and Oxford University Press)

associated with the prospect of formulating an Extended Synthesis or assessing whether one is needed. Philosophers must shuttle between the empirical content of evolutionary theory and abstract considerations of theory structure, always with an eye to specific goals of inquiry (Love 2008a). Here the motivation was to achieve a better understanding of the place of EvoDevo in evolutionary theorizing and its consequences for an Extended Synthesis, but it also has an impact on our understanding of scientific knowledge.

(b)

A

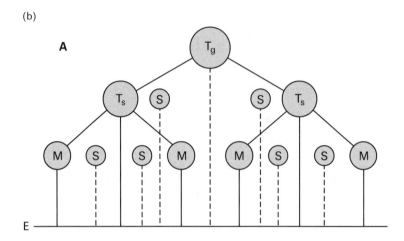

B

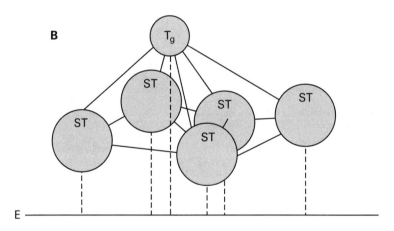

Figure 16.4
(Continued)

The physiologist Claude Bernard remarked that "every science is characterized by the nature of its problems and by the variety of phenomena that it studies" (Bernard 1957: 14). I have argued that a broad representation of evolutionary theory is usefully characterized in terms of the composition and organization of its multiple problem agendas, which encompass the biological phenomena relevant to comprehending evolutionary patterns and processes. The compositional element provides for how the content is circumscribed and the organization among the problem agendas is secured using constraining principles. When the criteria of explanatory adequacy are made explicit for each problem

agenda, the necessary contributions to the synthesis from diverse biological disciplines are more fully specified.

Quo Vadis? Extended Syntheses, Unification, and Coherence

A pluralist stance is an appropriate posture for the question of whether
we need an extended evolutionary synthesis. It directs our attention to
the reasons why we want to secure a revised synthesis. A focus on structure rather than content alongside the goal of understanding how the
contributions from EvoDevo and other fields articulate with our theoretical framework for evolution both respects the current controversy
and provides epistemic resources for a potential resolution. A bottom-up
methodology keeps the details of evolutionary theory unambiguously in
view. The choice of a broad representation resonates with textbook
treatments and the goal of evolutionary theorists, now and then, to synthesize different biological disciplines to produce an integrated body of
knowledge. An erotetic structure for evolutionary theory in terms of an
organization of problem agendas with diverse disciplinary contributors
provides a standpoint for probing the prospects and possibilities of an
Extended Synthesis. It also facilitates ongoing research, generates a
systematic conception of our knowledge of evolution, illuminates the
historical development of evolutionary theory, and yields resources for
responding to various challenges.[30]

An erotetic structure is also responsive to the concern that the Modern
Synthesis was in some sense too reductionist (Gould 2002; Woese 2004;
see chapter 17 in this volume). If evolutionary theory is composed of
multiple problem agendas that require contributions from diverse disciplinary perspectives, there is no "fundamental" viewpoint or level to
which we can reduce our picture of the evolutionary process. This holds
within each individual problem agenda as well (Love 2008c). My account
also meshes with the recognition that a fully unified view of evolutionary
processes may be out of reach, even though we seek integrated explanations of phenomena in different domains (Mitchell and Dietrich 2006).
The proliferation and heterogeneity of life science disciplines and methodologies following the advent of molecular biology has led to a cen-
trifugal force within evolutionary research, making it difficult to recover
a single big-picture or "grand unified theory." This plurality of biological
disciplines, offering complementary and competing contributions to
multidisciplinary explanations of phenomena related to problem agendas,
is suitably matched with a pluralism about structure in evolutionary

theory, which emphasizes a diversity of explanatory aims in biology through the utilization of erotetic units of analysis.

The nineteenth-century philosopher of science William Whewell argued that a scientific theory should not only successfully predict new phenomena and explain different kinds of cases than were used in formulating the theory ("consilience"), but also exhibit "coherence" or consilience *over time* (Snyder 2008). The coherence of a scientific theory is a function of whether substantial modifications to the theoretical framework, such as the introduction of new concepts or principles, are required when new types of phenomena are encountered. Arguably, one stimulus for an extended evolutionary synthesis stems from a perceived lack of coherence in the midst of new empirical findings, methodologies, concepts, and disciplinary approaches in evolutionary biology. The philosophical analysis offered here regarding the structure of evolutionary theory assists attempts to recover coherence through the vehicle of an Extended Synthesis. There is much work to be done, with respect to both structure and content, if we are to capture the breadth and depth of our knowledge about evolution while simultaneously yielding a coherent framework for prosecuting future research. The Modern Synthesis was both a process and a product, and the same is true today as we countenance various formulas of expansion and revision. We forget at our own peril that the work is never really finished, the process is ongoing, and the product is dynamically changing—theoretical coherence waxes and wanes. This is not a unique aspect of evolutionary research, as it can be observed in other scientific theories, but our vigorous attempts to address present manifestations of this dynamic change may blind us to its perpetuity.

Acknowledgments

I am grateful to Gerd Müller and Massimo Pigliucci for the invitation to contribute to this volume. In addition to the useful comments and suggestions of workshop participants ("Towards an Extended Evolutionary Synthesis," July 2008), I also received feedback from the Department of Ecology and Evolutionary Biology at SUNY–Stony Brook, where an early version of this material was presented in their seminar series (March 2008). Ricardo Azevedo, Mark Borrello, Ingo Brigandt, Gerd Müller, and Massimo Pigliucci provided numerous critical suggestions and insightful feedback on an earlier version of this chapter.

Notes

1. Whether this plurality is resolvable to a single comprehensive representation at some future point or reflects a perennial state of affairs is an open question (see Kellert et al. 2006). These representations can also have histories or be "conceptual lineages" (*sensu* Hull 1988).

2. Most of the essays in this volume exemplify this strategy for different facets of content. Even though I will speak of "content" somewhat monolithically, theory content is diverse, and includes empirical findings, dynamical models, and key concepts, among other items.

3. "The central core of Darwinian logic [remains] sufficiently intact to maintain continuity as the centerpiece of the entire field, but with enough important changes (to all major branches extending from this core) to alter the structure of evolutionary theory into something truly different by expansion, addition, and redefinition" (Gould 2002: 6).

4. Two other examples are worth mentioning explicitly. Mary Jane West-Eberhard has defended the need to include phenotypic plasticity in the content of evolutionary theory: "How does developmental plasticity fit within a genetic theory of evolution? This question remains largely unanswered . . . The result is a theory of evolution fraught with contradictions" (West-Eberhard 2003: 3). Marc Kirschner and John Gerhart have articulated a viewpoint labeled "facilitated phenotypic variation" that is meant to solve "Darwin's dilemma" of being unable to explain variation, and thus the origin of novelty: "The Modern Synthesis of 1940 was not so much wrong as it was incomplete . . . On the question of how the altered genotype caused an altered phenotype, the Modern Synthesis was silent" (Kirschner and Gerhart 2005: 29; cf. chapter 10 in this volume). Both of these books were negatively reviewed in major journals (Erwin 2005; De Jong and Crozier 2003).

5. Note that these controversies include substantial agreements that might be questioned. For example, Carroll holds that "everyone knew that genes must be at the center of the mysteries of both development and evolution" (SB Carroll, 2005: 8). Coyne does not necessarily disagree. But this claim might be challenged on separate grounds from the question of whether we need an extended evolutionary synthesis (see Love 2006b, 2008b).

6. Other "content" controversies can serve a similar purpose, such as rates of evolution or microbial evolution (RL Carroll 2000; Kutschera and Niklas 2004; Woese 2004).

7. "A theory is a relatively small body of general laws that work together to explain a large number of empirical generalizations, often by describing an underlying mechanism common to them all" (Rosenberg 1985: 121).

8. The "lawlessness" (and thus alleged disreputability) of evolutionary biology has stimulated many analyses of the structure of evolutionary theory (Beatty 1981; Caplan 1978; Tuomi 1981; Wasserman 1981).

9. Lisa Lloyd offered the following criterion for a philosophical reconstruction of evolutionary theory: "an adequate account of the structure of evolutionary theory will be capable of handling the range of theories already included under the name 'evolutionary theory'"(Lloyd 1988: 8). The pluralist stance implies that the range of theories included in evolutionary theory is not fixed, but can vary in kind or number, depending on our goals of analysis.

10. Textbooks at the level of secondary education often err on the side of inclusiveness. Higher-level textbooks (e.g., for graduate students) cannot simply include everything because they treat the subject matter with a higher degree of sophistication, often with an eye to being used in actual investigation.

11. E.g., "The origination of morphological structures, body plans, and forms should be regarded as a problem distinct from that of the variation and diversification of such entities (the central theme of current neo-Darwinian theory)" (Müller and Newman 2003: preface). With respect to external critics, I have in mind intelligent design proponents or others not participating in ongoing research. Sahotra Sarkar describes how different "narrow" representations are labeled and then used as rhetorical devices for criticizing evolutionary theory (Sarkar 2007: chap. 4).

12. Recovering this more robust explanatory power comes at a cost. It seemingly violates the distinction between pattern and process. If evolutionary theory is understood only in terms of mechanisms of change, then the inclusion of features such as classification will seem odd. At the same time, including only population-level mechanisms leaves the so-called pattern elements orphaned. If phylogenetic relationships between clades or biogeographical patterns of vicariance are not a part of evolutionary theory, then where do these inferences belong? Notice that the pattern/process distinction does not remove the controversy surrounding an extended evolutionary synthesis. For example, since variation is a critical component of the mechanism of natural selection, we need to comprehend the developmental production of variation in order to have an adequate evolutionary theory.

13. This does not mean the content is coherently integrated into a single comprehensive theoretical picture. There is also an issue about whether the material is actually taught, since textbooks are often selectively utilized.

14. Explicit justifications for adopting this narrow representation have been offered. Robert Brandon focused on the theory of natural selection alone to eliminate potential confusion: "The term 'evolutionary theory' is often used as a generic term standing for a whole cluster of interrelated theories in evolutionary biology, including theories of speciation, population genetics, evolutionary ecology and others" (Brandon 1981: 427). Michael Ruse recognized that systematics, morphology, embryology, and paleontology were part of evolutionary biology but argued that "all the different disciplines are unified in that they presuppose a background knowledge of genetics, particularly population genetics" (Ruse 1973: 48). Some have worried that too much attention has been given to population genetics: "many biologists and nearly all philosophers have overinvested in population genetics . . . it has led biologists and philosophers alike to misrepresent the structure and aim of evolutionary explanations" (Glymour 2006: 388).

15. George Simpson offered a similar perspective: "The basic problems of evolution are so broad that they cannot hopefully be attacked from the point of view of a single scientific discipline. Synthesis has become both more necessary and more difficult as evolutionary studies have become more diffuse and more specialized" (Simpson 1944: xv).

16. The third and fourth editions segregate the discussion of novelty to chapters on macroevolution but still treat various developmental phenomena as potentially relevant.

17. Some comments indicate a territoriality about integrating development with evolution: "Geneticists and developmental biologists are fully occupied with the enormously difficult problems that are their proper province; the evolutionary aspects of development must be studied by evolutionary biologists who understand and apply the techniques of molecular and developmental biology" (Futuyma 1986: 425).

18. "[T]he neo-Darwinian theory that emerged from the Modern Synthesis has been variously criticized as incomplete, inadequate as an explanation of evolution or just plain wrong. . . . Neo-Darwinian theory might well be incomplete. . . . It is true that classical neo-Darwinism emphasized some modes of explanation, levels of analysis, and questions to which inquiry was directed, at the expense of others. Evolutionary theory is currently being expanded . . . in all these respects . . . there is now a greater emphasis on higher levels of biological organization such as development and the constraints imposed by phylogenetic history. The importance of development and historical contingency has always been recognized, but until recently it has been the kind of formal, polite, recognition accorded to a stranger at an otherwise intimate party" (Futuyma 1986: 440).

19. The 1975 preface reiterates this: "My aim is still the same. I want to cover all important aspects of the theory of evolution" (Maynard Smith 1993: xxi).

20. "Synthesis denotes a blending of one or more parts to produce a new entity where the individuality of the original parts is not dissolved, though potentially transformed" (Love 2003: 312).

21. These components are sometimes characterized as theories (e.g., a theory of speciation) or disciplines (e.g., systematics or phylogeny), but neither is helpful in understanding

how multiple disciplines come together within evolutionary theory or why the components remain stable despite changes in theory (e.g., whether allopatry or sympatry is the preferred account of speciation).

22. "A unique combination of disciplines is emerging ... which focuses on the genes that affect ecological success and evolutionary fitness in natural environments and populations. Already this approach has provided new insights that were not available from its disciplinary components in isolation" (Feder and Mitchell-Olds 2003: 649). "An emerging synthesis of evolutionary biology and experimental molecular biology is providing much stronger and deeper inferences about the dynamics and mechanisms of evolution than were possible in the past. ... [It] combines the techniques of evolutionary and phylogenetic analysis with those of molecular biology, biochemistry and structural biology" (Dean and Thornton 2007: 675).

23. This pattern of reasoning is observable in Darwin's *On the Origin of Species*, where he claimed that an adequate theory of descent with modification needed to address "mutual affinities of organic beings," "embryological relations," "geographical distribution," "geological succession," and the exquisite nature of adaptation and "coadaptation" (Darwin 1964).

24. Erotetic units have received some attention from philosophers (e.g., Laudan 1977; Nickles 1981), though much less than other epistemic units and for reasons that go beyond the scope of the present chapter.

25. Problem agendas are distinct from other erotetic units such as individual research questions, which is pertinent to how stable they are across time. Individual questions, such as "How does the expression of *Dlx* genes make a causal difference in the origin of vertebrate jaws?," are part of the problem agenda of innovation and novelty, but they do not exhaust it. Problem agendas contain multiple, interrelated questions (both empirical and conceptual) that are not answerable by a few experiments or methods (e.g., "What causal interactions are needed to explain the origin of morphological structures?"), and therefore tend to exhibit more stability.

26. "Relations" are understood as material inferences, that is, inferential relations based on the specific subject matter in view and the concepts that represent this content. Material inferences are not licensed by their form (as is the case for deductive arguments), and include inductive, explanatory, and investigative modes of scientific reasoning. Thus the erotetic structure of evolutionary theory preserves the interrelation of content and structure.

27. My account is qualitatively different from Philip Kitcher's (1993). Kitcher argues that theories are answers to "significant" questions, where the answers conform to general patterns (abstract, explanatory schemata). This places the stress on the form of the answers rather than the structure of the questions. The latter is central to my erotetic account, where there is no requirement that multiple disciplinary contributions to problem agenda questions instantiate a set of abstract answer schemes.

28. Tuomi understood theoretical subunits to have associated problems, but did not equate the two: "each branch emphasizes the specific problems of that part of biological reality which is the primary object of study" (Tuomi 1981: 24). Caplan saw a role for problems in unifying the theoretical subunits but did not conceptualize these as problems: "the set of evolutionary disciplines and theories which constitute the synthetic theory are [sic] unified by their common concern with a particular set of biological phenomena. ... It is the commonness of problems and interests that lends theoretical unity and systematic power to the modern synthetic theory" (Caplan 1978: 274).

29. In some cases this was connected to the goal of axiomatizing evolutionary theory from the perspective of a syntactic view of theories (Ruse 1973: chap. 4).

30. For example, an erotetic structure leads to a novel way of interpreting the claim that "nothing makes sense in biology except in the light of evolution" (Dobzhansky 1973). Biological problem agendas take on their organizational relations to one another in the context of evolutionary theory, and we are able to contextualize the major topics of biology within an explicitly "evolutionary" organization.

References

Barton NH, Briggs DEG, Eisen JA, Goldstein DB, Patel NH (2007) Evolution. Cold Spring Harbor, NY: Cold Spring Harbor Laboratory Press.

Beatty J (1981) What's wrong with the received view of evolutionary theory? In PSA 1980: Vol. 2. PD Asquith, RN Giere eds.: 397–426. East Lansing, MI: Philosophy of Science Association.

Beatty, J (1986) The synthesis and the synthetic theory. In Integrating Scientific Disciplines. W Bechtel ed.: 125–135. Dordrecht: Martinus Nijhoff.

Beckner M (1959) The Biological Way of Thought. New York: Columbia University Press.

Bernard C (1957 [1865]) An Introduction to the Study of Experimental Medicine. HC Greene, trans. New York: Dover.

Brandon RN (1981) A structural description of evolutionary theory. In PSA 1980: Vol. 2. PD Asquith, RN Giere eds.: 427–439. East Lansing, MI: Philosophy of Science Association.

Campbell NA, Reece JB (2002) Biology. 6th ed. San Francisco: Benjamin Cummings.

Caplan AL (1978) Testability, disreputability, and the structure of the modern synthetic theory of evolution. Erkenntnis 13: 261–278.

Carroll RL (2000) Towards a new evolutionary synthesis. Trends in Ecology and Evolution 15: 27–32.

Carroll SB (2005) Endless Forms Most Beautiful: The New Science of Evodevo and the Making of the Animal Kingdom. New York: W.W. Norton.

Comte A (1988 [1830]) Introduction to Positive Philosophy. F Ferré ed. Indianapolis, IN: Hackett.

Coyne JA (2005) Switching on evolution: How does evo-devo explain the huge diversity of life on Earth? Nature 435: 1029–1030.

Darden L (1986) Relations among fields in the evolutionary synthesis. In Integrating Scientific Disciplines. W Bechtel ed.: 113–123. Dordrecht: Martinus Nijhoff.

Darwin C (1964 [1859]) On the Origin of Species. Cambridge, MA: Harvard University Press.

De Jong G, Crozier RH (2003) A flexible theory of evolution. Nature 424: 16–17.

Dean AM, Thornton JW (2007) Mechanistic approaches to the study of evolution: The functional synthesis. Nature Reviews Genetics 8: 675–688.

Dobzhansky T (1949) Towards a modern synthesis. Evolution 3: 376–377.

Dobzhansky T (1973) Nothing in biology makes sense except in the light of evolution. American Biology Teacher 35: 125–129.

Erwin DH (2005) A variable look at evolution. Cell 123: 177–179.

Feder ME, Mitchell-Olds T (2003) Evolutionary and ecological functional genomics. Nature Reviews Genetics 4: 649–655.

Freeman S (2002) Biological Science. Upper Saddle River, NJ: Prentice Hall.

Futuyma DJ (1979) Evolutionary Biology. Sunderland, MA: Sinauer.

Futuyma DJ (1986) Evolutionary Biology. 2nd ed. Sunderland, MA: Sinauer.

Futuyma DJ (1998) Evolutionary Biology. 3rd ed. Sunderland, MA: Sinauer.

Futuyma DJ (2005) Evolution. 4th ed. Sunderland, MA: Sinauer.

Gayon J (1990) Critics and criticisms of the modern synthesis: The viewpoint of a philosopher. In Evolutionary Biology: Vol. 24. MK Hecht, B Wallace, RJ Macintyre eds.: 1–49. New York and London: Plenum Press.

Glymour B (2006) Wayward modeling: Population genetics and natural selection. Philosophy of Science 73: 369–389.

Gould SJ (2002) The Structure of Evolutionary Theory. Cambridge, MA: Belknap Press of Harvard University Press.

Griesemer J (1984) Presentations and the status of theories. In PSA 1984: Vol. 1. PD Asquith ed.: 102–114. East Lansing, MI: Philosophy of Science Association.

Hansson B (1999) Interdisciplinarity: For what purpose? Policy Sciences 32: 339–343.

Hoekstra HE, Coyne JA (2007) The locus of evolution: Evo-devo and the genetics of adaptation. Evolution 61: 995–1016.

Hull D (1976) Informal aspects of theory reduction. In PSA 1974: Vol. 2. RS Cohen ed.: 653–670. Dordrecht: Reidel.

Hull D (1988) Science as a Process: An Evolutionary Account of the Social and Conceptual Development of Science. Chicago: University of Chicago Press.

Kellert SH, Longino HE, Waters CK (2006) Introduction: The pluralist stance. In Scientific Pluralism. SH Kellert, HE Longino, CK Waters eds.: vii–xxix. Minneapolis: University of Minnesota Press.

Kirschner MW, Gerhart JC (2005) The Plausibility of Life: Resolving Darwin's Dilemma. New Haven, CT : Yale University Press.

Kitcher P (1993) The Advancement of Science: Science Without Legend, Objectivity Without Illusions. New York: Oxford University Press.

Kuhn TS (1996 [1962/1970]) The Structure of Scientific Revolutions. 3rd ed. Chicago: University of Chicago Press.

Kutschera U, Niklas KJ (2004) The modern theory of biological evolution: An expanded synthesis. Naturwissenschaften 91: 255–276.

Laudan L (1977) Progress and Its Problems: Towards a Theory of Scientific Growth. Berkeley: University of California Press.

Lewontin R (1970) The units of selection. Annual Review of Ecology and Systematics 1: 1–14.

Lloyd EA (1988) The Structure and Confirmation of Evolutionary Theory. Westport, CT: Greenwood Press.

Love AC (2003) Evolutionary morphology, innovation, and the synthesis of evolutionary and developmental biology. Biology and Philosophy 18: 309–345.

Love AC (2006a) Evolutionary morphology and Evo-devo: Hierarchy and novelty. Theory in Biosciences 124: 317–333.

Love AC (2006b) Reflections on the middle stages of Evo-devo's ontogeny. Biological Theory 1: 94–97.

Love AC (2007) Morphological and paleontological perspectives for a history of Evo-devo. In From Embryology to Evo-devo: A History of Developmental Evolution. M Laubichler, J Maienschein eds.: 267–307. Cambridge, MA: MIT Press.

Love AC (2008a) From philosophy to science (to natural philosophy): Evolutionary developmental perspectives. Quarterly Review of Biology 83: 65–76.

Love AC (2008b) Explaining the ontogeny of form: Philosophical issues. In A Companion to Philosophy of Biology. S Sarkar, A Plutynski eds.: 223–247. Malden, MA: Blackwell.

Love AC (2008c) Explaining evolutionary innovation and novelty: Criteria of explanatory adequacy and epistemological prerequisites. Philosophy of Science 75: 874–886.

Maynard Smith J (1993 [1958, 1966, 1975]) The Theory of Evolution. Cambridge: Cambridge University Press.

McPeek MA (2006) What hypotheses are you willing to entertain? American Naturalist 168 (suppl.): S1–S3.

Mitchell SD, Dietrich MR (2006) Integration without unification: An argument for pluralism in the biological sciences. American Naturalist 168 (suppl.): S73–S79.

Monod JL (1975) On the molecular theory of evolution. In Problems of Scientific Evolution: Progress and Obstacles to Progress in the Sciences. R. Harré, ed.: 11–24. Oxford: Clarendon Press.

Morrison M (2007) Where have all the theories gone? Philosophy of Science 74: 195–228.

Müller GB (2007) Evo-devo: Extending the evolutionary synthesis. Nature Reviews Genetics 8: 943–949.

Müller GB, Newman SA (2005) The innovation triad: An EvoDevo agenda. Journal of Experimental Zoology 304B: 487–503.

Müller GB, Newman SA, eds. (2003) Origination of Organismal Form: Beyond the Gene in Developmental and Evolutionary Biology. Cambridge, MA: MIT Press.

Nickles T (1981) What is a problem that we may solve it? Synthese 47: 85–118.

Pennisi E (2008) Deciphering the genetics of evolution. Science 321: 760–763.

Pigliucci M (2003) The new evolutionary synthesis: Around the corner, or impossible chimaera? Quarterly Review of Biology 78: 449–453.

Pigliucci M (2007) Do we need an extended evolutionary synthesis? Evolution 61: 2743–2749.

Pigliucci M, Kaplan J (2006) Making Sense of Evolution. Chicago: University of Chicago Press.

Reif W-E, Junker T, Hoßfeld U (2000) The synthetic theory of evolution: General problems and the German contribution to the synthesis. Theory in Biosciences 119: 41–91.

Rice SH (1998) The evolution of canalization and the breaking of Von Baer's laws: Modeling the evolution of development with epistasis. Evolution 52: 647–656.

Rice SH (2004) Evolutionary Theory: Mathematical and Conceptual Foundations. Sunderland, MA: Sinauer.

Ridley M (1993) Evolution. Cambridge, MA: Blackwell Science.

Ridley M (1996) Evolution. 2nd ed. Cambridge, MA: Blackwell Science.

Ridley M (2004) Evolution. 3rd ed. Cambridge, MA: Blackwell Science.

Rosenberg A (1985) The Structure of Biological Science. Cambridge: Cambridge University Press.

Ruse M (1973) Philosophy of Biology. London: Hutchinson University Library.

Sarkar S (2007) Doubting Darwin: Creationist Designs on Evolution. Malden, MA: Blackwell.

Shapere D (1980) The meaning of the evolutionary synthesis. In The Evolutionary Synthesis: Perspectives on the Unification of Biology. E Mayr, WB Provine eds.: 388–398. Cambridge, MA: Harvard University Press.

Simpson GG (1944) Tempo and Mode in Evolution. New York: Columbia University Press.

Snyder LJ (2008) William Whewell. In The Stanford Encyclopedia of Philosophy. EN Zalta ed. http://plato.stanford.edu/archives/sum2008/entries/whewell/.

Sober E (1988) Reconstructing the Past: Parsimony, Evolution and Inference. Cambridge, MA: MIT Press.

Suppe F (2000) Understanding scientific theories: An assessment of developments, 1969–1998. Philosophy of Science 67: S102–S115.

Suppe F ed (1977) The Structure of Scientific Theories. 2nd ed. Urbana: University of Illinois Press.

Thompson P (1989) The Structure of Biological Theories. Albany, NY: SUNY Press.

Tuomi J (1981) Structure and dynamics of Darwinian evolutionary theory. Systematic Zoology 30: 22–31.

Wasserman GD (1981) On the nature of the theory of evolution. Philosophy of Science 48: 416–437.

West-Eberhard MJ (2003) Developmental Plasticity and Evolution. New York: Oxford University Press.

Winther RG (2006) Parts and theories in compositional biology. Biology and Philosophy 21: 471–499.

Woese CR (2004) A new biology for a new century. Microbiology and Molecular Biology Reviews 68: 173–186.

Zimmerman WF (2003) Stephen Jay Gould's final view of evolution. Quarterly Review of Biology 78: 454–459.

17 The Dialectics of Dis/Unity in the Evolutionary Synthesis and Its Extensions

Werner Callebaut

I believe one can divide men into two principal categories; those who suffer the tormenting desire for unity, and those who do not.
—George Sarton (quoted in Sarton 1959: 40–41)

The thesis of my contribution is that the evolution of evolutionary thinking (Hull 1988) since the making of the Modern Synthesis has been characterized by simultaneous unifying and disunifying tendencies, with no end in sight. This should be unsurprising, if not trivial, were it not for the claims of some postmodernists that unity is an ideological aberration, whereas disunity is welcome and real. This chapter elaborates on this theme and proposes a dialectical solution.

"The work of defending, expanding, challenging, and, perhaps, replacing the Modern Synthesis," Grene and Depew (2004: 248) observe, "has tended to bring out the philosopher in many evolutionary biologists." The "Darwin industry" has also attracted to the scene professional philosophers, whose efforts to clarify key concepts and inferential patterns have stimulated the formation of a professional philosophy of biology, which is now booming (Hull 2000). Are there genuine laws (i.e., universal, nomically necessary generalizations) in biology? How can evolutionary biology be explanatory if it is not usually predictive?[1] Answers to these and related questions have a direct bearing on any synthetic project. Beyond such philosophical concerns, calls for an extended or expanded Evolutionary Synthesis or "theory" (e.g., Gould 1982; Wicken 1987; Depew and Weber 1994; Odling-Smee et al. 2003; Kutschera and Niklas 2004; Jablonka and Lamb 2007; Müller 2007; Pigliucci 2007), if not for a "new," or even a "post-Darwinian," Synthesis (e.g., Wilson 1975; Carroll 2000; Johnson and Porter 2001; West-Eberhard 2003; Gilbert 2006; Rose and Oakley 2007), invite critical historical and sociological reflection on the rhetorical functions of discourse invoking "synthesis," "extension," and related terms.

The Rhetoric of Unification

In a review of recent work on niche construction (see chapter 8 in this volume), Okasha (2005: 10) writes approvingly that unlike certain authors who have called for "a major reorientation or re-structuring of evolutionary theory," Odling-Smee et al. provide "positive and practical suggestions" for how other researchers can put their ideas into practice. The would-be revolutionary whom Okasha targets is Stephen Jay Gould, whose initial dissenting attitude vis-à-vis the Synthesis, it should be noted, grew milder over the years: Gould (2002: 1003) contended only that "the synthesis can no longer assert full sufficiency to explain evolution at all scales" (see also Eldredge 1985), specifying that punctuated equilibrium was formulated as "the expected macroevolutionary extension of conventional allopatric speciation." Perhaps already forgotten is that Edward O. Wilson was, in comparison, a far greater heretic, or at least tried hard to be perceived as such. His *Sociobiology: The New Synthesis* (1975: 64) envisioned the development of a "stoichiometry" of social evolution—a theory that can predict particular biological events in ecological and evolutionary time. This task Wilson regarded as so formidable that it would require "such profound changes in attitude and working methods that it can rightfully be called *post-Darwinism*" (my emphasis).

Julie Klein (1990) pointed out that metaphors used to describe relations between scientific areas are typically drawn from geopolitics. Disciplines, domains, or fields are characterized as "territories" with "boundaries" that are to be "protected" and "crossed." Development in the early decades of the previous century has been characterized as an "uncharted swamp" that was to be "bypassed" (Hull 1998a: 89). Lindley Darden, in a meta-analysis of "fields" that originated as part of an effort to promote "unifying science without reduction" (Maull 1977), defines a scientific field in terms of a central problem; a domain to be explained; techniques and methods, unique to it or shared with other fields; concepts, laws, and theories; special vocabulary; and more general assumptions and goals more or less shared by those scientists using the techniques in trying to solve the central problem (Darden 1991: 19).

I have found Darden's and Maull's notion of *interfield* a useful tool to come to grips with the convergence of evolutionary and developmental biology resulting in EvoDevo. EcoDevo (Gilbert and Epel 2009), evolutionary economics (Nelson and Winter 1982), and evolutionary or "Darwinian" medicine (e.g., Stearns and Koella 2008), can be regarded as similar mergers in which the concepts and methods of one field are borrowed by the other to tackle its own problems, sometimes, as in DevoEvo and EvoDevo (Callebaut et al. 2007: 45), also the other way around. Darden's and Maull's account catalogs the kinds of interrelations that can hold between different bodies of knowledge, such as identity, part–whole, structure–function, and causal. Analogical relations provide a much weaker form of interconnection. On the interfield account, the Synthetic Theory is

a multifield, multilevel theory that postulates a well-integrated causal process [evolution by natural selection—W.C.] combining mechanisms from the different fields to solve the old problem of the origin of species. It established the study of isolating mechanisms as an important new area of study, while relating the already developed fields of Mendelian genetics, mathematical population genetics, and experimental and field studies of populations. (Darden 2006: 180)

The solution of the old problem not only required relating existing fields of knowledge but also necessitated the emergence of the study of isolating mechanisms as a new field. And because of the hierarchical, causal relations among the fields, their interrelations were *not* reciprocal: Mendelian genetics could proceed relatively independently of population genetics, but population genetics was dependent on new findings of mutational processes in genetics. Darden (1991, 2006) has applied her analysis to a number of historical cases. I suggest that, in combination with rhetorical accounts, it may help us clarify what is going on in trans- and interdisciplinary negotiations, of which I offer some examples.

In debates concerning biochemistry and molecular biology in the 1960s, molecular biology has been declared a subfield that biochemistry ought to naturally engulf, and vice versa (Abir-Am 1992). While the biochemical establishment resisted a loss of authority as a result of the rediscovery of nucleic acids by molecular biologists, who accounted for them in terms of information, organismic biologists contested molecular biologists' claim to have displaced evolution as the central problem of biology. Pnina Abir-Am (1992: 166) writes about this post-Synthesis development, which was to upset the prevailing unity of the biological order:

Theoretical, developmental, and population geneticists, such as Conrad H. Waddington, Theodosius Dobzhansky, and Richard Lewontin, aimed at restricting the epistemological and ontological claims of molecular biology and retaining the supremacy of evolutionary theory. Systematists and evolutionary theorists, such as George Gaylord Simpson and Ernst Mayr, tended to emphasize the limited biological relevance of molecular biology, which they referred to as "mere" chemistry.

Wilson's *Sociobiology* was another instance of scientific "territorial aggression" at work, to borrow an expression he coined in a biological context (Wilson 1971). Wilson (1975: 4–5) illustrated his proclamation that the social sciences were "the last branches of biology waiting to be included in the Modern Synthesis" with a figure in which sociobiology and behavioral ecology progressively engulf surrounding fields the way an amoeba engulfs its prey. In his *Consilience* (1998), named after Whewell's concept that for Wilson contains the key to unification,[2] he advocates the extension of the synthetic project to the entirety of knowledge. He imagines future analysts tracing an Ariadne's thread of "causal explanations from historical phenomena to the brain sciences and genetics," thus bridging the divide between the social and natural sciences. And to the question of how to extend the scale of time and space even further, he responds that there are "many potential entries across the whole range of human behavior, including those entailing art and ethics" (214). Despite Wilson's occasional characterization of interdisciplinarity as mutually beneficial "trade" across the boundaries that separate disciplinary domains, the majority of his metaphors again establish "an image of one territory dominating another through an expansionist war" (Ceccarelli 2001: 129).

Brooks and Wiley's attempt to displace "neo-Darwinism" by a unified theory of biology grounded in thermodynamics provides a third (failed) instance of "unfriendly takeover." They claimed that population biology could be "accommodated" within their own, "more general" theory:

Until now, the success of population biology has been taken by some to indicate de facto the superiority of neo-Darwinism. But if population biological phenomena can fit equally well within two different theories, the criteria for choosing between the theories will not be population biological ones. (Brooks and Wiley 1988: 255)

Ever since its inception, the Synthesis has been prone to similar territorial expansionist attacks. It has resisted them quite successfully until now.

Evolutionary biology before about 1935 was a "badly split field" (Mayr 1988: 325), if it could be called a field at all (Bowler 1983; Gould 2002). Leah Ceccarelli's comparative rhetorical analysis of Dobzhansky's *Genetics and the Origin of Species* (1937), Schrödinger's *What Is Life?* (1944), and Wilson's *Consilience* (1998) shows that Dobzhansky's book, arguably *the* foundational text for the emergence of the Modern Synthesis,[3] succeeded in simultaneously persuading the warring camps of geneticists and naturalists–systematists that "the other side" had something worthwhile to contribute. To accomplish this, Dobzhansky used metaphor, prolepsis (the anticipation of an objection), Aesopian form (the insertion of one meaning for a dominant audience and another, hidden meaning for a subordinate audience), and other rhetorical strategies. Schrödinger's navigation, a decade later, between the prestigious and well-financed discipline of physics and the comparatively less successful discipline of biology was similarly instrumental in bringing about the molecular biological revolution. In contrast, Wilson's more recent call for Enlightenment in a postmodern Dark Age (my characterization) "negatively influenced the outcome of his appeal for interdisciplinarity," as it was rhetorically designed to "fuel interdisciplinary hostilities" (Ceccarelli 2001: 128).

Whatever visionaries biology harbors today are likely to be found in systems biology, where the ultimate goal is "to understand every detail and principle of biological systems" (Kitano 2001: 1), or in synthetic biology, which has been characterized as down-to-earth intelligent design[4] (Craig Venter's "synthetic genomics on an absolutely massive scale followed by a process of selection"; Nicholls 2008). But *intellectually* speaking, our time—and here I am not thinking of biology alone—is certainly far less revolutionary-minded than, say, the late 1960s and early 1970s. Rather than revolutionary upheaval, the language of extension propounded in the present volume suggests to me "accommodation" in the Piagetian sense (see Gruber and Vonèche 1977) of accommodating an old "schema," the Synthesis, to new "objects," namely, the areas of biology that it left out or "blackboxed."

In the next section I will reconsider the history of the Synthesis with the aim of drawing some morals to inform our current debate about its extension. But first, three caveats. First, Cecciarelli (2001: 21), in reference to the secondary literature surrounding the Synthesis, notes correctly that scholars do not define the Synthesis as a "theory." Some *do* argue that developments in theory were co-responsible for this interdisciplinary "treaty" (Depew and Weber).[5] But much more than theory (change) was at stake in the making of the Synthesis. Indeed, many factors—conceptional, theoretical, social, and political—were involved (Ruse 1979; Mayr and Provine 1980; Depew and Weber 1994; Smocovitis 1996; Cecciarelli 2001; Gould 2002).

Second, *the Synthesis was not a "scientific revolution" in the Kuhnian sense.* There are general and specific reasons for caution with renderings of the Synthesis (or, for that matter, of the Darwinian Revolution itself) in the language of "paradigms" and related notions such as "normal science," "anomaly," and "crisis." Philosophers and historians of science reacting to the inflationary use of "paradigm" that followed the popularization of Kuhn's view have long become wary of the notion. John Greene, for one, concluded his essay "The Kuhnian Paradigm and the Darwinian Revolution in Natural History" by suggesting that Kuhn's "paradigm of paradigms can be made to fit certain aspects of the development of natural history from Ray to Darwin, but its adequacy as a conceptual model for that development seems doubtful" (Greene 1981: 54). In the case of the Synthesis, "there was no missing piece of information that needed to be found, nor was there a neat, elegant, critical experiment that settled the controversy" (Cecciarelli 2001: 21). Will Provine (1992: 173) concurs: "The synthesis is not characterized by startling or extraordinary new discoveries, concepts, or theories." Moreover, as a "hypertheory," "supertheory," or "metatheory" (Grene 1983; Burian 1988) that ties together a variety of scientific disciplines within biology, Darwinism differs considerably from the physical paradigms that Kuhn originally had in mind.

Third, following Ernst Mayr, I take it that not only biology itself but also its *history* is not so much a history of lawful theories (let alone facts) as it is one of concepts or "principles," which are taken to be more flexible and heuristically more fruitful than laws (Mayr 1982: 43). This is relevant to my story to the extent that evolutionary concepts such as adaptation (or an EvoDevo concept such as modularity) "have a way of expanding by cannibalizing other concepts that ought also to have a role in the whole explanatory framework" (Grene 1983: 7). This peculiarity of biology among the natural sciences commands particular philosophical attention (Callebaut et al. 2007: 30–32).

Reculer pour Mieux Sauter: Morals from the Modern Synthesis

The Modern Synthesis has been hailed as "a historical event that appeared to fulfill a project at least as deep as the Enlightenment project (or even deeper still) of unifying the branches of knowledge" (Smocovitis 1996: 7). Betty Smocovitis describes her book, *Unifying Biology*, as telling

... a story of the emergence, unification, and maturation of the central science of life—biology—within the positivist ordering of knowledge; and ... a story of the emergence of the central unifying discipline of evolutionary biology (complete with textbooks, rituals, problems, a discursive community, and a collective historical memory to delineate its boundaries). (Smocovitis 1996: 7)

In the last four decades of the twentieth century, the "positivist ordering of knowledge" gave way to philosophical naturalism and scientific realism, on one hand, and to a variety of social constructivist schemes in the history and social studies of science, on the other (Callebaut 1993). In some scientific quarters, positivism still lingers on; see, for example, Max Dresden's (1998) critique of the "myth of fundamentality" in physics, to which I will return below. We will have to investigate how this change in zeitgeist impinges on debates on "synthesis" and its "extension."

What the Modern Synthesis Was (Not)

The evolutionary consensus that was to become a bulwark of orthodoxy by the late 1950s was mostly set forth in a quartet of books by Dobzhansky, Mayr, Simpson, and Stebbins, published between 1937 and 1950, that molded Darwin's evolution by natural selection within the framework of rapidly advancing genetic knowledge (Ayala et al. 2000). It received no recognized name until Julian Huxley published his *Evolution: The Modern Synthesis* (1942), in which he extolled the virtues of unity and synthesis, comparing biology to physics—a recurrent theme in the history of biology. One chief result of the new consolidation, Huxley emphasized, was a "rebirth of Darwinism," which had been slow in recovering from its "eclipse"—the term is his—around 1900 (see Bowler 1983). Historians have devoted little attention to the significance of "modern" in "Modern Synthesis." Provine (1992: 169) is the exception when he expresses his suspicion that "when anyone describes his or her views as the 'new' or 'modern' way of seeing things, to be sharply distinguished from the 'old' inferior ways, the 'new' billing is often little more than scholarly overstatement, an attempt to attract attention." I note in passing that if "postmodernism," beyond "carrying modernist styles or practices to extremes" (American Heritage Dictionary), stands also for locality, disconnection, and the like, then "Postmodern Synthesis" is a contradiction in terms. But quite likely, biologists who refer to it (e.g., Luttikhuizen and Drent 2004; Butlin 2006) are not aware of the postmodernist movement and simply want to say "after the Modern Synthesis."

"Synthesis" refers to the process of putting together (*syntithenai*, in Greek) two or more things, concepts, elements, and such, to form a whole, and to the complex so formed. Chemical synthesis may be the exemplar here. In the laboratory, the preparation of a complicated organic substance is a very difficult task, for "every group, every atom must be placed in its proper position and this should be taken in its most literal sense" (Nicolaou et al. 2000: 46). Chemists take the practice of *total synthesis*—the "flagship" of organic synthesis—to require ingenuity, artistic taste, experimental skill, persistence, and character, which is why total synthesis is said to be an exact science and a fine art at the same time. But this dual nature also provides excitement and rewards of rare heights in the form of discoveries/inventions that impact not only other areas of chemistry but also, most significantly, materials science, biology, and medicine.

What prima facie distinguishes material synthesis from an intellectual construction such as the Evolutionary Synthesis is that the multifarious chemical architectures follow only the rules of chemical bonding, whereas the Modern Synthesis is typically described as having developed around a "core," whether Darwinian (e.g., Gould 2002: 508ff.) or population genetic (e.g., Futuyma 1988: 223; Grene and Depew 2004: chap. 9). This may suggest that the core is more fundamental, whereas other parts of the complex are (merely) peripheral and special (e.g., Lynch 2007). Yet Lewontin (2000: 192) argues against any foundationalist longings that "it is *not* within the problematic of population geneticists to discover the basic biological phenomena that govern evolutionary change, as it was for nuclear physics to discover universal forces between nuclear particles." I also note that although Mayr, Gould, and other evolutionists were self-declared "population thinkers" regarding *biological* evolution, they were more than reluctant to extend their anti-essentialism to the evolution of scientific theories, especially their own (Hull 2006). Gould (2002: 1), for one, argued that "Theories need both essences and histories." Anti-essentialism with respect to theories (etc.) is the view that scientific change can be accounted for *without* invoking an unchanging "hard core," a set of common ideas shared synchronically and through time by all researchers in the field (Hull 1988). The 16 "principal claims of the evolutionary synthesis" listed by Futuyma (2005: 10–11), which he presents as "the foundations of modern evolutionary biology" (see also the summary in chapter 1 of this volume) may be regarded as an attempt to characterize such a hard core. Futuyma's claim that most(!) evolutionary biologists today(!) "accept them as fundamentally valid" is qualified by the admission that "some of these principles have been extended . . . or modified since the 1940s."

What, then, was the Synthesis? I can offer only a nutshell view here. In the first decade of the twentieth century, three major alternatives to Darwinism, Lamarckism, saltationism, and orthogenesis "enjoyed substantial support, probably equal in extent . . . to the popularity of Darwinism itself" (Gould 2002: 506). (Beatty, in chapter 2 of this volume, discusses the roots of the elimination of directionality from variation and the continued subordination of chance variation throughout the Evolutionary Synthesis in Darwin's own work.) The rediscovery of Mendel's laws around 1900 had revivified the saltational alternative. The first, "Fisher–Haldane–Wright" phase of the Synthesis restricted the space of possibilities in three major moves:

1. By choosing the Darwinian central core as a proper and fundamental theory

2. By reading Mendelism in a different way to validate, rather than to confute, this central core

3. By utilizing this fusion to ban the three alternatives of Lamarckism, saltation, and orthogenesis. (Gould 2002: 506)

The resulting account of evolution (which is often referred to, somewhat misleadingly, as "neo-Darwinism"; contrast chapter 1 in this volume) "remained open and pluralistic in welcoming all notions consistent with the new formulation of Mendelism" (Gould 2002: 508), as is still reflected in the title of Haldane's book, *The Causes of Evolution* (1932). The second phase of the Synthesis is often depicted as a gathering of traditional subdisciplines and/or observations and experimental results under the umbrella constructed during the first phase (Waters 1992; Ceccarelli 2001). But this was "not so much a synthesis as it was a vast cut-down of variables considered important in the evolutionary process" (Provine 1992: 176). Provine has called this the *evolutionary constriction*:

Evolutionists after 1930 might disagree intensely about effective population size, population structure, random genetic drift, levels of heterozygosity, mutation rates, migration rates, and so on, but all could agree that these variables were or could be important in evolution in nature. . . . So the agreement was on the set of variables, and the disagreement concerned differences in evaluating relative influences of the agreed-upon variables. (Provine 1992: 177)

Gould's (1983) by now legendary "hardening" of the Modern Synthesis toward a selectionist interpretation occurred later, during the late 1940s and 1950s.

On Provine's account, which concentrates on theory, the Synthesis barely deserves its name. A quantitative synthesis of Mendelian heredity and various factors that can change gene frequencies in populations was accomplished by Fisher, Haldane, Wright, Hogben, Chetverikov, and others, although "they disagreed, often intensely, about actual processes of evolution in nature, even when their models were mathematically equivalent." But the rest of the so-called Synthesis "mostly consisted of exercises in removing barriers, consistency arguments, and forging a consensus. A lot of hand waving was also involved, especially at the time Huxley coined the name 'the modern synthesis'" (Provine 1992: 172).

Focusing on theory as well, E. O. Wilson (1975), with his background in biogeography, complements the conventional picture of the Synthesis by stressing that its first phase, the marriage of natural selection theory and population genetics, coincided with the creation of the foundations of mathematical population ecology by Lotka, Volterra, and others (cf. chapter 1 in this volume, on ecology's missing out on the Synthesis; and Scheiner and Willig [2008] on the prospect for a general theory of ecology). When the publication of Fisher's, Wright's, and Haldane's books closed this pioneering decade, "a respectable number of new ideas had been generated that constituted an extensive, albeit untested, framework on which a mature science might have been built" (Wilson 1975: 63). Alas, according to Wilson, evolutionary biology "did not and could not proceed in this straightforward manner." It first had to pass through a period of about 30 years of consolidation of information, innovation in empirical research, and slow progress. Wilson belittles the achievements of the Synthesis, and criticizes it for having created "very little theory in the strict sense" between 1930 and 1960, beyond that already laid down in the 1920s. Most of the branches of evolutionary biology were reformulated in the language of early population genetics. The greatest accomplishment of this period, he grants, was "the elucidation, through excellent empirical research, of the nature of genetic variation within species and of the means by which species multiply" (63). Other topics were clarified and extended as well. And yet,

some of the apparent new understanding of the Modern Synthesis was false illumination created by the too-facile use of a bastardized genetic lexicon: "fitness," "genetic drift," "gene migration," "mutation pressure," and the like. So many problems seemed to be solved by invoking these concepts, and so few really were. Stagnation inevitably followed. (Wilson 1975: 63–64)

Wilson's account of a generation of young evolutionists (roughly, those maturing in 1945–1960) cutting themselves off from "the central theory" is sobering indeed: "Having never grasped the true relation between theory and empiricism in the first place, they were willing to submit to authority rather than to advance the science by altering the central theory" (64). The dogmatism suggested here adds another dimension to Gould's "hardening" of the Synthesis. Discussing dissenting voices at the time of the fortieth anniversary of *Evolution*, the journal of the Synthesis movement, Douglas Futuyma diagnosed "sociological" deficiencies:

The Synthesis was largely intellectual rather than sociological. By and large, naturalists and systematists absorbed rather little of the population-genetic theory that is the core of neo-Darwinism; but it must be granted that, for decades after the Synthesis, few if any comprehensive treatments of population genetics explained the implications of the equations for evolutionary studies, especially in terms that a nonmathematical biologist could understand. (Futuyma 1988: 223)

The Synthesis also failed to establish a truly mutualistic relationship between historical studies (paleontology and systematics above the species level) and the synchronic analysis of processes at and near the population level, in contemporary populations. Provine (1992: 169–171), Gould (2002: 519), and Hull (2006) indirectly confirm this diagnosis of "too weak social bonds" when they point out that—literally—all participants in the Synthesis complained that their own contribution was not given enough credit and that even those working in the same institution often would not interact with each other.

    Given the relevance of national styles in science (see, e.g., Harwood 1993) and the notoriously different destinies of Darwinism in different countries and continents (see, e.g., Mayr and Provine 1980, part 2), it is tempting to ask also *where* the Synthesis took place. Grene and Depew (2004: 257) write that the Synthesis, although most intensively and institutionally pursued in the United States, was "an international accomplishment." But elsewhere, Grene (1983: 4) noted that Rensch was the only major German contributor who remained in Germany. France is altogether absent from the picture (Buican 1984). If one thinks of the importance of the work of the ecological geneticist Sergei Chetverikov and his school of population genetics, which successfully combined the methods of geneticists and naturalists by developing mathematical models and testing them through field observations, and of the influence of this school on Dobzhansky, who blended the field research practices that he brought from Russia, the laboratory techniques he had learned in Morgan's "fly room" lab at Columbia, and population genetic analysis, one is tempted to give more weight to the Russian contribution than is usually done (Adams 1970). See, for instance, Krementsov (2007) on the "particular synthesis" of biogeographical, taxonomic, genetic, ecological, and behavioral studies forged by Aleksandr Promptov, a student of

Chetverikov, which was ignored by the architects of the Synthesis, including Dobzhansky.

At this juncture, attempts to "decenter" the Synthesis, and, further back in time, the Darwinian Revolution itself, may become relevant to proponents of an Extended Synthesis. Consider the meaning of the very word "evolution." Throughout his career, the historian of biology Robert J. Richards has argued for a broader definition of "evolution" than the one currently used, a definition that embraces its older, embryological meanings (Richards 1992, 2002, 2008; see also Gilbert 1994 on the prominent place of embryology in Russian evolutionary thought). This "takes him to earlier and non-Anglocentric, Germanic locations for understanding the origins of evolutionary thought" (Smocovitis 2005: 43). Granting that there is much of value in Richards' approach, Smocovitis (2005: 43) asks: "but does it radically alter our understanding of something called the 'Darwinian Revolution'?" My answer is: maybe not of the Darwinian Revolution, but it certainly opens new perspectives on a richness of ideas that the "hardening of the constriction" (Provine) had hidden from our view. In their recent book *Ecological Developmental Biology*, Gilbert and Epel (2009) point the way by including a coda, "Philosophical Concerns Raised by Ecological Developmental Biology," in which they discuss, among other subjects, diverse integrative philosophical traditions that hitherto have been little explored by biologists, historians, and philosophers.

Discontents

Hardly any tenet of the Synthesis has remained unchallenged during the 1970s. In their excellent discussion, Grene and Depew (2004: chap. 9) refer to the disputes in the 1970s about how to bring behavior, especially cooperative behavior, into the Darwinian fold. These debates seem largely settled today, with kinship and direct and indirect reciprocity being agreed on as the explanatory mechanisms (Nowak 2006). (However, E. O. Wilson [2008] recently caused a stir by suggesting that eusociality can evolve if pre-adaptations provide the phenotypic flexibility required for eusociality, thus downplaying collateral kin selection and upgrading group selection in the process—a double insult to orthodoxy.) In the 1980s, macroevolution came to the fore when some paleontologists challenged Simpson's assumption that macroevolution is but microevolution writ large—the result of accumulated microevolutionary change at the level of populations (see chapter 13 in this volume). This debate has also reached anthropology (contrast Tattersall 2000 and Foley 2001). The

1990s saw "the consolidation of a long-simmering challenge from developmental biology" (Grene and Depew 2004: 248), which had been blackboxed by the Synthesis at its formation but has come back with a vengeance in the guise of EvoDevo. As these debates are discussed elsewhere in this volume (see chapters 10, 11, 12, and 14), I will not pursue them here. Instead, I focus on two more enduring trends, the challenges from molecular biology and systematics.

Molecular Biology   The twentieth century was to be the century of the gene (Keller 2000). The laboriously negotiated "treaty" did not establish the supremacy of the Synthesis for long. The struggle of its champions to accumulate intellectual credit and secure funding encountered increasing competition from their "physicalist" colleagues bent on applying physical and biochemical concepts and methods to biological issues. When the chemistry of DNA was identified with the classical concept of the gene, the molecular era begun (Allen 1978), dealing a body blow to the naturalist–systematist tradition. This blow was so severe, Greene (1999: 107) observes, that "Ernst Mayr began to devote increasing time and energy to historical and philosophical studies aimed at discrediting physics-oriented history and philosophy of science and fostering a rival 'organicist' view of science and nature."

Depew and Weber (1994: 359) suggest that the current atmosphere of "challenge, defensiveness, reconsideration, and sheer excitement" should be understood largely as a "magnification" of the old conflicts that the hardened Synthesis had swept under the carpet, under the new conditions created by molecular genetics. They focus on three debates in which Dobzhansky's student, Richard Lewontin, intervened prominently: Kimura's neutralism, the "selfish gene" hypothesis (Williams 1966), and the suggestion to expand Darwinian selection to include a variety of levels and entities, including groups and species, and at different rates (Eldredge 1985; Gould 2002; Vrba and Eldredge 1984; Vrba and Gould 1986). Chapters 4 and 13 in this volume update these debates.

Convinced as they are that biology at the start of the twenty-first century is achieving a substantive maturity of theory, experimental tools, and fundamental findings (due to the relatively secure foundations in genomics) that now necessitates an "intellectual transformation," Rose and Oakley (2007) are much less optimistic about the future of the Synthesis. In fact, they propose to remove a number of "dead parts" from it, a partial listing of which includes (1) "The genome is always a well-organized library of genes"; (2) "Genes usually have simple functions that have been specifically honed by powerful natural selection"; (3) "Species are finely adjusted to their ecological circumstances due to efficient adaptive adjustment of biochemical functions"; (4) "The durable units of evolution are species, and within them the organisms, organs, cells, and molecules, which are characteristic of the species"; and (5) "Given the adaptive nature of each organism and cell, their machinery can be modeled using principles of efficient design." (I note that it is far from obvious how these presumed "parts" can be made to correspond to some of Futuyma's aforementioned 16 tenets.) Will the Synthesis prove capable of coping with these and related challenges? One conceivable scenario is that, given the drastically changed scientific socialization of new cohorts of biologists, the Synthesis will fade away and eventually vanish into oblivion, except for some minority of biologists working in evolution. The "epistemologist's paradise" (Lewontin 2000: 192) could then become the historian's cemetery. Or does such a pessimistic view do injustice to the remarkable resilience of the Synthesis (see below) and the current vigor of evolutionary biology (see chapter 15 in this volume)? As Yogi Berra told us, it's tough to make predictions, especially about the future.

Systematics   In his cleverly balanced defense of the Synthesis against all kinds of dissenters, Futuyma was most discombobulated by the persistent attacks from the systematists' camp:

I find the animus of some systematists against the Synthesis most distressing of all. Even as systematics has grown in strength and rigor, its reputation among other evolutionary biologists has been diminished by the intemperance of some of its more vocal figures. Allowing for hyperbole, it is nonetheless disconcerting to read such statements by prominent and highly competent systematists as "Frankly, it does not matter to me who thought up the theory of evolution we are about to discard" [Daniel Brooks—W.C.] . . . or "There is no need to placate the ghost of neo-Darwinism; it will not haunt evolutionary theory much longer" [Donn Rosen—W.C.]. (Futuyma 1988: 222)

The most radical among the systematists, the pattern cladists, disavow any evolutionary content in their work. What began, with Hennig, as phylogenetic taxonomy in opposition to idealistic morphology has turned, in its pattern-cladistic form, against evolution itself (see Hull 1988 for a reconstruction of the debates in systematics). The key to understanding this deep-seated disagreement seems to be the perception of influential systematists such as Rosen that evolutionary theory can explain any phylogenetic history that might be postulated, and so prohibit no imaginable phylogenies. This challenge to the Synthesis has become less threatening to the extent that pattern cladism has become less prominent in systematics.

A Remarkable Resilience

Both Darwin's theory (which was a synthesis of its own) and the Evolutionary Synthesis were shaped in part by ignorance of features of life that were simply unknown or unknowable at that time (Rose and Oakley 2007; see chapter 3 in this volume). Even a summary investigation of the history of discontents with the Synthesis suggests that it has been remarkably adaptable to new challenges. This seems to support the view that it is first and foremost *a loosely and flexibly structured network of concepts and models* rather than a "theory" according to old-style hypothetico-deductivism. A rather extreme illustration of this flexibility is the fate of the ecologist Barry Commoner's long-standing attack on the "central dogma" of molecular biology (the molecular successor of Weismann's doctrine), which included the claim that DNA is not self-replicating (Commoner 1961, 1968). In a "News and Views" section in *Nature* (Anonymous 1968) we read about Commoner's "now familiar and, as far as most people are concerned, experimentally refuted arguments that DNA is not self-replicating." Today, the insight that DNA (or anything else) is self-replicating is increasingly rejected, the "central dogma" is contested (see already Buss 1983; see also chapter 7 in this volume), and nobody cries "Fire!" In this specific sense, the optimism of Kutschera and Niklas seems warranted when they write that the Modern Synthesis

continues to cast a bright light on what may be called the post-modern synthesis, one that continues to expand and elaborate our understanding of evolution as the result of the continuous and tireless exploration of virtually every branch of science, from paleobiology/geology to natural history and cell/molecular biology. ... Indeed, the "evolution" of evolutionary theory remains as vibrant and robust today as it ever was. (Kutschera and Niklas 2004: 273)

Many processes, patterns, and so on that were previously unknown are now coming into biologists' reach. Does this new knowledge threaten the Evolutionary Synthesis and call for its extension? Gavrilets (chapter 3 in this volume) suggests that the answer to this question is "a very subjective matter." Debates on the need (or not) to extend the Synthesis somewhat resemble the question of whether the glass is half full or half empty. Both the Darwinian Revolution and the Modern Synthesis were built on huge black boxes. "Darwin could never have written the *Origin of Species* if he had not wisely bracketed the mechanism of inheritance" (Amundson 1994: 576). Whereas pre–Synthesis Darwinians at least realized the need for a theory of inheritance, many proponents of the "hardened" Synthesis held that developmental considerations are irrelevant to our understanding of evolution. Today, someone like Michael Lynch remains unimpressed by much of the recent work on complexity (Depew and Weber 1994), epigenetic innovation (chapter 12 in this volume), evolvability (chapter 15 in this volume; see also chapter 9 in this volume), modularity (chapter 11 in this volume), robustness, and other topics in EvoDevo, cell biology (chapter 10 in this volume), and genomics (chapters 5 and 6 in this volume). For Lynch (2007: 8598), population genetics is the "essential framework for understanding how evolution occurs," and hence the ultimate arbiter in matters of genomic and phenotypic evolution (see Bromham 2009 and Pigliucci 2009 for critical discussion). He rhetorically asks: "Have evolutionary biologists developed a giant blind spot; are scientists from outside of the field reinventing a lot of bad wheels; or both?" (Lynch 2007: 8597). Here a presumed essence is rhetorically invoked to justify intellectual neophobia.

Alternatively, one can suggest continuity with the Synthesis by arguing that a novel idea such as the evolution of evolvability must not imply as radical a break from the population genetic tradition as is sometimes thought (chapter 15 in this volume), or that the need to incorporate cellular and developmental findings into evolutionary theory was foreshadowed by Sewall Wright (chapter 10 in this volume), who also pioneered the concept of fitness landscape (chapter 3 in this volume). Or one can invoke methodological biases of the mainstream—suggesting the possibility of incremental improvement—as do Jablonka and Lamb (chapter 7 in this volume) regarding model organisms, Jablonski (chapter 13 in this volume) regarding the fossil record, and Pigliucci (chapter 14 in this volume) regarding the difficulty of finding natural examples of genetic assimilation (the "tale-telling signs" may be gone from a natural population quickly). As far as I can see, such debates can go on forever. If, as seems obvious to me, the Synthesis has no essence, its extensions are negotiable. I consider this a Good Thing. In fact, as the editors of this volume suggest, we are in the midst of this negotiation.

Varieties of Dis/Unity

From the Unity of Science Movement to the Anti-Reductionist Consensus in Biology

Step back in time again. Gould (2002: 503) endorsed Smocovitis' (1996) view that the Evolutionary Synthesis owed much to, or at least reso-

nated, with the unity of science movement expounded by positivist philosophers of the Vienna Circle and "supported by biological pundits like
J. H. Woodger"[6] (Galison 1998 and Reisch 2005 discuss the fate of the
originally "Viennese" unity of science movement in the United States).
As an example, Gould mentioned that Mayr

> strongly supported the unity of science movement early in his career, but changed
> his mind when he began to fear that misplaced claims for grander synthesis could
> bury natural history in a reductionist scheme to uphold the primacy of physics
> and chemistry. (Gould 2002: 503)

Although there were some interactions between some of the architects
of the Synthesis and the positivists (Haldane, Dobzhansky, and Huxley
attended some of the meetings of the unity of science movement, first in
Europe and later in the United States), the positivists' influence on the
Synthesis should not be overrated. Rather, at the time when the Synthesis
came about, "unification of science" was in the air. Around 1946, when
the following sentences were written, it may have been the most popular
slogan in philosophy:

> This phrase [unification of science—W.C.] has not only become the cry of a
> specific group of philosophers, but it is now accepted as one of the aims of phi-
> losophy by most of the contemporary philosophic schools, with few exceptions.
> Each particular school believes that it has found *the* way of effecting such a uni-
> fication, implicitly assuming that it knows the conditions for a unified science.
> One who concerns himself with the literature of the movement soon becomes
> aware of current confusion in the meaning of the expression, "Unification of
> Science." The observer begins to wonder whether "Unity of Science" has not
> become a philosophic stereotype, designed to evoke feeling rather than thought.
> (Churchman and Ackoff 1946: 287)

And yet, Churchman and Ackoff insisted, "*The choice of unification is
not a free one; if science fails to unite, it will fail to exist*" (298; italics in
original).

Somewhat later, the physicist Max Delbrück (1949: 173) wrote that
"The history of biology ... records one great unifying theory, bringing
together separate fields: the theory of evolution." One should never
underestimate the "physics envy" of biologists—then or now. Hadn't
Paul Dirac, the "theorist with the purest soul" according to Niels Bohr,
stated confidently, at a time when some biologists were beginning to
think seriously about the unification of their own disciplines, that "the
underlying laws necessary for the mathematical theory of a large part of
physics and the whole of chemistry are ... completely known, and the

difficulty is only that exact applications of these laws lead to equations that are too complicated to be soluble" (Dirac 1929: 714)? For the logical positivists, unity was not so much an ontological as a *linguistic* or "logical" issue (Hempel 1969; Galison 1996; Morrison 2000; Frost-Arnold 2005). Their original *phenomenalism* aimed to reduce the vocabularies of the sciences to "sense data" terms. Its failure led them to articulate *physicalism*, a program to reduce scientific vocabulary to predicates referring only to observable, intersubjectively accessible qualities of physical things ("cows and calves" language—Otto Neurath). Reductionism as regards *intertheory* relations and developed by Ernest Nagel, Paul Oppenheim, Hilary Putnam, Kenneth Schaffner, and others is represented schematically in figure 17.1.

Four complications for the reductionist unification program ought to be mentioned briefly here. I don't have the space to develop any of these arguments; my only aim is to orient the interested reader to the pertinent literature.

Multiple Realizability (Hooker 2004; Richardson 2009; Bregant et al. 2009) Interesting lawlike statements can often be made about higher-level events whose physical descriptions have nothing in common. Whether the physical descriptions of the events subsumed by such generalizations have anything in common or not is irrelevant to the interestingness, truth, and so on of the former (see figure 17.1.b.1). This may (Bregant 2009) or may not (Sober 1999) undermine the reductionist program. The "special sciences" formulate generalizations of this kind. *Supervenience* has recently become a philosophers' way to capture cases of multiple realizability. Whether one talks about entities, phenomena, properties, or concepts, the basic notion of As supervening on Bs appears to subsume *covariance*, where variations in As are correlated with variations in Bs; *dependence* of As on Bs (or, if these are different, the determination of As by Bs); and the *nonreducibility* of As to Bs (Savellos and Yalçin 1995: 2).

Downward Causation (Van Gulick 1993; Robinson 2005) Although the events and objects picked out by the special sciences are composites of physical constituents, the causal powers of such objects are determined not solely by the physical properties of their constituents and the laws of physics, but also by the organization of those constituents within the composite (figure 17.1.b.2). In developmental biology, for instance, claims about a more fundamental level (e.g., gene expression) may have to be explained in terms of claims about a less fundamental level (e.g., descriptions of the relative positions of pertinent cells). Contrast Craver and Bechtel (2007), on whose construal of levels downward causation cannot occur.

Emergentism (e.g., Hooker 2004; Deacon 2006; Morrison 2006; Ricard 2006) According to emergentists, higher levels of organization "fulgurate" indeterminably out of lower-level ones and then causally feed back downward. Examples of chaotic systems and systems that undergo phase transitions are *not* instances of emergence: the inability of finite observers to predict macrobehavior on the basis of micro-behavior does not imply the metaphysical failure of determination (Hütteman 2004). Ladyman et al. (2007) deny that physics suggests that the world is structured in levels at all.

The Myth of Fundamentality (Dresden 1998; Wimsatt 2007; Callebaut 2010; see also Schaffer 2003, and Brown and Ladyman 2009) Structural features within physics (and science more generally) make a purely deductive "theory of everything" extremely unlikely and probably impossible:

While recognizing the significance of general principles, to construct a detailed theory of all physics, of all effects and phenomena, both in principle and in operational practice, new and different concepts and principles have to be introduced. These are over and above the ideas included in the fundamental triple [particles, interactions, dynamics—W.C.], nor can they be derived from them. This set of concepts is collectively denoted as "scales" or levels. Thus, . . . to have a truly comprehensive description of physics, one needs not only general concepts and principles (possibly irreducible) but also a classification and organization of physics in terms of a set of scales and levels at best remotely related to the general *a priori* principles (and, most likely, independent of them . . .). (Dresden 1998: 471)

Mainly as the result of insights gained from failed attempts to reduce transmission genetics to molecular genetics (Rosenberg 1994, 2006), philosophers of biology, with very few exceptions (Waters 1990; Rosenberg 2006), zeroed in on an *anti-reductionist consensus* more than two decades ago. A philosopher of chemistry has recently issued a call for a moratorium on the use of "reduction" and related words:

Given the enormous variety of possible intertheoretical relations, the proliferation of definitions of reduction, supervenience, emergence, unification, and so on, as well as the fact that empirical studies have provided support for almost any

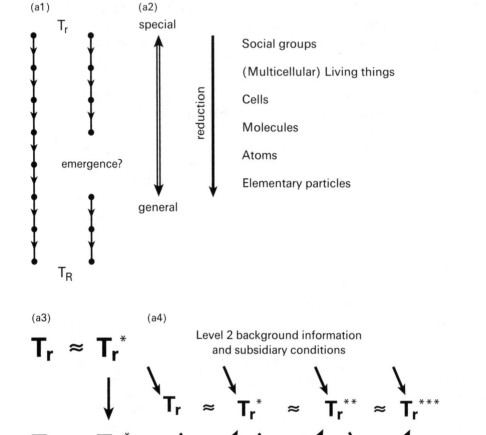

(b1)   (b2)

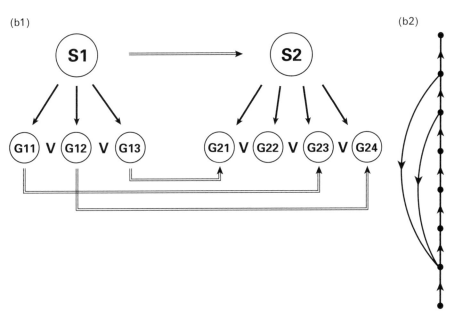

Figure 17.1
Rise and decline of reductionism. (a) Main stages in the reductionist program ($\rightarrow$ stands for "reduces to"). (a1) *Theory reduction through connectability*: only if the reducing theory T_R possesses the resources to refer to the same properties as the reduced science T_r is the latter reducible (Nagel 1961); failing connectability may (but need not) indicate emergence. Klein (2009) argues that this provides "reduction without reductionism." (a2) *Microreduction*: mereological (whole to parts) reduction of a special branch of science to a more fundamental branch that deals with the parts of the objects studied by the special branch (Oppenheim and Putnam 1958). Oppenheim and Putnam's "unity of science as a working hypothesis" was grounded in evolutionary considerations (Schlesinger 1961). (a3) *General reduction model*: a "corrected" theory T_r^* can be reduced to a "corrected" theory T_R^*, where the successor relation "$\approx$" is one of strong analogy (Schaffner 1993). (a4) *Co-evolution of theories at different levels* in Wimsatt's (2007) functional account of reduction. (b) Some complications: (b1) *Multiple realizability*, which inspired Fodor's (1974) "disunity of science as a working hypothesis": interesting lawlike statements at a higher level may be connectable to several or many lower-level (e.g., physical) descriptions that have nothing in common ($\rightarrow$ stands for "reduces to"; $\Rightarrow$ indicates a lawlike connection). (b2) *Downward causation* ($\rightarrow$ refers to causal relations). (Figures a1, a3, a4, and b2 after Wimsatt 2007).

metaphysical option, or, alternatively have shown rather conclusively that empirically the case is inconclusive, I suggest a moratorium on the use of words such as
"reduction," "supervenience" and "unification. . . . (Van Brakel 2003: 30)

Many perceptive philosophers of biology had drawn this moral before,
which had as an effect that the anti-reductionist consensus lingers on in
some kind of conceptual vacuum. As Sterelny puts it in a blurb on the
cover of Rosenbergs' 2006 book, "defenders of the consensus must slip
one way or the other: into spookiness about the biological, or into a
reduction program for the biological." But anti-reductionists like Mayr
could live comfortably with that.

The Charms of Disunity

In parallel with the growing disenchantment with reductionist unification
that was largely due to the *internal* difficulties of the program (it ended
up generating "epicycles upon epicycles"—David Hull), there has been
a postmodernist "call for the disunity of science" (Galison 1996: 2) in
recent years, focused mostly on what came to be seen as the "rhetoric"
of unity. While the organismic biologists and any remaining naturalists
in evolutionary biology were fighting to safeguard their own academic
existence, "the autonomy of condensed matter physics from the guiding
principles of particle physics has been debated from laboratories all the
way to the halls of Congress" (2)—Do we really need the supercollider?
On the other end of the academic spectrum, disunity was discussed in
the context of local history, feminism, multiculturalism, scientific relativ-
ism, social constructivism, and what have you.

One striking characteristic of this kind of disunification discourse is
that it catches scientific and political issues "from (pre-)Plato to (post-)
NATO" (Peter Galison) in the same, seamless web. For instance, a book
(Cartwright et al. 1996) devoted to Otto Neurath, the most colorful philosopher of the Vienna Circle, is subtitled *Philosophy Between Science
and Politics*, and its contents reflect the same polarity in multifarious
ways ("A Life Between Science and Politics," "The Primacy of Practical
Reason," etc.). The aim of the *International Encyclopedia of Unified
Science*, Neurath's brainchild, was to synthesize scientific activities such
as observation, experimentation, and reasoning, and show how they
helped to promote a unified science. "Encylopedism," his version of
unity opposing "pyramidism" (Reisch 1997), is discussed at some length,
but a clear picture of what it really meant has to be concocted by the
readers themselves from the bits and pieces that are offered—the post-
modern patchwork.

Another feature, which is most blatant in the less sophisticated appearances of the genre, is its normative orientation. Disunity often seems to be dictated by the categorical imperative of political correctness (cf. Reisch 2005). Interesting exceptions to this rule include Nancy Cartwright and John Dupré, whose arguments I should like to analyze in some depth at another occasion. Cartwright's (1999) analysis of our "dappled world" (see also Cartwright et al. 2002) implies that the laws of nature are limited in their range; in regions that seem to overlap, the separate laws may be helpful in calculating what happens, but the overall outcome may be highly context-dependent, or there may be no rules at all for composing the separate effects; and some situations may not be subject to law at all ("what happens, happens by hap"). Based on her knowledge of physics and economics, Cartwright's account of disunity is by far the profoundest I am aware of, but it is not clear to me that her arguments can be extended usefully to the biological realm, which according to some "knows only one law: that there aren't any" (Richard Lewontin, personal communication, 7 April 2007). Dupré presents himself as someone who devoted a career to promoting the idea of knowledge as disconnected and local, and whose view of the disunity of science grew out of an appreciation of the complexities and local specificities of classification. He is convinced that if science were unified, then the legitimate projects of inquiry would be (only) those that formed part of that unified whole. On his view, only a society with homogeneous, or at least hegemonic, political commitments and shared assumptions could expect a unified science. Unified science, he concludes, would require Utopia or totalitarianism (Dupré 1993, 1996; see Reisch 2005 for a critical appreciation). Reisch (1997) discusses the differences between Neurath's methodological rejection of reductionism and Dupré's metaphysical arguments in favor of disunity.

The Dialectics of Dis/Unity

Although I find many of Dupré's metaphysical and epistemological arguments against reduction, causal completeness, and the like valuable and quite compelling, I have a problem with his question "whether science is disunified simply because it has not yet been unified, or rather because disunity is its inevitable and appropriate condition" (Dupré 1996: 102). If this is a dilemma, then I must opt for its first horn, because I see unification happening around me all the time! And this movement started with Newton himself (see, e.g., Delbrück 1949; Grantham 2004). If physics and chemistry hadn't been united, where would we stand today?

The literature against unity and in favor of disunity seems to me hopelessly one-sided. Isn't it obvious that the increasing specialization of science (or any other subsystem of society, for that matter) cries out for some sort of (re-)unification, if only to restore the balance, however minimally? Modern, postmodern, amodern (Latour 1990)—who cares? Churchman and Ackoff seem to me to have been right on target when claiming that the choice of unification is not a free one, and that "if science fails to unite, it will fail to exist." For sure, we would have other interesting cognitive enterprises going on in the absence of any unifying tendencies worth the name, but would this be full-fledged science as it originated with the Scientific Revolution? Hacking (1996), despite his title ("The Disunities of the Sciences"), hints in the same direction. The friends of disunity have only replaced the ideology of positivist unification with an equally one-sided ideology of locality and disunity; the point is to let the dialectic continue.

Pragmatic Unifiers   This includes taking into account the unifying tendencies that are a consequence of regimes and practices of experimentation and instrumentation. These, as well as mathematical tools, "have been more powerful as a source of unity among diverse sciences than have grand unified theories" (Hacking 1996: 69). A nice contemporary example is the steady expansion of game theory since its inception in the 1940s, from a theory of human decision making to a tool for modeling the evolution of animal behavior and now also the evolution of biochemical systems. Callebaut et al. (2007: 69) discuss such biological instances.

A Unified Science does not yet Imply a Unified Nature   Arguments against unification, Morrison (2000: 140) points out, are often attempts to counter the ways in which philosophers (ab)use facts about unified theories to motivate a metaphysical view that extends to a unified natural order—a dangerous temptation that Kant was already aware of. This metaphysics may then be employed to argue for the "epistemological priority" of unified theories. Once it is understood how unity is produced, what its implications for a metaphysics of nature are (possibly none), and what its role in theory construction and confirmation is, it should cease to occupy the undesirable role attributed to it by advocates of disunity.

From Reductive to Synthetic Unity   Whereas in reductive unification two phenomena are being *identified* as of the same kind, synthetic unification only seeks to "integrate" two separate processes or phenomena under one theory. But what does "integration" mean? Although some biologists (e.g., Ricard 2006: 2–3) have provided a formal account, the question remains open in debates on the Synthesis. For instance, Dudley Shapere talked about modification and "supplementation":

Philosophers (and sometimes scientists) have limited themselves by thinking that the only kinds of relations of unification between theories are relations of reduction, where an old theory is absorbed into (its concepts defined in terms of, and its postulates deduced from) a newer or more general one. In the evolutionary synthesis, however, there was a mutual modification or supplementation of different theories (and, more generally, of the concepts, techniques, problems, and so forth, of different areas). (Shapere 1980: 396)

(Note that in the more sophisticated accounts of reduction, both the reducing and the reduced theory are also adjusted to each other; see figure 17.1.) For Provine, Shapere's view of synthesis boils down to mere consistency, which he finds too weak (necessary, but not sufficient) to deserve the name "synthesis." He prefers to call these developments "what they are: consistency arguments and removal of barriers," and "to reserve for 'synthesis' that which is actually synthesized" (Provine 1992: 172–173).

As I suggested before, Darden's "interfield" apparatus allows treating the kinds of interrelations that are at stake here in a more differentiated way (see also Grantham 2004; and Wimsatt's 2007 "coevolutionary" account of reduction). But independently of this, the heuristic importance of coherence or consistency should not be underestimated in our "omic" age, for no feature of life can contravene the findings of chemistry. When there is such incoherence, it has to be rectified by correcting either the biology or the chemistry, or both (Rose and Oakley 2007). Burian (2004: chap. 3) argues that coherence and interdisciplinary integration are middle-range norms that play a guiding role in the longer-term treatment of biological problems (see also Bechtel and Hamilton 2007).

Unification and Explanation   To conclude this section, let us try to clarify some questions regarding the relation between unification and explanation. Deductive derivation is *not* explanation (yet). True explanations ought to provide us with the *mechanism*, the how and why at work in any particular pattern or process we attempt to explain (Morrison 2000; Darden 2006; Callebaut et al. 2007). The Fisher vs. Wright debate resulted in some formal or structural unity at the level of the mathematical models of population genetics, but according to Morrison (2000, 2002) this was insufficient to achieve explanatory unity. Plutynski (2005) disagrees:

often scientists seek the "big picture," a theoretical framework and reasons for adopting it. This quest often appeals to mathematical argument or demonstration. In this context, "How possibly?" and "Why necessarily?" questions are important, and there is no point in denying explanatory relevance to their answers. Plutynski's more liberal view is reminiscent of Kitcher and Salmon's (1989) conciliatory stance, according to which bottom-up explanation (the causal–mechanistic account of how the explanandum comes about) and top-down explanation (showing how the explanandum belongs to a class of phenomena that are derivable from a "common argument pattern") may complement each other usefully.

Further Extensions of the Evolutionary Synthesis

In this section I briefly review some currently envisaged "lateral extensions" of the Synthesis (*sensu* Fernando and Szathmáry; see chapter 9 in this volume) that are little or not discussed elsewhere in this book, focusing on their interrelations (see also chapter 4 in this volume).

There was one important area of biology whose problems Maynard Smith hardly touched on in his *Problems of Biology* (1986): ecology. This was because in addition to evolution by natural selection, "there are other ideas in ecology that have more in common with economics than with anything in contemporary biology" (vii). *Bioeconomics* is one attempt to bridge these differences. In the inaugural issue of the *Journal of Bioeconomics*, Landa and Ghiselin (1999: 5) describe the aims of bioeconomics as the integration or "consilience" of economics and biology for the purpose of enriching both disciplines by substantially enlarging the theoretical and empirical bases which ultimately contribute to the building of new hypotheses, theorems, theories, and paradigms. Unfortunately, the examples they adduce of how biology can contribute to economics and vice versa are quite shallow compared to the foundational publications that economists such as Gordon Tullock, Gary Becker, and Jack Hirshleifer, and the evolutionary biologist Michael Ghiselin contributed to this interfield some decades earlier (Tullock 1971; Ghiselin 1974; Hirshleifer 1977, 1978). Their common thread was that ultimately, evolutionary biology and economics share the very same material object: "the economy of nature." Although Landa and Ghiselin refer to the "new institutional economics" as having much to offer to biologists in explaining aspects of nonhuman societies, including differences in organizational forms, the general flavor is adaptationist and selectionist. In fact, E. O. Wilson's sociobiology was an obvious inspiration to these bioeconomists. Bioeconomics represents the kind of merger Marjorie Grene had in mind when she wrote that

> a good case can be made for the existence of something about the Victorian state of mind that makes "utility," whether in Bentham's social thought or in the adaptationist bent of selection theory, seem self-explanatory. Adaptations, like utility generally, are means and explain what they explain not in themselves but in relation to ends. But is the end of survival really adequate for human as well as organic history? Whatever the answer, the point here is just: there is something teasing about the way in which both utilitarianism and selection theory (in the strictest, or perhaps most reductive form) appeal to thinkers of a certain cast of mind. (Grene 1983: 3–4)

In comparison to bioeconomics, *evolutionary economics* as pioneered by Nelson and Winter (1982) is much more pluralistic in its reception of ideas coming from evolutionary biology, which it tends to cautiously treat as "heuristically useful," if not "merely analogical." Evolutionary economics is also deeply entrenched in Simon's "bounded rationality" tradition, which as a naturalistic framework—in the philosophical sense—seems much better suited to a conversation with "real life" biology than the orthodox rational actor model of neoclassical provenience (Callebaut 2007). If one puts the conceptual and historical efforts of evolutionary economists (e.g., Hodgson and Knudsen 2006) into the perspective of the Synthesis, one is inclined to say that they are just now entering the first phase of restriction (rejection of Lamarckism and teleology in general).

Proposals for "a unified science of cultural evolution" (Mesoudi et al. 2006) or for "the unification of the behavioral sciences" (Gintis 2007) appear ever more frequently in the literature, which is a sure sign that cultural evolution is finally being recognized as "a proper player in the now rearticulating Darwinian and human sciences" (Wimsatt 2006: 340). Whereas Mesoudi et al. start from evolutionary biology—broadly conceived so as to include EvoDevo and other recent developments—and cautiously try to adjust elements and aspects of culture to its mold, Gintis rather imperialistically pushes his *beliefs, preferences, and constraints (BPC) model*, which is the good old rational actor model of consistent preferences deprived of its name. Like the foundational work of the aforementioned pioneers of bioeconomics, his enterprise looks very much like an update of Wilson's sociobiology, now fortified by the lessons from evolutionary psychology and neuroeconomics.

Conclusions

"Evolutionary everything is hot right now" (Hull 1998b: 513). "Universal Darwinism" (Dawkins 1998) is not only conquering the social sciences successfully, but already extends its arms to the humanities and the normative sphere of ethics (e.g., Casebeer 2003), where it meets with considerable resistance (e.g., Sheets-Johnstone 2008). Understanding the evolution of language may remain the pièce de résistance for some time to come (Christiansen and Kirby 2003; see also chapter 9 in this volume).[7] It remains to be seen if, in a broader social and cultural embedding, "evolution" (or at least its "gene's eye" variety) will retain the aura of austerity that someone like George Williams (1966) cast on the theme almost half a century ago. Or will it, rather, function as ersatz religion, not unlike Huxley's and E. O. Wilson's secular humanism, but now for the masses (the "new atheism")?

Evolutionary everything is hot—or is it? While evolution remains a theme in origins of life research and synthetic biology (Rasmussen et al. 2008), it is virtually absent from the considerations of systems biology, the present-day incarnation of physiology (Kitano 2001; Alon 2006). Thus, in a volume on the philosophical foundations of systems biology, we read that it is important to realize that systems biology tries to understand "life as it is now," and does not focus on evolution. It may use reasoning derived from evolutionary biology (homologies), but it does "not yet" claim to be able to explain the evolution of biological systems. This preference, the authors explain,

reflects the conviction that *life should be understandable without reference to the histories of all life forms.* Systems biology aims at acquiring a molecular mechanistic understanding of biological systems subject to challenges to their existence from the outside world. For systems biology, the properties of molecules are considered as important as the systemic behavior, and understanding is considered to have been achieved when it can be shown how the latter emerges from the former when their nonlinear interactions and arrangements are taken into account. Because of the complicated organization of cells and nonlinearity, hence condition dependence of the interactions, detailed models are required to "calculate the emergence of life from its dead molecular constitution." (Boogerd et al. 2007: 325; my emphasis)

The emphasis on emergence may be seen as a welcome departure from the reductionism of Williams's (1966) gene-selectionist view (although many reductionists would claim that they can incorporate nonlinear interactions into their framework as well). But Williams rejoined Gould when it came to the issue of the contingency of evolutionary history:

A biological explanation should invoke no factors other than the laws of physics, natural selection, and the contingencies of history. The idea that an organism has a complex history through which natural selection has been in constant operation imposes a special constraint on evolutionary theorizing. (Williams 1985: 1–2)

Violating this constraint to meet the demands of modeling is one thing, explaining it away quite another. The claim that evolvability is "functional" (Boogerd et al. 2007: 325) is equally problematic from an epistemological point of view. The trend toward physicalization is at odds with the expectation that "omics" approaches will find biological "meaning" (Krohs and Callebaut 2007). Mayr's case for biology as an autonomous science co-equal with physics (and, in fact, superior to it as a bridge to the social sciences and humanities) is definitely lost in these endeavors.

Before closing, I want to turn briefly to EvoDevo because of its potential implications for the future of population thinking, which for Mayr was the hallmark of the Synthesis. Noting that the "German-language strain" of morphological thinking, which at least some proponents of EvoDevo tap for inspiration, "clearly remains on the typological side," Grene (1983: 5) wondered if in the hands of Ghiselin and Hull, population thinking had not been taken too far:

The development of the view that "species are individuals" has lately characterized population thinking to the point of denying the need ever to notice *any* similarities whatever among living things. Taxa are to be taken purely as lineage taxa, and that, said Ghiselin and Hull, is that. . . . Yet, carried to its logical conclusion, this view seems to undercut the very starting point of any biological science, including the theory of evolution. How does one tell which "individuals" (in the everyday sense) are parts of which larger "species-individuals" except by noting some kind of likeness among some and not others?

Hull has suggested that here, as in other areas, theory simply has to go against common sense. But organisms *do* "seem to behave in quite common-sensical ways" (Lewontin 2000: 192). Grene is unconvinced:

What Riedl calls our "hereditary common sense" about natural kinds does seem, at some level, to underlie the practice of even the most theoretical biology. Doesn't one need some judgment of what something is *like* in order to notice either homologies or analogies? . . . Perhaps judgments of "this such" are only everyday starting points, to be left behind, like physicists' everyday judgments, in the more sophisticated statements of a developed evolutionary science. (5–6)

But then, how could Mayr have been so persuasive in the debate over numerical taxonomy by insisting on the importance of the weighting of taxonomic traits by the experienced taxonomist? Grene concludes that "there seems to be some minimal, almost foundational contribution here that some arguments drawn from a rigorous selection theory want us not only to overlook but to abandon" (6). Pursuing this line of thought, I suggest that advocates of EvoDevo, or at least their philosophical companions, should not rest with further exploding the "myth of essentialism" (Winsor 2003; Stamos 2005; Walsh 2006), but also investigate the functional origins of our psychological essentialism (Barrett 2001) to specify the case for this "foundational contribution" of evolutionary epistemology.

Let me conclude. Ever since its inception, the Synthesis has been characterized by "lateral" and "vertical" extensions *and* by simultaneous disunifying tendencies, with no end in sight. I call this double movement "dialectical." The inexorable "vertical" expansion of molecular biology since, roughly, the 1960s may seem to threaten the Synthesis's dominance by making it more and more irrelevant to the real concerns of most practicing biologists (Rose and Oakley 2007), although the chapters in this volume that are representative of the molecular trend rather suggest a supplementation of Synthesis ideas. On the other end of the ontological spectrum, starting with sociobiology in the mid-1970s, a highly successful, narrowly selectionist and adaptationist "horizontal" expansion toward the social sciences and even the humanities reinvigorated a "hardened" version of the Synthesis. Located between these poles, EvoDevo (broadly conceived so as to include epigenetics, innovation studies, and macroevolution) and niche construction theory will most likely be accommodated by the Synthesis as further extensions, albeit ones that may require major conceptual reshuffling.

As the historian of physics I. Bernard Cohen (1985: 297) said some decades ago, it should not be thought that the Darwinian Revolution is over.

Acknowledgments

I would like to thank the participants in the 18th Altenberg Workshop in Theoretical Biology for very useful feedback and exciting discussions generally, and Gerd Müller and Massimo Pigliucci for their comments on and patience with the long gestation of this chapter. I am especially grateful to Elena Aronova for her insightful comments on many of the issues discussed here. Error clause as usual.

Notes

1. See, e.g., Lorenzano (2006) for a recent review. Although biologists seem to have paid little attention to the issue of biological laws (Mayr 1982: 37), it has important implications for the autonomy of biology. Gould (see McIntyre 1997), Mayr (e.g., 1982: chap. 2), and others have argued that there are factors unique to biological theorizing that prevent the formulation of laws (but see Brooks and Wiley 1988: 3–11). One argument Mayr (2000: 69) used to fend off physicalism is that in evolutionary biology "laws give way to concepts," and that concepts and the theories based on them cannot be reduced to the laws and theories of the physical sciences. I will consider the issue of biological explanation briefly in the section on dis/unity.

2. Consilience ("jumping together") involves the explanation and prediction of facts of a kind different from those considered in the formation of a hypothesis or law. It contributes to unity insofar as it demonstrates that facts that once appeared to be of different kinds are in fact the same. This in turn results in simpler theories as the number of hypotheses and laws that are required is reduced. For Whewell, unity was a step in the direction of "ultimate simplicity," in which all knowledge within a particular branch of science will follow from one basic principle (see Morrison 2000: 16). A complication here is that the "domains" of the sciences—the "physical," the "chemical," the "biological," and so on—do not exist *out there* but are "theory-laden" and, insofar, *constructed* (Hempel 1969); for instance, "movement," which for us is a purely physical notion, was an eminently biological concept for Aristotelians.

3. Darden (2006: 170) writes: "the synthetic theory of evolution as proposed by Theodosius Dobzhansky in 1937." According to Gould, when the founding fathers of the Synthesis could be drawn back into their pasts, one distinct memory emerged:

The great works of the synthesis had not been a set of independent volumes, each drawing its separate inspiration from Fisher, Haldane, or Wright. Instead, all the gathered authors looked at Dobzhansky ... and said that they had drawn primary inspiration from *Genetics and the Origin of Species*.... Dobzhansky had not simply been first, by good fortune, in an inevitable line; his book had been the direct instigator of all volumes that followed. (Gould 1982; quoted in Cecciarelli 2001: 25)

Dobzhansky's student Lewontin has suggested that Dobzhansky persuaded "by the sheer force of his personality and reemphasis of things that were in the literature as information but not really in people's heads as knowledge" (quoted in Ceccarelli 2001: 30). See also Mayr (1982: 569); Ayala et al. (2000: 6941); Ceccarelli (2001: 24–29); and Grene and Depew (2004: chap. 9).

4. The *other* intelligent design, of creationist fame, may be seen as an unintended waste product of a hardened Synthesis:

Leaving developmental biology out of evolutionary biology has left evolutionary biology open to attacks from promoters of "intelligent design.... But once development is added to the evolutionary synthesis, it is straightforward to see how the eye can develop through induction, and that the concepts of modularity and correlated progression can readily explain such a phenomenon.... (Gilbert 2006: 749)

5. For the architects of the Synthesis, progress in solving the problems of evolutionary biology would be guaranteed only under certain circumstances. By "treaty," Depew and Weber (1994: 300) refer to "a set of interdisciplinary agreements, some tacit, some explicit, that generated the conceptual space within which a whole raft of specific research programs became possible, while other approaches were ruled out by common consent as a waste

of time." Population thinking thus ruled out the macromutationist Mendelism of Goldschmidt as well as any "regression" to the old morphological–developmental model of evolution invoking bauplans.

6. John H. Woodger, a Cambridge embryologist-turned-logician and participant in the "Biotheoretical Gathering" in the early 1930s (Abir-Am 1987), tried to force biological theories into the positivist deductive mold. He has become a privileged target of criticisms by logicians and philosophers of biology alike. One reviewer called his enterprise "altogether archaic, indeed antediluvian," considering the development of logic in recent decades (Dawson 1974: 353), adding that "Woodger's logical works add nothing to logic or mathematics; whether they provide a stimulus for biological research is, to this reviewer, doubtful" (354). Hull (2000: 68) complained that he was forced to work his way through all of Woodger's publications, and has "never gotten over it." Very few biologists or philosophers give any indication of ever having read Woodger, who got no mention in Mayr's monumental *The Growth of Biological Thought* (1982). As late as 1997, Griffiths invoked Woodger's negative authority to put Mahner and Bunge's (1997) use of formal logic in a bad light, although present-day philosophy of biology could arguably benefit from more formal approaches.

7. A fringe extension that cannot be considered here is Hoffmeyer's (1997) call to unify biology on the basis of a semiotic theory of life.

References

Abir-Am PG (1987) The Biotheoretical Gathering, trans-disciplinary authority and the incipient legitimation of molecular biology in the 1930s: New perspective on the historical sociology of science. History of Science 25: 1–70.

Abir-Am PG (1992) The politics of macromolecules: Molecular biologists, biochemists, and rhetoric. Osiris 7: 164–191.

Adams MB (1970) Towards a Synthesis: Population concepts in Russian evolutionary thought, 1925—1935. Journal of the History of Biology 3: 107–129.

Allen GE (1978) Life Sciences in the Twentieth Century. Cambridge: Cambridge University Press.

Alon U, ed. (2006) An Introduction to Systems Biology: Design Principles of Biological Circuits. London: Chapman and Hall.

Amundson R (1994) Two concepts of constraint: Adaptationism and the challenge from developmental biology. Philosophy of Science 61: 556–578.

Anonymous (1968) Central dogma, right or wrong? Nature 218: 317.

Ayala FJ, Fitch WM, Clegg MT (2000) Variation and evolution in plants and microorganisms: Toward a new synthesis 50 years after Stebbins. Proceedings of the National Academy of Sciences of the USA 97: 6941–6944.

Barrett HC (2001) On the functional origins of essentialism. Mind and Society 3: 1–30.

Bechtel W, Hamilton A (2007) Reduction, integration, and the unity of science: Natural, behavioral, and social sciences and the humanities. In General Philosophy of Science: Focal Issues. T. Kuipers, ed.: 337–430. Amsterdam: Elsevier.

Boogerd FC, Bruggeman FJ, Hofmeyr J-HS, Westerhoff HV (2007) Afterthoughts as foundations for systems theory. In Systems Biology: Philosophical Foundations. F.C. Boogerd, F.J. Bruggeman, J-H.S. Hofmeyr, H.V. Westerhoff, eds.: 321–336. Amsterdam: Elsevier.

Bowler PJ (1983) The Eclipse of Darwinism: Anti-Darwinian Evolution Theories in the Decades around 1900. Baltimore: Johns Hopkins University Press.

Bregant J, Stozer A, Cerkvenik M (2009) Molecular reduction: Reality or fiction? Synthese (in press).

Bromham L (2009) Does nothing in evolution make sense except in the light of population genetics? Biology and Philosophy (forthcoming).

Brooks DR, Wiley EO (1988) Evolution as Entropy: Toward a Unified Theory of Biology. 2nd ed. Chicago: University of Chicago Press.

Brown R, Ladyman J (2009) Physicalism, supervenience and the fundamental level. Philosophical Quarterly 59: 21–38.

Buican D (1984) Histoire de la génétique et de l'évolutionisme en France. Paris: Presses Universitaires de France.

Burian RM (1988) Challenges to the evolutionary synthesis. Evolutionary Biology 23: 247–259.

Burian RM (2004) Epistemology of Development, Evolution, and Genetics. Cambridge: Cambridge University Press.

Buss LW (1983) Evolution, development, and the units of selection. Proceedings of the National Academy of Sciences of the USA 80: 1387–1391.

Butlin R (2006) Post-modern synthesis. Trends in Ecology and Evolution 21: 536.

Callebaut W (2007) Herbert Simon's silent revolution. Biological Theory 2: 76–86.

Callebaut W (2010) Multiscale phenomena in biology and scientific perspectivism. In Multiscale Phenomena in Biology. RM Sinclair and KM Stiefel, eds. Melville, NY: Springer.

Callebaut W (1993) Taking the Naturalistic Turn, or How Real Philosophy of Science Is Done. Chicago: University of Chicago Press.

Callebaut W, Müller GB, Newman S (2007) The Organismic Systems Approach: Evo-Devo and the streamlining of the naturalistic agenda. In Integrating Evolution and Development: From Theory to Practice. R Sansom and RN Brandon, eds.: 25–92. Cambridge, MA: MIT Press.

Carroll RL (2000) Towards a new evolutionary synthesis. Trends in Ecology and Evolution 15: 27–32.

Cartwright N (1999) The Dappled World: A Study of the Boundaries of Science. Cambridge: Cambridge University Press.

Cartwright N et al. (2002) Book Symposium, The Dappled World. Philosophical Books 43: 241–278.

Cartwright N, Cat J, Fleck L, Uebel TE (1996) Otto Neurath: Philosophy Between Science and Politics. Cambridge: Cambridge University Press.

Casebeer WD (2003) Natural Ethical Facts: Evolution, Connectionism, and Moral Cognition. Cambridge, MA: MIT Press.

Ceccarelli L (2001) Shaping Science with Rhetoric: The Cases of Dobzhansky, Schrödinger, and Wilson. Chicago: University of Chicago Press.

Christiansen MH, Kirby S (2003) Language evolution: The hardest problem in science? In Language Evolution. MHChristiansen and S Kirby, eds.: 1–15. Oxford: Oxford University Press.

Churchman CW, Ackoff RL (1946) Varieties of unification. Philosophy of Science 13: 287–300.

Cohen IB (1985) Revolution in Science. Cambridge, MA: Belknap Press.

Commoner B (1961) In defense of biology. Science 133: 1745–1748.

Commoner B (1968) Failure of the Watson–Crick theory as a chemical explanation of inheritance. Nature 220: 334–340.

Craver CF, Bechtel W (2007) Top-down causation without top-down causes. Biology and Philosophy 22: 547–563.

Darden L (1991) Theory Change in Science: Strategies from Mendelian Genetics. New York: Oxford University Press.

Darden L (2006) Reasoning in Biological Discoveries: Essays on Mechanisms, Interfield Relations, and Anomaly Resolution. Cambridge: Cambridge University Press.

Dawkins R (1998[1983]) Universal Darwinism. In The Philosophy of Biology. DL Hull and M Ruse, eds.: 15–37. Oxford: Oxford University Press.

Dawson EE (1974) From biology to mathematics. Review of Woodger (1952–53). Journal of Symbolic Logic 39: 353–354.

Deacon TW (2006) Emergence: The hole at the wheel's hub. In The Re-Emergence of Emergence: The Emergentist Hypothesis from Science to Religion. P Clayton and P Davies, eds.: 111–150. New York: Oxford University Press.

Delbrück M (1949) A physicist looks at biology. Transactions of the Connecticut Academy of Arts and Sciences 38: 173–190.

Depew DJ, Weber BH (1994) Darwinism Evolving: Systems Dynamics and the Genealogy of Natural Selection. Cambridge, MA: MIT Press.

Dirac PAM (1929) Quantum mechanics of many-electron systems. Proceedings of the Royal Society of London A123: 714–733.

Dobzhansky T (1937) Genetics and the Origin of Species. New York: Columbia University Press. Reprint, 1982.

Dresden M (1998) The Klopsteg Memorial Lecture. Fundamentality and numerical scales: Diversity and the structure of physics. American Journal of Physics 66: 468–482.

Dupré J (1993) The Disorder of Things. Cambridge, MA: Harvard University Press.

Dupré J (1996) Metaphysical disorder and scientific disunity. In The Disunity of Science. PGalison and DJ Stump, eds.: 101–117. Stanford, CA: Stanford University Press.

Eldredge N (1985) Unfinished Synthesis: Biological Hierarchies and Modern Evolutionary Thought. New York: Oxford University Press.

Fodor JA (1974) Special sciences (Or: the disunity of science as a working hypothesis). Synthese 28: 97–115.

Foley R (2001) In the shadow of the Modern Synthesis? Alternative perspectives on the last fifty years of paleoanthropology. Evolutionary Anthropology 10: 5–14.

Frost-Arnold G (2005) The large-scale structure of logical empiricism: Unity of science and the elimination of metaphysics. Philosophy of Science 72: 826–838.

Futuyma DJ (1988) *Sturm und Drang* and the evolutionary synthesis. Evolution 42: 217–226.

Futuyma DJ (2005) Evolution. Sunderland, MA: Sinauer.

Galison P (1996) Introduction: The context of disunity. In The Disunity of Science. P Galison and DJ Stump, eds.: 1–33. Stanford, CA: Stanford University Press.

Galison P (1998) The Americanization of unity. Daedalus 127: 45–71.

Ghiselin MT (1974) The Economy of Nature and the Evolution of Sex. Berkeley: University of California Press.

Gilbert SF (1994) Dobzhansky, Waddington and Schmalhausen: Embryology and the Modern Synthesis. In The Evolution of Theodosius Dobzhansky. MB Adams, ed.: 143–154. Princeton, NJ: Princeton University Press.

Gilbert SF (2006) Developmental Biology. 8th ed. Sunderland, MA: Sinauer.

Gilbert SF, Epel D (2009) Ecological Developmental Biology: Integrating Epigenetics, Medicine, and Evolution. Sunderland, MA: Sinauer.

Gintis H (2007) A framework for the unification of the behavioral sciences. Behavioral and Brain Sciences 30: 1–61.

Gould SJ (1982) Darwinism and the expansion of evolutionary theory. Science 216: 380–387.

Gould SJ (1983) The hardening of the Modern Synthesis. In Dimensions of Darwinism. M Grene, ed.: 71–93. Cambridge: Cambridge University Press.

Gould SJ (2002) The Structure of Evolutionary Theory. Cambridge, MA: Belknap Press.

Grantham TA (2004) Conceptualizing the (dis)unity of science. Philosophy of Science 71: 133–155.

Greene JC (1981) Science, Ideology, and World View: Essays in the History of Evolutionary Ideas. Berkeley: University of California Press.

Greene JC (1999) Reflections on Ernst Mayr's *This is Biology*. Biology and Philosophy 14: 103–116.

Grene M (1983) Introduction. In Dimensions of Darwinism: Themes and Counterthemes in Twentieth-Century Evolutionary Theory. M Grene, ed.: 1–15. Cambridge: Cambridge University Press.

Grene M, Depew D (2004) The Philosophy of Biology: An Episodic History. Cambridge: Cambridge University Press.

Griffiths PE (1997) Axioms for biology [review of Mahner and Bunge 1997]. Nature 389: 250.

Gruber HE, Vonèche J, eds. (1977) The Essential Piaget: An Interpretive Reference and Guide. New York: Basic Books.

Hacking I (1996) The disunities of the sciences. In The Disunity of Science. P Galison and DJ Stump, eds.: 37–74. Stanford, CA: Stanford University Press.

Haldane JBS (1932) The Causes of Evolution. London: Longmans Green.

Harwood J (1993) Styles of Scientific Thought: The German Genetics Community, 1900–1933. Chicago: University of Chicago Press.

Hempel CG (1969) Reduction: Ontological and linguistic facts. In Philosophy, Science, and Method: Essays in Honor of E. Nagel. S Morgenbesser, P Suppes, M White, eds.: 179–199. New York: St. Martin's Press.

Hirshleifer J (1977) Economics from a biological point of view. Journal of Law and Economics 20: 1–52.

Hirshleifer J (1978) Competition, cooperation, and conflict in economics and biology. American Economic Review 68: 238–243.

Hodgson GM, Knudsen T (2006) Dismantling Lamarckism: Why descriptions of socio-economic evolution as Lamarckian are misleading. Journal of Evolutionary Economics 16: 343–366.

Hoffmeyer J (1997) Biosemiotics: Towards a new synthesis in biology. European Journal for Semiotic Studies 9: 355–376.

Hooker CA (2004) Asymptotics, reduction and emergence. British Journal for the Philosophy of Science 55: 435–479.

Hull DL (1988) Science as a Process: An Evolutionary Account of the Social and Conceptual Development of Science. Chicago: University of Chicago Press.

Hull DL (1998a) Introduction to Part II. In The Philosophy of Biology. DL Hull and M Ruse, eds.: 89–92. Oxford: Oxford University Press.

Hull DL (1998b) Review of Anthony O'Hear, *Beyond Evolution: Human Nature and the Limits of Evolutionary Explanation* (Oxford: Clarendon Press, 1997). British Journal for the Philosophy of Science 49: 511–514.

Hull DL (2000) The professionalization of science studies: Cutting some slack. Biology and Philosophy 15: 61–91.

Hull DL (2006) The essence of scientific theories. Biological Theory 1: 17–19.

Hüttemann A (2004) What's Wrong with Microphysicalism? London: Routledge.

Huxley J (1942) Evolution: The Modern Synthesis. London: Allen and Unwin.

Jablonka E, Lamb MJ (2007) The expanded evolutionary synthesis—A response to Godfrey-Smith, Haig, and West-Eberhard. Biology and Philosophy 22: 453–472.

Johnson NA, Porter AH (2001) Toward a new synthesis: Population genetics and evolutionary developmental biology. Genetica 112–113: 45–58.

Keller EF (2000) The Century of the Gene. Cambridge, MA: Harvard University Press.

Kitano H, ed. (2001) Foundations of Systems Biology. Cambridge, MA: MIT Press.

Kitcher P, Salmon WC, eds. (1989) Scientific Explanation. Minnesota Studies in the Philosophy of Science, vol. 13. Minneapolis: University of Minnesota Press.

Klein C (2009) Reduction without reductionism: A defence of Nagel on connectability. Philosophical Quarterly 59: 39–53.

Klein JT (1990) Interdisciplinarity: History, Theory, and Practice. Detroit: Wayne State University Press.

Krementsov N (2007) A particular synthesis: Aleksandr Promptov and speciation in birds. Journal of the History of Biology 40: 637–682.

Krohs U, Callebaut W (2007) Data without models merging with models without data. In Systems Biology: Philosophical Foundations. FC Boogerd, FJ Bruggeman, J-HS Hofmeyr, HV Westerhoff, eds.: 181–213. Amsterdam: Elsevier.

Kutschera U, Niklas KJ (2004) The modern theory of biological evolution: An expanded synthesis. Naturwissenschaften 91: 255–276.

Ladyman J, Ross D, Spurrett D, Collier J (2007) Every Thing Must Go: Metaphysics Naturalized. Oxford: Oxford University Press.

Landa JT, Ghiselin MT (1999) The emerging discipine of bioeconomics: Aims and scope of the Journal of Bioeconomics. Journal of Bioeconomics 1: 5–12.

Latour B (1990) Postmodern? No, simply amodern! Steps towards an anthropology of science. Studies in the History of the Philosophy of Science 21: 145–171.

Lewontin RC (2000) What do population geneticists know and how do they know it? In Biology and Epistemology. R Creath and J Maienschein, eds.: 191–214. Cambridge: Cambridge University Press.

Lorenzano P (2006) Fundamental laws and laws of biology. In Philosophie der Wissenschaft—Wissenschaft der Philosophie. (G Erst and GK-H Niebergall, eds.: 129–155. Paderborn: Mentis Verlag.

Luttikhuizen PC, Drent J (2004) Post-modern synthesis? Review of West-Eberhard 2003. Heredity 92: 596–597.

Lynch M (2007) The frailty of adaptive hypothesis for the origins of organismal complexity. Proceedings of the National Academy of Sciences of the USA 104(suppl. 1): 8597–8604.

Mahner M, Bunge M (1997) Foundations of Biophilosophy. Berlin: Springer.

Maull NL (1977) Unifying science without reduction. Studies in the History and Philosophy of Science 8: 143–162.

Maynard Smith J (1986) The Problems of Biology. Oxford: Oxford University Press.

Mayr E (1982) The Growth of Biological Thought: Diversity, Evolution, and Inheritance. Cambridge, MA: Belknap Press.

Mayr E (1988) Toward a New Philosophy of Biology: Observations of an Evolutionist. Cambridge, MA: Belknap Press.

Mayr E (2000) Darwin's influence on modern thought. Scientific American (July): 66–71.

Mayr E, Provine WB, eds. (1980) The Evolutionary Synthesis: Perspectives on the Unification of Biology. Cambridge, MA: Harvard University Press.

McIntyre L (1997) Gould on laws in biological science. Biology and Philosophy 12: 357–367.

Mesoudi A, Whiten A, Laland KN (2006) Towards a unified science of cultural evolution. Behavioral and Brain Sciences 29: 329–383.

Morrison M (2000) Unifying Scientific Theories: Physical Concepts and Mathematical Structures. Cambridge: Cambridge University Press.

Morrison M (2002) Modelling populations: Pearson and Fisher on Mendelism and Biometry. British Journal for the Philosophy of Science 53: 39–68.

Morrison M (2006) Unification, explanation, and explaining unity: The Fisher–Wright controversy. British Journal for the Philosophy of Science 57: 233–245.

Müller GB (2007) Evo-Devo: Extending the evolutionary synthesis. Nature Reviews Genetics 8: 943–949.

Nagel E (1961) The Structure of Science. New York: Harcourt, Brace, and World.

Nelson RR, Winter SG (1982) An Evolutionary Theory of Economic Change. Cambridge, MA: Belknap Press.

Nicholls H (2008) Synthetic biology. Lancet 372(suppl. 1): S45–S49.

Nicolaou KC, Vourloumis D, Winssinger N, Baran PS (2000) The art and science of total synthesis at the dawn of the twenty-first century. Angewandte Chemie (international ed.) 39: 44–122.

Nowak MA (2006) Evolutionary Dynamics: Exploring the Equations of Life. Cambridge, MA: Harvard University Press.

Odling-Smee FJ, Laland KN, Feldman MW (2003) Niche Construction: The Neglected Process in Evolution. Princeton, NJ: Princeton University Press.

Okasha S (2005) On niche construction and extended evolutionary theory. Biology and Philosophy 20: 1–10.

Oppenheim P, Putnam H (1958) Unity of science as a working hypothesis. In Minnesota Studies in the Philosophy of Science. H Feigl, M Scriven, G Maxwell, eds.: vol. 2: 3–36. Minneapolis: University of Minnesota Press.

Pigliucci M (2007) Do we need an extended evolutionary synthesis? Evolution 61: 2743–2749.

Pigliucci M (2008) The proper role of population genetics in modern evolutionary theory. Biological Theory 3: 316–324.

Plutynski A (2005) Explanatory unification and the early synthesis. British Journal for the Philosophy of Science 56: 595–609.

Provine WB (1992) Progress in evolution and meaning in life. In Julian Huxley: Biologist and Statesman of Science. CK Waters and A Van Helden, eds.: 165–180. Houston, TX: Rice University Press.

Rasmussen S, Bedau MA, Chen L, Deamer D, Krakauer DC, Packard NH, Stadler PF, eds. (2008) Protocells: Bridging Nonliving and Living Matter. Cambridge, MA: MIT Press.

Reisch GA (1997) How postmodern was Neurath's idea of unity of science? Studies in the History and Philosophy of Science 28: 439–451.

Reisch GA (2005) How the Cold War Transformed Philosophy of Science. Cambridge: Cambridge University Press.

Ricard J (2006) Emergent Collective Properties, Networks and Information in Biology. Amsterdam: Elsevier.

Richards RJ (1992) Evolution. In Keywords in Evolutionary Biology. (EF Keller and EA Lloyd, eds.: 95–105. Cambridge, MA: Harvard University Press.

Richards RJ (2002) The Romantic Conception of Life: Science and Philosophy in the Age of Goethe. Chicago: University of Chicago Press.

Richards RJ (2008) The Tragic Sense of Life: Ernst Haeckel and the Struggle over Evolutionary Thought. Chicago: University of Chicago Press.

Richardson RC (2009) Multiple realization and methodological pluralism. Synthese 167: 473–492.

Robinson WS (2005) Zooming in on downward causation. Biology and Philosophy 20: 117–136.

Rose MR, Oakley TH (2007) The new biology: Beyond the Modern Synthesis. Biology Direct 2: 30.

Rosenberg A (1994) Instrumental Biology, or The Disunity of Science. Chicago: University of Chicago Press.

Rosenberg A (2006) Darwinian Reductionism Or, How to Stop Worrying and Love Molecular Biology. Chicago: University of Chicago Press.

Ruse M (1979) The Darwinian Revolution: Science Red in Tooth and Claw. Chicago: University of Chicago Press. 2nd ed., 1999.

Sarton M (1959) I Knew a Phoenix. New York: Norton.

Savellos EE, Yalçin ÜD (1995) Introduction. In Supervenience: New Essays. EE Savellos and ÜD Yalçin, eds.: 1–15. Cambridge: Cambridge University Press.

Schaffer J (2003) Is there a fundamental level? Noûs 37: 498–517.

Schaffner KF (1993) Discovery and Explanation in Biology and Medicine. Chicago: University of Chicago Press.

Scheiner SM, Willig MR (2008) A general theory of ecology. Theoretical Ecology 1: 21–28.

Schlesinger G (1961) The prejudice of micro-reduction. British Journal for the Philosophy of Science 12: 215–224.

Schrödinger E (1944) What Is Life? The Physical Aspect of the Living Cell. Cambridge: Cambridge University Press.

Shapere D (1980) The meaning of the evolutionary synthesis. In The Evolutionary Synthesis: Perspectives on the Unification of Biology. E Mayr and WB Provine, eds.: 388–398. Cambridge, MA: Harvard University Press.

Sheets-Johnstone M (2008) The Roots of Morality. University Park, PA: Pennsyvania State University Press.

Smocovitis VB (1996) Unifying Biology: The Evolutionary Synthesis and Evolutionary Biology. Princeton, NJ: Princeton University Press.

Smocovitis VB (2005) "It Ain't Over 'til it's Over": Rethinking the Darwinian Revolution. Journal of the History of Biology 38: 33–49.

Sober E (1999) The multiple realizability argument against reductionism. Philosophy of Science 66: 542–564.

Stamos DN (2005) Pre-Darwinian taxonomy and essentialism: A reply to Mary Winsor. Biology and Philosophy 20: 79–96.

Stearns SC, Koella JC, eds. (2008) Evolution in Health and Disease. 2nd ed. New York: Oxford University Press.

Tattersall I (2000) Paleoanthropology: The last half-century. Evolutionary Anthropology 9: 2–16.

Tullock G (1971) Biological externalities. Journal of Theoretical Biology 33: 565–576.

Van Brakel J (2003) The *ignis fatuus* of reduction and unification. Annals of the New York Academy of Sciences 988: 30–43.

Van Gulick R (1993) Who's in charge here? And who's doing all the work? In Mental Causation. J Heil and A Mele, eds.: 233–256. Oxford: Oxford University Press.

Vrba ES, Eldredge N (1984) Individuals, hierarchies and processes: Towards a more complete evolutionary theory. Paleobiology 10: 217–228.

Vrba ES, Gould SJ (1986) The hierarchical expansion of sorting and selection: Sorting and selection cannot be equated Paleobiology 12: 217–228.

Walsh D (2006) Evolutionary essentialism. British Journal for the Philosophy of Science 57: 425–448.

Waters CK (1990) Why the anti-reductionist consensus won't survive the case of classical Mendelian genetics. In: PSA 1990. Vol. 1: 125–139.

Waters CK (1992) Introduction: Revising our picture of Julian Huxley. In Julian Huxley: Biologist and Statesman of Science. CK Waters and A Van Helden, eds.: 1–27. Houston, TX: Rice University Press.

West-Eberhard MJ (2003) Developmental Plasticity and Evolution. Oxford: Oxford University Press.

Wicken JS (1987) Evolution, Thermodynamics and Information: Extending the Darwinian Paradigm. New York: Oxford University Press.

Williams GC (1966) Adaptation and Natural Selection. Princeton, NJ: Princeton University Press.

Williams GC (1985) A defense of reductionism in evolutionary biology. Oxford Surveys in Evolutionary Biology 2: 1–27.

Wilson EO (1971) Competitive and aggressive behavior. In Man and Beast: Competitive Social Behavior. JF Eisenberg and WS Dillon, eds.: 183–217. Washington, DC: Smithsonian Institution Press.

Wilson EO (1975) Sociobiology: The New Synthesis. Cambridge, MA: Harvard University Press.

Wilson EO (1998) Consilience: The Unity of Knowledge. London: Little, Brown.

Wilson EO (2008) One giant leap: How insects achieved altruism and colonial life. BioScience 58: 17–25.

Wimsatt WC (2006) Reengineering the Darwinian sciences in social context. Biological Theory 1: 338–341.

Wimsatt WC (2007) Re-engineering Philosophy for Limited Beings: Piecewise Approximations to Reality. Cambridge, MA: Harvard University Press.

Winsor MP (2003) Non-essentialist methods in pre-Darwinian taxonomy. Biology and Philosophy 18: 387–400.

Contributors

John Beatty Department of Philosophy, University of British Columbia, Vancouver BC, Canada

Werner Callebaut Faculty of Sciences, Hasselt University, Diepenbeek, Belgium, Konrad Lorenz Institute for Evolution and Cognition Research, Altenberg, Austria

Jeremy Draghi Yale University, Department of Ecology and Evolutionary Biology, New Haven CT, USA

Chrisantha Fernando MRC National Institute for Medical Research, Mill Hill, London, UK

Sergey Gavrilets Departments of Ecology & Evolutionary Biology and Mathematics, National Institute for Mathematical and Biological Synthesis, University of Tennessee, Knoxville TN, USA

John C. Gerhart Department of Molecular and Cell Biology, University of California, Berkeley, Berkeley CA, USA

Eva Jablonka Cohn Institute for the History and Philosophy of Science and Ideas, Tel Aviv University, Tel Aviv, Israel

David Jablonski Department of Geophysical Sciences, University of Chicago, Chicago IL, USA

Marc W. Kirschner Department of Systems Biology, Harvard Medical School, Boston MA, USA

Marion J. Lamb Independent scholar

Alan C. Love Department of Philosophy, University of Minnesota, Minneapolis MN, USA

Gerd B. Müller Department of Theoretical Biology, University of Vienna, Konrad Lorenz Institute for Evolution and Cognition Research, Altenberg, Austria

Stuart A. Newman Department of Cell Biology and Anatomy, New York Medical College, New York NY, USA

Massimo Pigliucci Department of Philosophy, City University of New York–Lehman College, New York NY, USA

John Odling-Smee School of Anthropology, University of Oxford, Oxford, UK

Eörs Szathmáry Collegium Budapest (Institute for Advanced Study), Budapest, Hungary; Parmenides Foundation, Munich, Germany; and Institute of Biology, Eötvös University, Budapest, Hungary

Michael Purugganan Department of Biology and Center for Genomics and Systems Biology, New York University, New York NY, USA

Günter P. Wagner Department of Ecology and Evolutionary Biology, Yale University, New Haven CT, USA

David Sloan Wilson Departments of Biology and Anthropology, Binghamton University, Binghamton NY, USA

Gregory A. Wray Department of Biology and Institute for Genome Sciences & Policy, Duke University, Durham NC, USA

Index